Annals of Mathematics Studies

Number 155

Markov Processes from K. Itô's Perspective

by

Daniel W. Stroock

PRINCETON UNIVERSITY PRESS

PRINCETON AND OXFORD

2003

Copyright © 2003 by Princeton University Press
Published by Princeton University Press, 41 William Street,
Princeton, New Jersey 08540
In the United Kingdom: Princeton University Press,
3 Market Place, Woodstock, Oxfordshire OX20 1SY

The Annals of Mathematics Studies are edited by
John N. Mather and Elias M. Stein

ISBN 0-691-11542-7 (cloth)
ISBN 0-691-11543-5 (pbk.)

British Library Cataloging-in-Publication Data is available.

The publisher would like to acknowledge the author of this
volume for providing the camera-ready copy from which this book was printed

Printed on acid-free paper ∞
www.pup.princeton.edu

Printed in the United States of America

1 3 5 7 9 10 8 6 4 2

1 3 5 7 9 10 8 6 4 2
(Pbk.)

This book is dedicated to the only person to whom it could be:

Kiyosi Itô

Contents

Preface

In spite of (or, maybe, because of) his having devoted much of his life to the study of probability theory, Kiyosi Itô is not a man to leave anything to chance. Thus, when Springer-Verlag asked S.R.S. Varadhan and me to edit a volume [14] of selected papers by him, Itô wanted to make sure that we would *get it right*. In particular, when he learned that I was to be the author of the introduction to that volume, Itô , who had spent the preceding academic year at the University of Minnesota, decided to interrupt his return to Japan with a stop at my summer place in Colorado so that he could spend a week tutoring me. In general, preparing a volume of selected papers is a pretty thankless task, but the opportunity to spend a week being tutored by Itô more than compensated for whatever drudgery the preparation of his volume cost me. In fact, that tutorial is the origin of this book.

Before turning to an explanation of what the present book contains, I cannot, and will not, resist the temptation to tell my readers a little more about Itô the man. Specifically, I hope that the following, somewhat trivial, anecdote from his visit to me in Colorado will help to convey a sense of Itô's enormous curiosity and his determination to understand the world in which he lives. On a day about midway through the week which Itô spent in Colorado, I informed him that one of my horses was to be shod on the next day and that, as a consequence, it might be best if we planned to suspend my tutorial for a day. Considering that it was he who had taken the trouble to visit me, I was somewhat embarrassed about asking Itô to waste a day. However, Itô's response was immediate and completely positive. He wanted to learn how we Americans put shoes on our horses and asked if I would pick him up at his hotel in time for him to watch. I have a vivid memory of Itô bending over right next to the farrier, all the time asking a stream of questions about the details of the procedure. It would not surprise me to learn that, after returning home, Itô not only explained what he had seen but also suggested a cleverer way of doing it. Nor would it surprise me to learn that the farrier has avoided such interrogation ever since.

The relevance of this anecdote is that it highlights a characteristic of Itô from which I, and all other probabilists in my generation, have profited immeasurably. Namely, Itô's incurable pedagogic itch. No matter what the topic, Itô is driven to master it in a way that enables him to share his insights with the rest of us. Certainly the most renowned example of Itô's skill is his introduction of stochastic differential equations to explain the Kolmogorov–Feller theory of Markov processes. However, the virtues of his theory were not immediately recognized. In fact, sometime during the week of my tutorial, Itô confided in me his disappointment with the initial reception which his theory received. It seems that J.L. Doob was the first person to fully appreciate what Itô had done. To wit, Doob not only played a crucial role in arranging for Itô's memoirs [11] to be published by the AMS, he devoted §5 of the ninth chatper in his own famous book [6] to explaining, extending, and improving Itô's theory. However, Doob's book is sufficiently challenging that hardly anyone who made it to §5 of Chapter IX was ready to absorb its content. Thus, Itô's theory did not receive the attention that it deserves until H.P. Mc Kean, Jr., published his lovely little book [23].

Like Lerey's theory of spectral sequences, Itô's theory was developed as a pedagogic device, and, like Leray's theory, Itô's theory eventually took on a life of its own. Both Itô's and Mc Kean's treatments of the theory concentrate on practical matters: the application of stochastic integration to the path-wise construction of Markov processes. However, around the same time that Mc Kean's book appeared, P.A. Meyer and his school completed the program which Doob had initiated in Chapter IX. In particular, Meyer realized that his own extension of Doob's decomposition theorem allowed him to carry out for essentially general martingales the program which Doob had begun in his book, and, as a result, Itô's theory became an indispensable tool for the Strasbourg School of Probability Theory. Finally, for those who found the Strasbourg version too arcane, the theory was brought back to earth in the beautiful article [21], where Kunita and Watanabe showed that, if one takes the best ideas out of the Strasbourg version and tempers them with a little reality, the resulting theory is easy, æsthetically pleasing, and remarkably useful.

In the years since the publication of Mc Kean's book, there have been lots of books written about various aspects of Itô's theory of stochastic integration. Among the most mathematically influential of these are the book by Delacherei and P.A. Meyer's [5], which delves more deeply than most will care into the intimate secrets of stochastic processes, N. Ikeda and S. Watanabe's [16], which is remarkable for both the breadth and depth of the material covered, and the book by D. Revuz and M. Yor's [27], which demonstrates the power of stochastic integration by showing how it can be used to elucidate the contents Itô's book with Mc Kean [15], which, ironically, itself contains *no* stochastic integration. Besides these, there are several excellent, more instruction oriented texts: the ones by B. Øksendal [25] and by K.L. Chung and R. Williams [3] each has devoted followers. In addition, the lure of mammon has encouraged several authors to explain stochastic integration to economists and to explain economics to stochastic integrationists.[1]

In view of the number of books which already exist on the subject, one can ask, with reason, whether yet another book about stochastic integration is really needed. In particular, what is my excuse for writing this one? My answer is that, whether or not it is needed, I have written this book to redress a distortion which has resulted from the success of Itô's theory of stochastic integration. Namely, Itô's stochastic integration theory is a *secondary theory*, the *primary theory* being the one which grew out of Itô's ideas about the structure of Markov processes. Because his primary theory is the topic of Chapters 1 through 4, I will restrict myself here to a few superficial comments. Namely, as is explained in the introduction to [14], when, as a student at Tokyo University, Itô was assigned the task of explaining the theory of Markov processes to his peers, he had the insight that Kolmogorov's equations arise naturally if one thinks of a Markov process as *the integral curve of a vector field on the space* $\mathbf{M}_1(\mathbb{R}^n)$ *of probability measures on* $\mathbb{R}^n$. Of course, in order to think about vector fields on $\mathbf{M}_1(\mathbb{R}^n)$, one has to specify a coordinate system, and Itô realized that the one which not only leads to Kolmogorov's equations but also explains their connection to infinitely divisible laws is the coordinate system determined by integrating $\mu \in \mathbf{M}_1(\mathbb{R}^n)$ against smooth functions. When these coordinates are adopted, the infinitely divisible flows play the role of "rays," and Kolmogorov's equations arise as the equations which determine the integral curve of a vector field composed of these rays.[2]

From a conventional standpoint, Chapter 1 is a somewhat peculiar introduction to the theory of continuous-time Markov processes on a finite state space. Namely, because it is the setting in which all technical difficulties disappear, Chapter 1 is devoted to the development of Itô's ideas in the setting of the n-point space $\mathbb{Z}_n \equiv \{0, \ldots, n-1\}$. To get started, I first give $\mathbf{M}_1(\mathbb{Z}_n)$ the differentiable structure that it inherits as a simplex in $\mathbb{R}^n$. I then think of $\mathbb{Z}_n$ as an

---

[1] Whatever the other economic benefits of this exercise have been, it certainly provided (at least for a while) a new niche in the job market for a generation of mathematicians.

[2] It should be mentioned that, although this perspective is enormously successful, it also accounts for one of the most serious shortcomings of Itô's theory. Namely, Itô's ideas apply only to Markov processes which are smooth when viewed in the coordinate system which he chose. Thus, for example, they cannot handle diffusions corresponding to divergence-form, strictly elliptic operators with rough coefficients, even though the distribution of such a diffusion is known to be, in many ways, remarkably like that of Brownian motion.

Abelian group (under addition modulo n), develop the corresponding theory of infinitely divisible laws, and show that the inherited differentiable structure has a natural description in terms of infinitely divisible flows, which now play the role that rays play in Euclidean differential geometry. Having put a differentiable structure on $\mathbf{M}_1(\mathbb{Z}_n)$, one knows what vector fields are there, and so I can discuss their integral curves. Because it most clearly highlights the analogy between the situation here and the one which is encountered in the classical theory of ordinary differential equations, the procedure which I use to integrate vector fields on $\mathbf{M}_1(\mathbb{Z}_n)$ is the Euler scheme: the one in which the integral curve is constructed as the limit of "polygonal" curves. Of course, in this setting, "polygonal" means "piecewise infinitely divisible." In any case, the integral curve of a general vector field on $\mathbf{M}_1(\mathbb{Z}_n)$ gives rise to a nonlinear Markov flow: one for which the transition mechanism depends not only on where the process presently is but also on what its present distribution is. In order to get usual Markov flows (i.e., linear ones with nice transition probability functions), it is necessary to restrict one's attention to vector fields which are affine with respect to the obvious convex structure of $\mathbf{M}_1(\mathbb{Z}_n)$, at which point one recovers the classical structure in terms of Kolmogorov's forward and backward equations.

The second half of Chapter 1 moves the considerations of the first half to a pathspace setting. That is, in the second half I show that the flows produced by integrating vector fields on $\mathbf{M}_1(\mathbb{Z}_n)$ can be realized in a canonical way as the one-time marginal distributions of a probability measure $\mathbb{P}$ on the space $D([0,\infty);\mathbb{Z}_n)$ of right-continuous paths, piecewise constant $p : [0,\infty) \longrightarrow \mathbb{Z}_n$. After outlining Kolmogorov's completely general approach to the construction of such measures, the procedure which I adopt and emphasize here is Itô's. Namely, I begin by constructing the $\mathbb{P}$, known as the Lévy process, which corresponds to an infinitely divisible flow. I then show how the paths of a Lévy process can be massaged into the paths of a general Markov process on $\mathbb{Z}_n$. More precisely, given a transition probability function on $\mathbb{Z}_n$, I show how to produce a mapping, the "Itô map," of $D([0,\infty);\mathbb{Z}_n)$ into itself in such a way that the $\mathbb{P}$ corresponding to the given transition probability function is the image under the Itô map of an appropriate Lévy process. The chapter ends by showing that the construction of the Itô map can be accomplished by the pathspace analog of the Euler approximation scheme used earlier to integrate vector fields on $\mathbf{M}_1(\mathbb{Z}_n)$.

The second and third chapters recycle the ideas introduced in Chapter 1, only this time when the state space is $\mathbb{R}^n$ instead of $\mathbb{Z}_n$. Thus, Chapter 2 develops the requisite machinery, especially the Lévy–Khinchine formula,[3] about

---

[3] In the hope that it will help to explain why these techniques apply to elliptic, parabolic,

infinitely divisible flows, and then shows[4] that, once again, the infinitely divisible flows are the rays in this differentiable structure. Once this structure has been put in place, I integrate vector fields on $\mathbf{M}_1(\mathbb{R}^n)$ using Euler's approximation scheme and, after specializing to the affine case, show how this leads back to Kolmogorov's description of Markov transition probability functions via his forward and backward equations.

Before I can proceed to the pathspace setting, I have to develop the basic theory of Lévy processes on $\mathbb{R}^n$. That is, I have to construct the pathspace measures corresponding to infinitely divisible flows in $\mathbf{M}_1(\mathbb{R}^n)$, and, for the sake of completeness, I carry this out in some detail in the last part of Chapter 2. Having laid the groundwork in Chapter 2, I introduce in Chapter 3 Itô's procedure for converting Lévy processes into more general Markov processes. Although the basic ideas are the same as those in the second half of Chapter 1, everything is more technically complicated here and demands greater care. In fact, it does not seem possible to carry out Itô's procedure in complete generality, and so Chapter 3 includes an inquiry into the circumstances in which his procedure does work. Finally, Chapter 3 concludes with a discussion of some examples which display both the virtues and the potential misinterpretation of Itô's theory.

Chapter 4 treats a number of matters which are connected with Itô's construction. In particular, it is shown that, under suitable conditions, his construction yields measures on pathspace which vary smoothly as a function of the starting point, and this observation leads to a statement of uniqueness which demonstrates that the pathspace measures at which he arrives are canonically connected to the affine vector fields from which they arise.

In some sense, Chapter 4 could, and maybe should, be the end of this book. Indeed, by the end of Chapter 4, the essence of Itô's theory has been introduced, and all that remains is to convince afficionados of Itô's theory of stochastic integration that the present book is about their subject. Thus, Chapter 5 develops Itô's theory of stochastic integration for Brownian motion and applies it to the constructions in Chapter 3 when Brownian motion is the underlying Lévy process. Of course, Itô's famous formula is the centerpiece of all this. Chapter 6 showcases several applications of Itô's stochastic integral theory, especially his formula. Included are Tanaka's treatment of local time, the Cameron–Martin formula and Girsanov's variation thereof, pinned Brownian motion, and

---

but not hyperbolic equations, I have chosen to base my proof of their famous formula on the minimum principle.

[4] Unfortunately, certain technical difficulties prevented me from making the the connection here as airtight as it is in the case of $\mathbf{M}_1(\mathbb{Z}_n)$.

Itô's treatment of Wiener's theory of homogeneous chaos.

Kunita and Watanabe's extension of Itô's theory to semimartingales is introduced in Chapter 7 and is used there to prove various representation theorems involving continuous martingales. Finally, Chapter 8 deals with Stratonovich's variant on Itô's integration theory. The approach which I have taken to this topic emphasizes the interpretation of Stratonovich's integral, as opposed to Itô's , as the one at which someone schooled in L. Schwartz's distribution theory would have arrived. In particular, I stress the fact that Stratonovich's integral enjoys coordinate inveriance properties which, as is most dramatically demonstrated by Itô's formula, Itô's integral does not. In any case, once I have introduced the basic ideas, the treatment here follows closely the one developed in [38]. Finally, the second half of this chapter is devoted to a proof of the "support theorem" (cf. [39] and [40]) for degenerate diffusions along the lines suggested in that article. Although applications to differential geometry are among the most compelling reasons for introducing Stratonovich's theory, I have chosen to not develop them in this book, partly in the hope that those who are interested will take a look at [37].

Before closing, I should say a few words about possible ways in which this book might be read. Of course, I hope that it will eventually be read by at least one person, besides myself, from cover to cover. However, I suspect that most readers will not to do so. In particular, because this book is really an introduction to continuous time stochastic processes, the reader who is looking for the most efficient way to learn how to do stochastic integration (or price an option) is going to be annoyed by Chapters 1 and 2. In fact, a reader who is already comfortable with Brownian motion and is seeking a no frills introduction to the most frequently used aspects of Itô's theory should probably start with Chapter 5 and dip into the earlier parts of the book only as needed for notation. On the other hand, for someone who already knows the nuts and bolts of stochastic integration theory and is looking to acquire a little "culture," a reasonable selection of material would be the contents of Chapter 1, the first half of Chapter 2, and Chapter 3. In addition, all readers may find some topics of interest in the later chapters.

Whatever is the course which they choose, I would be gratified to learn that some of my readers have derived as much pleasure out of reading this book as I have had writing it.

Daniel W. Stroock, August 2002

Markov Processess from K. Itô's Perspective

Finite State Space, a Trial Run

In his famous article [19], Kolmogorov based his theory of Markov processes on what became known as Kolmogorov's *forward*[1] and *backward* equations. In an attempt to explain Kolmogorov's ideas, K. Itô [14] took a crucial step when he suggested that Kolmogorov's forward equation can be thought of as describing the "flow of a vector field on the space of probability measures." The purpose of this chapter is to develop Itô's suggestion in the particularly easy case when the state space E is finite. The rest of the book is devoted to carrying out the analogous program when $E = \mathbb{R}^n$.

1.1 An Extrinsic Perspective

Suppose that $\mathbb{Z}_n = \{0, \ldots, n-1\}$ where $n \geq 1$. Then $\mathbf{M}_1(\mathbb{Z}_n)$ is the set of probability measures $\mu_{\boldsymbol{\theta}} = \sum_{m=0}^{n-1} \theta_m \delta_m$, where each δ_m is the unit point mass at m and $\boldsymbol{\theta}$ is an element of the simplex $\Theta_n \subseteq \mathbb{R}^n$ consisting of vectors whose coordinates are non-negative and add to 1. Furthermore, if Θ_n is given the topology which it inherits as a subset of $\mathbb{R}^n$ and $\mathbf{M}_1(\mathbb{Z}_n)$ is given the topology of weak convergence (i.e., the topology in which convergence is tested in terms of integration against bounded, continuous functions), then the map $\boldsymbol{\theta} \in \Theta_n \longmapsto \mu_{\boldsymbol{\theta}} \in \mathbf{M}_1(\mathbb{Z}_n)$ is a homeomorphism. Thus it is reasonable to attempt using this homeomorphism to gain an understanding of the differentiable structure on $\mathbf{M}_1(\mathbb{Z}_n)$.

§1.1.1. **The Structure of** Θ_n: It is important to recognize that the differentiable structure which Θ_n inherits as a subset of $\mathbb{R}^n$ is not entirely trivial. Indeed, even when $n = 2$, Θ_n is a submanifold with boundary, and it is far worse when $n \geq 3$. More precisely, its interior $\mathring{\Theta}_n$ is a nice $(n-1)$-dimensional submanifold of $\mathbb{R}^n$, but the boundary $\partial\Theta_n$ of Θ_n breaks into faces which are disconnected submanifolds of dimensions $(n-2)$ through 0. For example,

$$(\theta_1, \theta_2) \in \left\{(\xi_1, \xi_2) \in (0,1)^2 : \xi_1 + \xi_2 < 1\right\} \longmapsto (1 - \theta_1 - \theta_2, \theta_1, \theta_2) \in \mathring{\Theta}_3$$

---

[1] In the physics literature, Kolmogorov's forward equation is usually called the Fokker-Planck equation.

is a diffeomorphism, and

$$\partial \Theta_3 = \mathring{I}_0 \cup \mathring{I}_1 \cup \mathring{I}_2 \cup \{(1,0,0)\} \cup \{(0,1,0)\} \cup \{(0,0,1)\}$$

where each of the maps $\theta_1 \in (0,1) \longmapsto (0, \theta_1, 1 - \theta_1) \in I_0$, $\theta_0 \in (0,1) \longmapsto (\theta_0, 0, 1 - \theta_0) \in I_1$, and $\theta_0 \in (0,1) \longmapsto (\theta_0, 1 - \theta_0, 0) \in I_2$ is a diffeomorphism. In particular, after a little thought it becomes clear that although, starting at each $\boldsymbol{\theta} \in \Theta_n$, motion in Θ_n can be performed in $(n-1)$ independent directions, movement in some, or all, those directions may have restricted parity. For example, if $\boldsymbol{\theta} \in I_0$, then movement from $\boldsymbol{\theta}$ will require that the 0th coordinate be initially nondecreasing. Similarly, if $\boldsymbol{\theta} = (1,0,0)$, at least initially, the 0th coordinate will have to be nonincreasing while both the 1st and 2nd coordinates will have to be nondecreasing.

For the reasons just discussed, it is best to describe the notion of a differentiable path in Θ_n in terms of one-sided paths. That is, we will say that $t \in [0, \infty) \longmapsto \boldsymbol{\theta}(t) \in \Theta_n$ is *differentiable* if[2] the derivative of

$$t \in [0, \infty) \longmapsto \dot{\boldsymbol{\theta}}(t) \equiv \lim_{h \searrow 0} \frac{\boldsymbol{\theta}(t+h) - \boldsymbol{\theta}(t)}{h} \in \mathbb{R}^n \quad \text{is a smooth curve,}$$

in which case we will call $\dot{\boldsymbol{\theta}}(0)$ a tangent to Θ_n at $\boldsymbol{\theta}(0)$. Further, given $\boldsymbol{\theta} \in \Theta_n$, we will use $T_{\boldsymbol{\theta}}(\Theta_n)$ to denote the set of all vectors $\dot{\boldsymbol{\theta}}(0)$ which arise as the (right) derivative at $t = 0$ of a smooth $t \in [0, \infty) \longmapsto \boldsymbol{\theta}(t) \in \Theta_n$ with $\boldsymbol{\theta}(0) = \boldsymbol{\theta}$. The following statement is essentially obvious.

LEMMA 1.1.1. *For each* $m \in \mathbb{Z}_n$, *let* $\mathbf{e}_m \in \mathbb{R}^n$ *be the vector whose mth coordinate is 1 and whose other coordinates are 0. Then* Θ_n *is the convex hull of* $\{\mathbf{e}_0, \ldots, \mathbf{e}_{n-1}\}$. *Moreover, if* $\boldsymbol{\theta} \in \Theta_n$, *then* $\mathbf{b} \in \mathbb{R}^n$ *is a tangent to* Θ_n *at* $\boldsymbol{\theta}$ *if and only if* $\sum_{m=0}^{n-1} (\mathbf{b}, \mathbf{e}_m)_{\mathbb{R}^n} = 0$, $(\boldsymbol{\theta}, \mathbf{e}_m)_{\mathbb{R}^n} = 0 \implies (\mathbf{b}, \mathbf{e}_m)_{\mathbb{R}^n} \geq 0$, *and* $(\boldsymbol{\theta}, \mathbf{e}_m)_{\mathbb{R}^n} = 1 \implies (\mathbf{b}, \mathbf{e}_m)_{\mathbb{R}^n} \leq 0$. *In particular,* $T_{\boldsymbol{\theta}}(\Theta_n)$ *is a closed, convex cone in* $\mathbb{R}^n$.

As we anticipated, the shape of $T_{\boldsymbol{\theta}}(\Theta_n)$ undergoes radical changes as $\boldsymbol{\theta}$ moves from $\mathring{\Theta}_n$ to the faces making up $\partial \Theta_n$. In particular, the preceding description makes it difficult to spot any structure which plays here the role of *model space* the way $\mathbb{R}^n$ does in the study of n-dimensional differentiable manifolds. On the other hand, as yet, we have not taken advantage of the basic property enjoyed by Θ_n: it is a simplex. In particular, we should expect that the structure of $T_{\boldsymbol{\theta}}(\Theta_n)$ for all $\boldsymbol{\theta} \in \Theta_n$ can be understood in terms of the structure of spaces $T_{\boldsymbol{\theta}}(\Theta_n)$ as $\boldsymbol{\theta}$ runs over the vertices $\mathbf{e}_0, \ldots, \mathbf{e}_{n-1}$ of Θ_n. This expectation is given substance by the following simple result.

---

[2] When differentiating a function f with respect to time, we will often use $\dot{f}$ to denote the derivative.

LEMMA 1.1.2. *For each* $0 \leq m \leq n - 1$, $\mathbf{a} \in T_{\mathbf{e}_m}(\Theta_n)$ *if and only if*

$$\left(\mathbf{a}, \mathbf{e}_\ell\right)_{\mathbb{R}^n} \geq 0 \quad \text{for each } \ell \neq m \text{ and } \left(\mathbf{a}, \mathbf{e}_m\right)_{\mathbb{R}^n} = -\sum_{\ell \neq m} \left(\mathbf{a}, \mathbf{e}_\ell\right)_{\mathbb{R}^n}.$$

Moreover, for any $\boldsymbol{\theta} \in \Theta_n$, $\mathbf{b} \in T_{\boldsymbol{\theta}}(\Theta_n)$ *if and only if there exist* $\mathbf{a}_\ell \in T_{\mathbf{e}_\ell}(\Theta_n)$, $0 \leq \ell \leq n - 1$, *such that*

$$(1.1.3) \qquad\qquad \mathbf{b} = \sum_{\ell=0}^{n-1} \left(\boldsymbol{\theta}, \mathbf{e}_\ell\right)_{\mathbb{R}^n} \mathbf{a}_\ell.$$

PROOF: The first assertion is a special case of Lemma 1.1.1. In proving the second part, we use the notation $\theta_\ell = \left(\boldsymbol{\theta}, \mathbf{e}_\ell\right)_{\mathbb{R}^n}$, $b_m = \left(\mathbf{b}, \mathbf{e}_m\right)_{\mathbb{R}^n}$, and $a_{\ell,m} = \left(\mathbf{a}_\ell, \mathbf{e}_m\right)_{\mathbb{R}^n}$.

If $\mathbf{b} = \sum_{\ell=0}^{n-1} \theta_\ell \mathbf{a}_\ell$, where $\mathbf{a}_\ell \in T_{\mathbf{e}_\ell}\Theta_n$ for each $\ell \in \mathbb{Z}_n$, then $b_m = \sum_{\ell=0}^{n-1} \theta_\ell a_{\ell,m}$, and so

$$\sum_{m=0}^{n-1} b_m = \sum_{\ell=0}^{n-1} \theta_\ell \left(\sum_{m=0}^{n-1} a_{\ell,m} \right) = 0,$$

$$\theta_k = 0 \implies b_k = \sum_{\ell \neq k} \theta_\ell a_{\ell,k} \geq 0, \quad \text{and } \theta_k = 1 \implies b_k = a_{k,k} \leq 0.$$

Hence, $\mathbf{b} \in T_{\boldsymbol{\theta}}(\Theta_n)$.

To go the other way, we work by induction on $n \geq 2$. Thus, suppose that $\boldsymbol{\theta} \in \Theta_2$ and $\mathbf{b} \in T_{\boldsymbol{\theta}}\Theta_2$ are given. Clearly, either $\mathbf{b} = \mathbf{0}$, and there is nothing to do, or $b_0 > 0$ and $1 - \theta_0 > 0$, or $b_1 > 0$ and $1 - \theta_1 > 0$. Thus, we may and will assume that $b_1 > 0$ and $1 - \theta_1 > 0$, in which case we take $\mathbf{a}_0 = \frac{1}{1-\theta_1}\mathbf{b}$ and $\mathbf{a}_1 = \mathbf{0}$. Next, let $n \geq 3$ and assume the result for $n - 1$. Reasoning as before, we reduce to the case when $b_{n-1} > 0$ and $1 - \theta_{n-1} > 0$. Set

$$\hat{\boldsymbol{\theta}} = \frac{(\theta_0, \ldots, \theta_{n-2})}{1 - \theta_{n-1}} \quad \text{and} \quad \hat{\mathbf{b}} = \frac{\left(b_0 + \hat{\theta}_0 b_{n-1}, \ldots, b_{n-2} + \hat{\theta}_{n-2} b_{n-1}\right)}{1 - \theta_{n-1}},$$

and observe that $\hat{\boldsymbol{\theta}} \in \Theta_{n-1}$ and $\hat{\mathbf{b}} \in T_{\hat{\boldsymbol{\theta}}}\Theta_{n-1}$. Using the induction hypothesis, find $\hat{\mathbf{a}}_\ell \in T_{\mathbf{e}_\ell}\Theta_{n-1}$ so that $\hat{\mathbf{b}} = \sum_{\ell=0}^{n-2} \hat{\theta}_\ell \hat{\mathbf{a}}_\ell$. Next, take $\mathbf{a}_\ell = \mathbf{0}$ if $\ell = n-1$ or $\theta_\ell = 0$, and, when $0 \leq \ell \leq n - 2$ and $\theta_\ell > 0$, take

$$\mathbf{a}_\ell = \left(\hat{\mathbf{a}}_\ell, 0\right) - \frac{b_{n-1}}{1 - \theta_{n-1}}\mathbf{e}_\ell + \frac{b_{n-1}}{1 - \theta_{n-1}}\mathbf{e}_{n-1}.$$

It is then an easy matter to check that $\mathbf{a}_\ell \in T_{\mathbf{e}_\ell}\Theta_n$ for each $0 \leq \ell \leq n - 1$ and $\mathbf{b} = \sum_{\ell=0}^{n-1} \theta_\ell \mathbf{a}_\ell$. $\square$

§1.1.2. Back to $\mathbf{M}_1(\mathbb{Z}_n)$: We now transfer the above analysis back to $\mathbf{M}_1(\mathbb{Z}_n)$. Namely, using $\mu \in \mathbf{M}_1(\mathbb{Z}_n) \longmapsto \boldsymbol{\theta}_\mu \in \Theta_n$ to denote the inverse of the map $\boldsymbol{\theta} \in \Theta_n \longmapsto \mu_{\boldsymbol{\theta}} \in \mathbf{M}_1(\mathbb{Z}_n)$, we say that the curve $t \in [0, \infty) \longmapsto \mu_t \in \mathbf{M}_1(\mathbb{Z}_n)$ is *differentiable* if $t \in [0, \infty) \longmapsto \boldsymbol{\theta}(t) \equiv \boldsymbol{\theta}_{\mu_t} \in \Theta_n$ is differentiable. In order to describe the tangent corresponding to such a curve, first note that, since $\boldsymbol{\theta}_\mu = \big(\mu(\{0\}), \ldots, \mu(\{n-1\})\big)$, an equivalent formulation of the preceding is to say that that, for each $\varphi \in C(\mathbb{Z}_n; \mathbb{R})$, the map[3] $t \in [0, \infty) \longmapsto \langle \varphi, \mu_t \rangle \in \mathbb{R}$ is differentiable. Thus, it is reasonable to identify the corresponding tangent $\dot\mu_0$ as the linear functional on $C(\mathbb{Z}_n; \mathbb{R})$ given by

$$(1.1.4) \qquad\qquad \dot\mu_0 \varphi = \lim_{h \searrow 0} \frac{\langle \varphi, \mu_h \rangle - \langle \varphi, \mu_0 \rangle}{h}.$$

Clearly, the relationship between $\dot\mu_0$ and $\dot{\boldsymbol{\theta}}(0)$ is

$$\dot\mu_0 \varphi = \sum_{m=0}^{n-1} \varphi(m)\big(\dot{\boldsymbol{\theta}}(0), \mathbf{e}_m\big)_{\mathbb{R}^n}.$$

Hence Λ_μ is an element of the space $T_\mu\big(\mathbf{M}_1(\mathbb{Z}_n)\big)$ of tangent vectors to $\mathbf{M}_1(\mathbb{Z}_n)$ at $\mu \in \mathbf{M}_1(\mathbb{Z}_n)$ if and only if (cf. Lemma 1.1.1)

$$(1.1.5) \quad \Lambda_\mu \varphi = \sum_{m=0}^{n-1} \varphi(m)\big(\mathbf{b}, \mathbf{e}_m\big)_{\mathbb{R}^n}, \quad \varphi \in C(\mathbb{Z}_n; \mathbb{R}), \text{ for some } \mathbf{b} \in T_{\boldsymbol{\theta}_\mu}(\Theta_n).$$

Of course, this means that, like $T_{\boldsymbol{\theta}_\mu}(\Theta_n)$, $T_\mu\big(\mathbf{M}_1(\mathbb{Z}_n)\big)$ is a closed, convex cone, this time of linear functionals on $C(\mathbb{Z}_n; \mathbb{R})$. It also means (cf. Lemma 1.1.2) that each $\Lambda_\mu \in T_\mu\big(\mathbf{M}_1(\mathbb{Z}_n)\big)$ admits the representation

$$(1.1.6) \quad \Lambda_\mu = \int_{\mathbb{Z}_n} \Lambda_{\delta_x}\, \mu(dx), \quad \text{where } x \in \mathbb{Z}_n \longmapsto \Lambda_{\delta_x} \in T_{\delta_x}\big(\mathbf{M}_1(\mathbb{Z}_n)\big).$$

On the other hand, except when μ is a point mass, the representation in (1.1.6) is far from unique! In fact, as the proof of Lemma 1.1.2 makes abundantly clear, except at vertices there are innumerable ways of decomposing $\mathbf{b} \in T_{\boldsymbol{\theta}}\Theta_n$ into $\mathbf{a}_\ell$'s, and each decomposition leads to a different representation in (1.1.6).

REMARKS 1.1.7. Our derivation of (1.1.6) is a straight forward computation which fails to reveal the reason why we should have expected its validity. To understand its intuitive origin, one needs to know the interpretation of

---

[3] $\langle \varphi, \mu \rangle$ denotes the integral of the function φ with respect to the measure μ.

convergence of measures in terms of *coupling*.[4] For our purposes, suffice it to say that when ν is close to μ, then there exists a map $x \in \mathbb{Z}_n \longmapsto P(x, \cdot) \in \mathbf{M}_1(\mathbb{Z}_n)$ such that $\nu = \int P(x, \cdot) \mu(dx)$ and $P(x, \cdot)$ is close to δ_x for most (as measured by μ) $x \in \mathbb{Z}_n$. Hence, if $t \rightsquigarrow \mu_t$ is differentiable at $t = 0$, then it is reasonable to hope that $\mu_t = \int P_t(x, \cdot) \mu(dx)$ where, for each x, $t \rightsquigarrow P_t(x, \cdot)$ is differentiable at $t = 0$ and determines an element $\dot{P}_0(x, \cdot) \in T_{\delta_x} \mathbf{M}_1(\mathbb{Z}_n)$, which is, obviously, a candidate for Λ_{δ_x}. In addition, this picture helps to explain the lack of uniqueness alluded to above. Namely, there are lots of couplings, and each one has the potential to give a different choice of $x \rightsquigarrow \Lambda_{\delta_x}$ in (1.1.6). Unfortunately, a proof of (1.1.6) based on this line of reasoning eludes me. Nonetheless, as we will see in the next subsection, we can give a more *intrinsic* and revealing way to understand (1.1.6).

A second, and more mundane, remark deals with our notion of a tangent vector. Namely, we have implicitly adopted a particular coordinate system in which to describe tangent vectors. To be precise, our coordinates are the maps $\mu \rightsquigarrow \langle \varphi, \mu \rangle$ as φ runs through $C(\mathbb{Z}_n; \mathbb{R})$, and we have described a tangent vector by saying how it acts, as a directional derivative, on these coordinate maps. Knowing this, the action of a tangent vector on a more general function is determined the chain rule. That is,[5]

$$\dot{\mu}_0 F = \lim_{h \searrow 0} \frac{F(\mu_t) - F(\mu_0)}{h} = \sum_{m=0}^{n-1} \partial_{\theta m} \tilde{F}(\boldsymbol{\theta}_{\mu_0}) \dot{\mu}_0 \mathbf{1}_{\{m\}},$$

where $\tilde{F}(\boldsymbol{\theta}) \equiv F(\mu_{\boldsymbol{\theta}})$.

1.2 A More Intrinsic Approach

So far, all of our considerations have rested heavily on the relationship between $\mathbf{M}_1(\mathbb{Z}_n)$ and Θ_n. In order to prepare for the analogous considerations when the state space $\mathbb{Z}_n$ is replaced by $\mathbb{R}^n$, it is important for us to develop another, more intrinsic, way of thinking about tangent vectors to $\mathbf{M}_1(\mathbb{Z}_n)$.

Taking a hint from (1.1.6) and the first part of Remark 1.1.7, we begin by trying to develop a deeper understanding of the tangent spaces $T_{\delta_x} \mathbf{M}_1(\mathbb{Z}_n)$ to the point masses. For this purpose, we need to take care of a few preparations.

§1.2.1. **The Semigroup Structure on $\mathbf{M}_1(\mathbb{Z}_n)$:** All the contents of this section derive from the simple observation that $\mathbb{Z}_n$ admits a natural Abelian group structure, the one corresponding to addition modulo n, which induces an Abelian semigroup structure, the one corresponding to convolution, on

---

[4] An excellent account of this topic is given in §11.6 of [7].
[5] We use $\mathbf{1}_\Gamma$ to denote the indicator (a.k.a. the characteristic) function of Γ. That is, $\mathbf{1}_\Gamma(x)$ equals 1 or 0 depending on whether x is or is not an element of Γ.

$\mathbf{M}_1(\mathbb{Z}_n)$. That is, using $x_1 + x_2$ to denote addition modulo n in $\mathbb{Z}_n$, we define the convolution product $M_1 \star M_2$ of measures M_1 and M_2 on $\mathbb{Z}_n$ so that

$$\langle \varphi, M_1 \star M_2 \rangle = \iint_{\mathbb{Z}_n \times \mathbb{Z}_n} \varphi(x_1 + x_2)\, M_1(dx_1) M_2(dx_2),$$

and note that the restriction of this product to $\mathbf{M}_1(\mathbb{Z}_n)$ induces an Abelian semigroup structure there. Clearly, δ_0 is the identity element for this semigroup, and $\delta_{x+y} = \delta_x \star \delta_y$ for all $x,\, y \in \mathbb{Z}_n$.

Analysis of convolution products is expedited by the introduction of Fourier analysis. Thus, we introduce the Fourier functions

$$\mathfrak{e}_{\xi,n}(x) \equiv \exp\left(\frac{\sqrt{-1}\, 2\pi \xi x}{n}\right), \quad \xi \in \mathbb{Z}_n,$$

and, for given $\mu \in \mathbf{M}_1(\mathbb{Z}_n)$, define its Fourier transform $\hat{\mu} : \mathbb{Z}_n \longrightarrow \mathbb{C}$ by $\hat{\mu}(\xi) = \langle \mathfrak{e}_{\xi,n}, \mu \rangle$. If $\lambda_{\mathbb{Z}_n}$ is the normalized Haar (a.k.a., counting) measure on $\mathbb{Z}_n$, then it is an elementary matter to check that $\{\mathfrak{e}_{\xi,n} : \xi \in \mathbb{Z}_n\}$ forms an orthonormal basis in $L^2(\lambda_{\mathbb{Z}_n}; \mathbb{C})$. Hence, μ can be recovered from $\hat{\mu}$ via the formula

$$\mu(\{x\}) = \int_{\mathbb{Z}_n} \overline{\mathfrak{e}_{\xi,n}(x)} \hat{\mu}(\xi)\, \lambda_{\mathbb{Z}_n}(d\xi).$$

Of course, the reason for introducing Fourier analysis when studying convolution products is the identity $\widehat{\mu_1 \star \mu_2} = \widehat{\mu_1}\widehat{\mu_2}$.

§1.2.2. Infinitely Divisible Flows: There are certain curves on $\mathbf{M}_1(\mathbb{Z}_n)$ which play the role for the semigroup structure there that straight lines play in Euclidean geometry. That is, a ray in $\mathbb{R}^n$ is a smooth curve $t \in [0, \infty) \longmapsto \ell_t \in \mathbb{R}^{n-1}$ with the property that $\ell_{s+t} = \ell_s + \ell_t$ for all $s, t \geq 0$, and a (one-sided) straight line is a curve $t \in [0, \infty) \longmapsto p_t \in \mathbb{R}^n$ which satisfies $p_{s+t} = p_s + \ell_t$, $s, t \geq 0$, for some ray $t \rightsquigarrow \ell_t$.

The analog of a *ray* in $\mathbf{M}_1(\mathbb{R}^n)$ is a smooth curve $t \in [0, \infty) \longmapsto \lambda_t \in \mathbf{M}_1(\mathbb{R}^n)$ which satisfies the property that $\lambda_{s+t} = \lambda_s \star \lambda_t$. Thus, the analog of a *straight line* is a curve $t \rightsquigarrow \mu_t$ such that $\mu_{s+t} = \lambda_t \star \mu_s$, $s, t \geq 0$. For historical reasons, this sort of ray is called an *infinitely divisible flow*.

As the conjunction of part (i) in Exercise 1.2.9 with the following result shows, it is quite easy to characterize all the infinitely divisible flows on $\mathbb{Z}_n$ which start from δ_0.

THEOREM 1.2.1. *Given a curve $t \in [0, \infty) \longmapsto \lambda_t \in \mathbf{M}_1(\mathbb{Z}_n)$ which is differentiable at $t = 0$ and starts from δ_0, $t \rightsquigarrow \lambda_t$ is an infinitely divisible flow if and only if there is a measure M on $\mathbb{Z}_n$ such that $M(\{0\}) = 0$ and,* for each $t \in [0, \infty)$,

$$(1.2.2) \qquad \lambda_t = e^{-tM(\mathbb{Z}_n)} \sum_{j=0}^{\infty} \frac{t^j}{j!} M^{\star j},$$

where $M^{\star 0} \equiv \delta_0$ and $M^{\star j} \equiv M \star M^{\star(j-1)}$ for $j \geq 1$. In fact, M is uniquely determined by $M(\{0\}) = 0$ and

$$(1.2.3) \qquad \dot{\lambda}_0 \varphi = \int_{\mathbb{Z}_n} (\varphi(y) - \varphi(0)) \, M(dy), \quad \varphi \in C(\mathbb{Z}_n; \mathbb{R}).$$

PROOF: For M with $M(\{0\}) = 0$, it is easy to check that the $t \rightsquigarrow \lambda_t$ given by (1.2.2) is an infinitely differentiable flow starting at δ_0. Moreover, it is clear that M is related to $\dot{\lambda}_0$ via (1.2.3). In particular, for $x \neq 0$, we find that $M(\{x\}) = \dot{\lambda}_0 \mathbf{1}_{\{x\}}$, which means that M is uniquely determined by (1.2.3) plus the condition $M(\{0\}) = 0$. Hence, all that remains to check is that every infinitely divisible flow $t \rightsquigarrow \lambda_t$ starting from δ_0 admits an M with $M(\{0\}) = 0$ for which the representation in (1.2.2) holds.

Suppose that $t \rightsquigarrow \lambda_t$ is an infinitely divisible flow which is differentiable at $t = 0$ and starts from δ_0. Note that, because $\widehat{\lambda_{s+t}} = \widehat{\lambda_s} \widehat{\lambda_t}$,

$$\frac{d}{dt} \widehat{\lambda_t}(\xi) = f(\xi) \widehat{\lambda_t}(\xi),$$

where $f(\xi) \equiv \dot{\lambda}_0 \mathbf{e}_{\xi, n}$. Hence, because $\widehat{\lambda_0} \equiv 1$, $\widehat{\lambda_t} = e^{tf}$. But, by definition,

$$f(\xi) = \lim_{h \searrow 0} \frac{1}{h} \int_{\mathbb{Z}_n} (\mathbf{e}_{\xi, n}(y) - 1) \, \lambda_h(dy) = \int_{\mathbb{Z}_n} (\mathbf{e}_{\xi, n}(y) - 1) \, M(dy),$$

where $M(\{x\}) \equiv \dot{\lambda}_0 \mathbf{1}_{\{x\}} \geq 0$ for $x \in \mathbb{Z}_n \setminus \{0\}$ and $M(\{0\}) \equiv 0$. Next observe that, either $f \equiv 0$, in which case $\lambda_t = \delta_0$ for all $t \geq 0$ and (1.2.3) holds with $M = 0$, or $M(\mathbb{Z}_n) > 0$ and we can write

$$e^{tf} = e^{-tM(\mathbb{Z}_n)} \sum_{j=0}^{\infty} \frac{(M(\mathbb{Z}_n)t)^j}{j!} \left(\frac{\widehat{M}}{M(\mathbb{Z}_n)} \right)^{\star j},$$

which shows that e^{tf} is the Fourier transform of the right hand side of (1.2.2). $\square$

We say that a (non-negative) measure M on $\mathbb{Z}_n$ with $M(\{0\}) = 0$ is a *Lévy measure*.

§1.2.3. An Intrinsic Description of $T_{\delta_x}(\mathbf{M}_1(\mathbb{Z}_n))$: We first note that the semigroup structure on $\mathbf{M}_1(\mathbb{Z}_n)$ allows us to concentrate our attention on $T_{\delta_0}\mathbf{M}_1(\mathbb{Z}_n)$. Indeed, for any $x \in \mathbb{Z}_n$, $t \rightsquigarrow \mu_t$ is a smooth curve starting from δ_x if and only if $t \rightsquigarrow \mu_t \star \delta_{-x}$ is a smooth curve starting from δ_0. Hence, if $\tau_x : \mathbb{Z}_n \longrightarrow \mathbb{Z}_n$ denotes translation by x, then $\Lambda_{\delta_x} \in T_{\delta_x}\mathbf{M}_1(\mathbb{Z}_n)$ if and only if $\Lambda_{\delta_x} = (\tau_x)_*\Lambda_{\delta_0}$,[6] where $\Lambda_{\delta_0} = (\tau_{-x})_*\Lambda_{\delta_x} \in tT_{\delta_0}\mathbf{M}_1(\mathbb{Z}_n)$. That is,

$$(1.2.4) \qquad T_{\delta_x}\mathbf{M}_1(\mathbb{Z}_n) = \big\{(\tau_x)_*\Lambda_{\delta_0} : \Lambda_{\delta_0} \in T_{\delta_0}\mathbf{M}_1(\mathbb{Z}_n)\big\}.$$

We now turn our attention to $T_{\delta_0}\mathbf{M}_1(\mathbb{Z}_n)$, and our idea is to use infinitely divisible flows as our canonical choice of curves to represent elements of $T_{\delta_0}\mathbf{M}_1(\mathbb{Z}_n)$. Namely, we want to check that

$$(1.2.5) \qquad T_{\delta_0}\mathbf{M}_1(\mathbb{Z}_n) = \big\{\dot{\lambda}_0 : t \rightsquigarrow \lambda_t \text{ infinitely divisible with } \lambda_0 = \delta_0\big\}.$$

Indeed, for each infinitely divisible flow $t \rightsquigarrow \lambda_t$ with $\lambda_0 = \delta_0$, $\dot{\lambda}_0$ is certainly an element of $T_{\delta_0}\mathbf{M}_1(\mathbb{Z}_n)$. Conversely, given $\Lambda_{\delta_0} \in T_{\delta_0}\mathbf{M}_1(\mathbb{Z}_n)$, define $M(\{x\}) = \Lambda_{\delta_0}\mathbf{1}_{\{x\}}$ for $x \neq 0$ and $M(\{0\}) = 0$, and determine $t \rightsquigarrow \lambda_t$ by (1.2.2). Clearly, $\dot{\lambda}_0 = \Lambda_{\delta_0}$.

Although the preceding is satisfactory, there is another, somewhat better, way to reach the same conclusion. To wit, suppose that $t \rightsquigarrow \mu_t$ is a differentiable curve starting from δ_0 with $\dot{\mu}_0 = \Lambda_{\delta_0}$. For each $N \geq 1$, determine the infinitely divisible flow $t \rightsquigarrow \lambda_{t,N}$ by

$$\lambda_{t,N} = e^{-tN}\sum_{j=0}^{\infty}\frac{t^j}{j!}\big(N\mu_{1/N}\big)^{\star j},$$

and notice that

$$\widehat{\lambda_{t,N}}(\xi) = \exp\left(tN\int_{\mathbb{Z}_n}\big(\mathbf{e}_{\xi,n}(y) - 1\big)\mu_{1/N}(dy)\right) \longrightarrow \exp\big(t\Lambda_{\delta_0}\mathbf{e}_{\xi,n}\big).$$

As a consequence, one sees that, if $f(\xi) \equiv \Lambda_{\delta_0}\mathbf{e}_\xi$, then, for each $t \geq 0$, e^{tf} is the Fourier transform of a $\lambda_t \in \mathbf{M}_1(\mathbb{Z}_n)$, and clearly $t \rightsquigarrow \lambda_t$ is an infinitely divisible flow for which $\Lambda_{\delta_0} = \dot{\lambda}_0$.

Whatever way one chooses to arrive at it, the preceding considerations (especially, the last part of Theorem 1.2.1 and (1.2.4)) lead us to the following one-to-one correspondence between the tangent vectors Λ_{δ_x} at a point mass δ_x and the cone of Lévy measures M:

$$(1.2.6) \qquad \Lambda_{\delta_x}\varphi = \int_{\mathbb{Z}_n}\big(\varphi(x + y) - \varphi(x)\big)M(dy).$$

---

[6] We adopt the differential geometric convention of using f_* to denote the "pushforward" of a vector under a map f. Thus, $(\tau_x)_*\Lambda_{\delta_0}\varphi = \Lambda_{\delta_0}(\varphi \circ \tau_x)$.

§**1.2.4. An Intrinsic Approach to (1.1.6):** To complete our program, we must find an intrinsic way to rationalize (1.1.6). The key to our[7] proof is contained in the following.

LEMMA 1.2.7. *If* $t \rightsquigarrow \mu_t$ *is a differentiable curve on* $\mathbf{M}_1(\mathbb{Z}_n)$*, then there is an* $x \in \mathbb{Z}_n$ *such that* $\mu_0(\{x\}) > 0$ *and either* $\mu_0 = \delta_x$ *or*

$$\dot{\mu}_0\varphi = \mu_0(\{x\})\Lambda_{\delta_x}\varphi + \big(1 - \mu_0(\{x\})\big)\Lambda_{\nu_0}\varphi,$$

where $\Lambda_{\delta_x} \in T_{\delta_x}\mathbf{M}_1(\mathbb{Z}_n)$*,* $\nu_0 \in \mathbf{M}_1(\mathbb{Z}_n)$ *satisfies* $\nu_0(\{x\}) = 0$*, and* $\Lambda_{\nu_0} \in T_{\nu_0}\mathbf{M}_1(\mathbb{Z}_n)$*.*

PROOF: If either $\dot{\mu}_0 = 0$ or $\mu_0(\{x\}) = 1$ for some $x \in \mathbb{Z}_n$, then there is nothing to do. Thus, we will assume that $\dot{\mu}_0 \neq 0$ and that $\mu_0(\{x\}) < 1$ for all $x \in \mathbb{Z}_n$.

Now, because $\sum_{x \in \mathbb{Z}_n} \dot{\mu}_0 \mathbf{1}_{\{x\}} = 0$, we know that there is an $x \in \mathbb{Z}_n$ such that $\dot{\mu}_0 \mathbf{1}_{\{x\}} < 0$. Hence, we can find a $T > 0$ such that $\mu_t(\{x\}) < \mu_0(\{x\})$ for all $t \in (0, T]$. In particular, we can define $\mu_t' \in \mathbf{M}_1(\mathbb{Z}_n)$ for $t \in [0, T)$ so that

$$\mu_0(\{x\})\mu_t'(\{y\}) = \begin{cases} \mu_t(\{x\}) & \text{if } y = x \\ \frac{\mu_0(\{x\}) - \mu_t(\{x\})}{1 - \mu_t(\{x\})}\mu_t(\{y\}) & \text{if } y \neq x. \end{cases}$$

Clearly, for each $t \in [0, T)$, $\mu_t - \mu_0(\{x\})\mu_t' = \big(1 - \mu_0(\{x\})\big)\nu_t$ where

$$\big(1 - \mu_t(\{x\})\big)\nu_t(\{y\}) = \begin{cases} 0 & \text{if } y = x \\ \mu_t(\{y\}) & \text{if } y \neq x. \end{cases}$$

Hence, $\dot{\mu}_0 = \mu_0(\{x\})\dot{\mu}_0' + \big(1 - \mu_0(\{x\})\big)\dot{\nu}_0$, which is a decomposition of the required sort. □

With Lemma 1.2.7 at hand, it is now clear how one can "peel off" the Λ_{δ_x}'s which enter (1.1.6).

§**1.2.5. Exercises**

EXERCISE 1.2.8. The major source of our problems with the tangent space to $\mathbf{M}_1(\mathbb{Z}_n)$ at a μ which is not a point mass is that $\mathbf{M}_1(\mathbb{Z}_n)$ is only a semigroup and not a group. Indeed, show that $\mu \in \mathbf{M}_1(\mathbb{Z}_n)$ admits an inverse (i.e., a $\nu \in \mathbf{M}_1(\mathbb{Z}_n)$ such that $\mu * \nu = \delta_0$) if and only if $\mu = \delta_x$ for some $x \in \mathbb{Z}_n$, in which case its inverse is δ_{-x}.

---

[7] The reasoning here profited from several conversations that I had with Balint Virag on the topic.

EXERCISE 1.2.9. Here are a few easy applications of Fourier analysis.

(i) Show that $\hat{\mu} : \mathbb{Z}_n \longrightarrow \{0,1\}$ if and only if μ is the normalized Haar measure for a subgroup of $\mathbb{Z}_n$. In particular, check that $\hat{\mu} \equiv 1$ if and only if $\mu = \delta_0$ and $\hat{\mu} = \mathbf{1}_{\{0\}}$ if and only if $\mu = \lambda_{\mathbb{Z}_n}$.

(ii) If $t \rightsquigarrow \lambda_t$ is an infinitely divisible flow, show that λ_0 must be the normalized Haar measure for some subgroup of $\mathbb{Z}_n$. If, in addition, $t \rightsquigarrow \lambda_t$ is differentiable at $t = 0$, show that $\lambda_t = \lambda_0 \star \lambda_t'$, where $t \rightsquigarrow \lambda_t'$ is an infinitely divisible flow starting at δ_0.

(iii) Say that μ admits a square root if $\mu = \nu * \nu$ for some $\nu \in \mathbf{M}_1(\mathbb{Z}_n)$. Show that $\mu \in \mathbf{M}_1(\mathbb{Z}_2)$ admits a square root if and only if $\mu = p\delta_0 + (1-p)\delta_1$, where $p \in \left[\frac{1}{2}, 1\right]$, in which case one can take $\nu = q\delta_0 + (1 - q)\delta_1$, where $2q - 1 = \sqrt{2p - 1}$.

EXERCISE 1.2.10. There is another, entirely different, way to see that every tangent to $\mathbf{M}_1(\mathbb{Z}_n)$ at δ_x is of the form in (1.2.6). Namely, show that if $\Lambda_{\delta_x} \in T_{\delta_x}\mathbf{M}_1(\mathbb{Z}_n)$, then Λ_{δ_x} annihilates constant functions and satisfies the *minimum principle* at x in the sense that $\Lambda_{\delta_x}\varphi \geq 0$ whenever φ takes its minimum value at x. Conclude from this that the action of Λ_{δ_x} on $C(\mathbb{Z}_n; \mathbb{R})$ must admit the representation given in (1.2.6). (See Lemma 2.1.8 for ideas.)

1.3 Vector Fields and Integral Curves on $\mathbf{M}_1(\mathbb{Z}_n)$

Obviously, a vector field on $\mathbf{M}_1(\mathbb{Z}_n)$ should be a map which takes a $\mu \in \mathbf{M}_1(\mathbb{Z}_n)$ to a $\Lambda_\mu \in T_\mu(\mathbf{M}_1(\mathbb{Z}_n))$, and, at the very least, we will require our vector fields to be continuous in the sense that $\mu \rightsquigarrow \Lambda_\mu \varphi$ should be continuous for each $\varphi \in C(\mathbb{Z}_n; \mathbb{R})$.

By combining (1.1.6) with (1.2.6), we know that each Λ_μ map must admit the representation

$$(1.3.1) \qquad \Lambda_\mu \varphi = \int_{\mathbb{Z}_n} \left(\int_{\mathbb{Z}_n} \left(\varphi(x + y) - \varphi(x)\right) M\big((x,\mu), dy\big) \right) \mu(dx)$$

in terms of a map $x \rightsquigarrow M\big((x,\mu), \cdot\big)$ from $\mathbb{Z}_n$ into Lévy measures. However, as we pointed out right after (1.1.6), $x \rightsquigarrow M\big((x,\mu), \cdot\big)$ will not, in general, be determinable from Λ_μ except when μ is a point mass. To avoid this ambiguity, we will adopt the convention that specifying a *vector field* on $\mathbf{M}_1(\mathbb{Z}_n)$ means giving a continuous map $(x,\mu) \rightsquigarrow M\big((x,\mu), \cdot\big)$ from $\mathbb{Z}_n \times \mathbf{M}_1(\mathbb{Z}_n)$ into Lévy measures. With such a map we associate the vector field $\mu \rightsquigarrow \Lambda_\mu$ given by (1.3.1).

§1.3.1. **Affine and Translation Invariant Vector Fields:** It is important to recognize that there is a significant subclass of vector fields on

$\mathbf{M}_1(\mathbb{Z}_n)$ for which the ambiguity alluded to above disappears. Namely, we say that the vector field Λ is *affine* if

$$\Lambda_{((1-\theta)\mu+\theta\nu)}\varphi = (1-\theta)\Lambda_\mu\varphi + \theta\Lambda_\nu\varphi$$
$$\text{for } \theta \in [0,1], \ \mu, \nu \in \mathbf{M}_1(\mathbb{Z}_n), \text{ and } \varphi \in C(\mathbb{Z}_n; \mathbb{R}).$$

Obviously, Λ is affine if and only if the associated $\mu \rightsquigarrow M\big((x, \mu), \cdot\big)$ in (1.3.1) can be taken independent of μ. That is, Λ is affine if and only if

$$(1.3.2) \qquad \Lambda_\mu\varphi = \int_{\mathbb{Z}_n} \left(\int_{\mathbb{Z}_n} \big(\varphi(x+y) - \varphi(x)\big) M(x, dy) \right) \mu(dx)$$

for some choice of $x \in \mathbb{Z}_n \rightsquigarrow M(x, \cdot)$. Specializing even further, notice that among the affine vector fields are the vector fields which are *translation invariant* in the sense that

$$(1.3.3) \qquad \Lambda_{\mu\star\nu} = \int_{\mathbb{Z}_n} (\tau_x)_*\Lambda_\nu\, \mu(dx), \quad \mu, \nu \in \mathbf{M}_1(\mathbb{Z}_n).$$

Indeed, because $\mu = \mu \star \delta_0$, (1.3.3) implies that $\Lambda_\mu = \int_{\mathbb{Z}_n} (\tau_x)_*\Lambda_{\delta_0}\, \mu(dx)$, from which it is clear that Λ is affine and the associated $x \rightsquigarrow M(x, \cdot)$ is constant. That is, we have shown that Λ is translation invariant if and only if there exists a unique Lévy measure M such that, for all $\mu \in \mathbf{M}_1(\mathbb{Z}_n)$,

$$\Lambda_\mu\varphi = \int_{\mathbb{Z}_n} \left(\int_{\mathbb{Z}_n} \big(\varphi(x+y) - \varphi(x)\big) M(dy) \right) \mu(dx), \quad \varphi \in C(\mathbb{Z}_n; \mathbb{R}).$$

§1.3.2. Existence of an Integral Curve: We will say that $t \in [0, \infty) \longmapsto$ $\mu_t \in \mathbf{M}_1(\mathbb{Z}_n)$ is an *integral curve* of the vector field Λ starting from μ if

$$\langle \varphi, \mu_t \rangle - \langle \varphi, \mu \rangle = \int_0^t \Lambda_{\mu_s}\varphi \, ds, \quad (t, \varphi) \in [0, \infty) \times C(\mathbb{Z}_n; \mathbb{R}).$$

In this subsection we will address the problem of proving that a vector field Λ admits at least one integral curve. To get started, notice that there is nothing to do when Λ is translation invariant. Indeed, simply observe if $t \rightsquigarrow \lambda_t$ is the infinitely divisible flow with $\dot{\lambda}_0 = \Lambda_{\delta_0}$, then, for each $\mu \in \mathbf{M}_1(\mathbb{Z}_n)$, "the straight line" $t \rightsquigarrow \lambda_t \star \mu = \int (\tau_x)_*\lambda_t\, \mu(dx)$ is an integral curve of Λ starting at μ.

To take the next step, we try the analog of the Euler approximation scheme for solving ordinary differential equations. Namely, Euler's scheme for integrating vector fields on $\mathbb{R}^n$ is to create a polygonal approximation by breaking time into short intervals, looking where you are at the beginning of each interval, and, during the interval, following the straight line whose tangent is the value which the vector field takes at the beginning of the interval. In the setting of $\mathbf{M}_1(\mathbb{Z}_n)$, we know that the role of rays is played by infinitely divisible flows. Thus, the only new complication here is the fact that, when μ is nondegenerate, one has to proceed along a randomly distributed straight line. However, as we are about to see, this complication is easily handled. To wit, let Λ be a general vector field, and suppose that $(x, \mu) \rightsquigarrow M((x, \mu), \cdot)$ determines Λ. For each (x, μ), use $t \rightsquigarrow (\lambda^x_\mu)_t$ to denote the infinitely divisible flow with $(\dot{\lambda}^x_\mu)_0 \varphi = \int (\varphi(y) - \varphi(x)) M((x, \mu), dy)$. Now let μ be given, and, for each $N \in \mathbb{N}$, define $t \rightsquigarrow \mu^N_t$ so that

$$(1.3.4) \qquad \mu^N_0 = \mu \text{ and } \mu^N_t = \int_{\mathbb{Z}_n} (\tau_x)_* (\lambda^x_{\mu^N_{m2^{-N}}})_{t-m2^{-N}} \mu^N_{m2^{-N}}(dx)$$

when $m2^{-N} \leq t < (m+1)2^{-N}$. Because $(x, \mu) \rightsquigarrow M((x, \mu), \cdot)$ is continuous and $\mathbb{Z}_n \times \mathbf{M}_1(\mathbb{Z}_n)$ is compact, it is an easy matter to check that, as a subset of $C([0, \infty); \mathbf{M}_1(\mathbb{Z}_n))$, the sequence $\{\mu^N : N \geq 0\}$ is equicontinuous and therefore relatively compact. Moreover, for each $\varphi \in C(\mathbb{Z}_n; \mathbb{R})$ and $m2^{-N} \leq t < (m+1)2^{-N}$, $\frac{d}{dt} \langle \varphi, \mu^N_t \rangle$ is equal to

$$\int_{\mathbb{Z}_n} \left(\int_{\mathbb{Z}_n} \left(\int_{\mathbb{Z}_n} \left(\varphi(x+y+z) - \varphi(x+y) \right) \right. \right.$$
$$\left. \left. \times M((x, \mu^N_{m2^{-N}}), dz) \right) (\lambda^x_{\mu^N_{m2^{-N}}})_{t-m2^{-N}}(dy) \right) \mu^N_{m2^{-N}}(dx),$$

from which it should be clear that any limit of $\{\mu^N : N \geq 0\}$ will have to be an integral curve of Λ. Hence, after putting this together with the preceding remarks about relative compactness, we have now proved the following statement.

LEMMA 1.3.5. *For each vector field Λ on $\mathbf{M}_1(\mathbb{Z}_n)$ and $\mu \in \mathbf{M}_1(\mathbb{Z}_n)$, there is an integral curve of Λ which starts at μ.*

§1.3.3. **Uniqueness for Affine Vector Fields:** Anyone familiar with the uniqueness problems which arise in the study of ordinary differential equations will realize that there is no general uniqueness statement to complement Lemma 1.3.5. On the other hand, because (cf. Exercise 1.3.15 below) all questions about integrating vector fields in this setting can be expressed in terms of a relatively simple system of ordinary differential equations, it should come as no surprise that uniqueness is guaranteed by the Lipschitz continuity condition[8]

$$(1.3.6) \qquad \max_{x \in \mathbb{Z}_n} \text{var}\left(M((x, \mu), \cdot) - M((x, \nu), \cdot) \right) \leq C \, \text{var}(\mu - \nu).$$

---

[8] Here, and elsewhere, var(μ) is the variation norm of the measure ν.

However, because the proof relies so heavily on the present setting, it has been relegated to an exercise so that we can concentrate here on an argument which will admit a ready analog when $\mathbb{Z}_n$ is replaced by $\mathbb{R}^n$. Namely, we will take advantage of the general principle that *uniqueness and existence are dual properties*. (Cf. Remark 1.3.14 below.) Unfortunately, this argument is useful only when the vector field is affine.

THEOREM 1.3.7. *Given a map $x \rightsquigarrow M(x, \cdot)$ from $\mathbb{Z}_n$ into Lévy measures, let Λ be the associated affine vector vector field on $\mathbf{M}_1(\mathbb{Z}_n)$. Then, for each $\varphi \in C(\mathbb{Z}_n; \mathbb{R})$ there is a $u \in C\big([0, \infty) \times \mathbb{Z}_n; \mathbb{R}\big)$ with the properties that*

$$\dot{u}(t, x) = \int \big(u(t, x + y) - u(t, x)\big) M(x, dy) \quad \text{and} \quad u(0, \cdot) = \varphi.$$

Moreover, if $t \rightsquigarrow \mu_t$ is an integral curve of Λ, then $\langle \varphi, \mu_T \rangle = \langle u(T, \cdot), \mu_0 \rangle$. In particular, there is only one integral curve of Λ starting from any given $\mu \in \mathbb{Z}_n$.

PROOF: To emphasize the basic duality principle alluded to above, we begin by showing why, if u exists for a given φ, $\langle \varphi, \mu_T \rangle = \langle u(T, \cdot), \mu_0 \rangle$ for any integral curve $t \rightsquigarrow \mu_t$. To this end, simply observe that $\frac{d}{dt}\langle \varphi_{T-t}, \mu_t \rangle = 0$ for $t \in (0, T)$, and so

$$\langle \varphi, \mu_T \rangle - \langle u(T), \mu_0 \rangle = \langle u(T - t), \mu_t \rangle \Big|_{t=0}^{T} = 0.$$

To produce u, one can proceed as follows. Define the $n \times n$ matrix Q so that

$$Q_{k\ell} = \begin{cases} M(k, \{\ell - k\}) & \text{if } \ell \neq k \\ -\sum_{\ell \neq k} Q_{k,\ell} & \text{if } \ell = k; \end{cases}$$

and, given $\varphi \in C(\mathbb{Z}_n; \mathbb{R})$, set

$$u(t, k) \equiv \sum_{\ell=0}^{n-1} \left(e^{tQ}\right)_{k\ell} \varphi(\ell).$$

Obviously, $u(0, \cdot) = \varphi$, and, because $\frac{d}{dt} e^{tQ} = Q e^{tQ}$, one can easily check that $\dot{u}(t, x) = \int \big(u(t, x + y) - u(t, x)\big) M(x, dy)$. $\square$

REMARK 1.3.8. It should be obvious that the duality principle works both ways. Namely, existence of an integral curve $t \rightsquigarrow \mu_t$ means that, for any solution u, $t \rightsquigarrow \langle u(t, \cdot), \mu_0 \rangle = \langle u(0, \cdot), \mu_t \rangle$.

§1.3.4. The Markov Property and Kolmogorov's Equations: Let Λ be an affine vector field on $\mathbf{M}_1(\mathbb{Z}_n)$. For each $x \in \mathbb{Z}_n$, let $t \rightsquigarrow P_t(x, \cdot)$ be the integral curve of Λ starting from δ_x. As a corollary of uniqueness, we get the *Markov property*

$$(1.3.9) \qquad \mu_{s+t} = \int_{\mathbb{Z}_n} P_t(y, \cdot)\, \mu_s(dy), \quad s, t \in [0, \infty)$$

for integral curves $t \rightsquigarrow \mu_t$ of Λ. Indeed, all that one has to do is note that, as functions of t, both sides are integral curves of Λ starting at μ_s. Of course, when $\mu_0 = \delta_x$, (1.3.9) becomes the *Chapman–Kolmogorov equation*:

$$(1.3.10) \qquad P_{s+t}(x, \cdot) = \int_{\mathbb{Z}_n} P_t(y, \cdot)\, P_s(x, dy), \quad s, t \geq 0 \text{ and } x \in \mathbb{Z}_n.$$

Because (1.3.9) determines how integral curves of Λ evolve, the map $(t, x) \rightsquigarrow P_t(x, \cdot)$ is called the *transition probability function determined by* Λ.

Of course, when Λ is translation invariant, uniqueness shows that, for any integral curve $t \rightsquigarrow \mu_t$ of Λ,

$$(1.3.11) \qquad \mu_{s+t} = \lambda_t \star \mu_s, \quad s, t \in [0, \infty),$$

where $t \rightsquigarrow \lambda_t$ is the infinitely divisible flow whose derivative at $t = 0$ is Λ_{δ_0}. In particular, $P_t(x, \cdot) = \lambda_t \star \delta_x$ when the vector field is translation invariant. Using the terminology introduced in §1.2.2, what we are saying is that *the integral curve of a translation invariant vector field is a straight line*, and in this sense translation invariant vector fields are the vector fields on $\mathbf{M}_1(\mathbb{R}^n)$ with *constant coefficients*. Notice that this way of thinking is completely consistent with our assertion in §1.3.2 that the proof of Lemma 1.3.5 is an Euler approximation scheme.

Having, in the preface to this chapter, advertised this whole development as an outgrowth of Itô's attempt to explain Kolmogorov's forward and backward equations, it is only right that we now show in what sense it does. For this purpose, suppose that $x \rightsquigarrow M(x, \cdot)$ determines the affine vector field Λ, and define the operator L on $C(\mathbb{Z}_n; \mathbb{C})$ so that

$$L\varphi(x) = \Lambda_{\delta_x}\varphi = \int_{\mathbb{Z}_n} \big(\varphi(x + y) - \varphi(x)\big)\, M(x, dy), \quad x \in \mathbb{Z}_n.$$

Then the integral curve $t \rightsquigarrow \mu_t$ of Λ starting at μ is characterized by the properties that $\mu_0 = \mu$ and $\partial_t \langle \varphi, \mu_t \rangle = \langle L\varphi, \mu_t \rangle$ for all $\varphi \in C(\mathbb{Z}_n; \mathbb{C})$. Thinking of $\mathbf{M}_1(\mathbb{Z}_n)$ as a subset of the dual space $C(\mathbb{Z}_n; \mathbb{C})^*$ of $C(\mathbb{Z}_n; \mathbb{C})$ and using L^*

to denote the adjoint of L on $C(\mathbb{Z}_n; \mathbb{C})^*$, one can summarize these properties in the statement that

$$(1.3.12) \qquad\qquad \dot{\mu}_t = L^*\mu_t \quad \text{with } \mu_0 = \mu,$$

which is *Kolmogorov's forward equation*. By taking $\varphi = \mathbf{1}_{\{y\}}$, one obtains a more explicit form of (1.3.12):

$$\partial_t\mu_t(\{y\}) = \sum_{z\in\mathbb{Z}_n}\Big(\mu_t(\{y-z\})M(y-z, \{z\}) - \mu_t(\{y\})M(y, \{z\})\Big),$$

which, when applied to $t \rightsquigarrow P_t(x, \cdot)$ explains the origin of the name "forward." Namely, if one thinks of $P_t(x, \{y\})$ as the conditional probability of being at y at time t having started from x at time 0, then y is the *forward variable*: the one which is observed when looking "forward" in time, whereas x is the *backward variable*, the one observed when looking "backward" in time. Since the forward equation describes the evolution of $t \rightsquigarrow P_t(x, \cdot)$ in terms of the forward variable, it is natural that it should be known as the forward equation.

In order to derive Kolmogorov's backward equation, remember that μ_t is given by (1.3.9) and apply (1.3.10) to arrive first at

$$\langle P_{t+h}(\cdot, \{y\}), \mu\rangle = \langle P_t(\cdot, \{y\}), \mu_h\rangle$$

and then at

$$\langle \partial_t P_t(\cdot, \{y\}), \mu\rangle = \langle L P_t(\cdot, \{y\}), \mu\rangle$$

for all $(t, \mu) \in [0, \infty) \times \mathbf{M}_1(\mathbb{Z}_n)$. Hence, we have now derived *Kolmogorov's backward equation*

$$(1.3.13) \qquad\qquad \partial_t P_t(\cdot, \{y\}) = L P_t(\cdot, \{y\})$$

which describes the evolution of $t \rightsquigarrow P_t(\cdot, \{y\})$ in terms of the backward variable.

REMARK 1.3.14. For anyone familiar with elementary functional analysis, it should be apparent that the distinction between the backward and forward equations is the distinction between the evolution equation determined by the operator L and its adjoint L^*. In particular, our proof of uniqueness in Theorem 1.3.7 is an example of the duality principle that existence for the L-evolution equation implies uniqueness for the L^*-evolution equation.

§1.3.5. Exercises

EXERCISE 1.3.15. The purpose of this exercise is to explain the remark made at the beginning of §1.3.3. That is, let Λ be the vector field on $\mathbf{M}_1(\mathbb{Z}_n)$ determined by the map $(x, \mu) \rightsquigarrow M\big((x, \mu), \cdot\big)$, and define the matrix valued function $\boldsymbol{\theta} \in \Theta_n \longmapsto Q(\boldsymbol{\theta}) \in \mathrm{Hom}(\mathbb{R}^n; \mathbb{R}^n)$ so that

$$Q(\boldsymbol{\theta})_{k\ell} = \begin{cases} M\big((k, \mu_{\boldsymbol{\theta}}), \{\ell - k\}\big) & \text{if } \ell \neq k \\ -M\big((k, \mu_{\boldsymbol{\theta}}, \mathbb{Z}_n\big) & \text{if } \ell = k. \end{cases}$$

Next, define $\mathbf{u} \in \mathbb{R}^n \setminus \{\mathbf{0}\} \longmapsto \boldsymbol{\theta}(\mathbf{u}) \in \Theta_n$ by

$$\boldsymbol{\theta}(\mathbf{u}) = \left(\frac{|u_0|}{\sum_0^{n-1} |u_m|}, \dots, \frac{|u_{n-1}|}{\sum_0^{n-1} |u_m|} \right),$$

and choose a function $\eta \in C^\infty\big(\mathbb{R}^n; [0, 1]\big)$ so that $\eta(\mathbf{u}) = 0$ if $\sum_0^{n-1} |u_m| \leq \frac{1}{2}$ and $\eta(\mathbf{u}) = 1$ if if $\sum_0^{n-1} |u_m| \geq \frac{3}{4}$. Finally, extend Q to the whole of $\mathbb{R}^n$ by taking $Q(\mathbf{0}) = 0$ and $Q(\mathbf{u}) = \eta(\mathbf{u})Q\big(\boldsymbol{\theta}(\mathbf{u})\big)$ when $\mathbf{u} \in \mathbb{R}^n \setminus \{\mathbf{0}\}$.

(i) If $t \rightsquigarrow \mu_t$ is an integral curve of Λ, let $\mathbf{u}(t)$ be the row vector whose kth entry is $\mu_t(\{k\})$ and show that

(*) $$\dot{\mathbf{u}}(t) = \mathbf{u}(t)Q\big(\mathbf{u}(t)\big), \quad t \in [0, \infty).$$

(ii) Suppose that $t \rightsquigarrow \mathbf{u}(t)$ is a solution to (*), and show that $\sum_0^{n-1} u(t)_k = \sum_0^{n-1} u(0)_k$ for all $t \geq 0$.

(iii) Again suppose that $t \rightsquigarrow \mathbf{u}(t)$ solves (*). If $\mathbf{u}(0) \in \Theta_n$, show that $\mathbf{u}(t) \in \Theta_n$ for all $t \geq 0$.

Hint: In view of part (ii), it suffices to prove that each entry $u(t)_k$ of $\mathbf{u}(t)$ stays non-negative. To this end, first note that

$$\dot{u}(t)_k = \sum_{\ell \neq k} Q(t)_{\ell,k}\big(u(t)_\ell - u(t)_k\big) + V(t)_k u(t)_k, \quad (t, k) \in [0, \infty) \times \mathbb{Z}_n,$$

where $Q(t) \equiv Q\big(\mathbf{u}(t)\big)$ and $V(t)_k \equiv \sum_{\ell \neq k}\big(Q(t)_{\ell,k} - Q(t)_{k,\ell}\big)$. Next, let $T > 0$ be given, set $1 + \beta_T = \min_{(t,k) \in [0,T] \times \mathbb{Z}_n} V(t)_k$, define $\mathbf{w}(t) = e^{-\beta_T t}\mathbf{u}(t)$, and observe that

$$\dot{w}(t)_k = \sum_{\ell \neq k} Q_{\ell,k}(t)\big(w(t)_\ell - w(t)_k\big) + \big(V(t) - \beta_T\big)w(t)_k \quad \text{for } (t, k) \in [0, T] \times \mathbb{Z}_n.$$

Now suppose that $u(t)_k < 0$ for some $t \in [0, T]$ and $k \in \mathbb{Z}_n$, and conclude that there would exist some $(t, k) \in (0, T] \times \mathbb{Z}_n$ such that $u(t)_k < 0$. Hence, there would exist a $(t_0, k_0) \in (0, T] \times \mathbb{Z}_n$ for which $w(t_0)_{k_0} = \min_{(t,k) \in [0,T] \times \mathbb{Z}_n} w(t)_k < 0$. Finally, show that this would lead to the contradiction that $0 \geq \dot{w}(t_0)_{k_0} \geq -w(t_0)_{k_0} > 0$.

(iv) Assuming that (1.3.6) holds, show that, for each $\mu \in \mathbf{M}_1(\mathbb{Z}_n)$, there is precisely one solution $t \rightsquigarrow \mathbf{u}(t)$ to (*) with $\mathbf{u}(0) = \boldsymbol{\theta}_\mu$. As a consequence, conclude $t \rightsquigarrow \mu_{\mathbf{u}(t)}$ is the one and only integral curve of Λ starting at μ.

(v) Sharpen the reasoning in the **Hint** for (ii) to see that, for each $T > 0$, $\mu_t(\{k\}) \geq e^{\beta_T t} \min_{\ell \in \mathbb{Z}_n} \mu_0(\{\ell\})$ for all $(t, k) \in [0, T] \times \mathbb{Z}_n$.

EXERCISE 1.3.16. As we will see in the next section, there are circumstances when it is important to deal with affine vector fields which are time dependent. That is, one is looking at a map taking $t \in [0, \infty)$ into an affine vector field $\Lambda(t)$ on $\mathbf{M}_1(\mathbb{Z}_n)$, or, equivalently, a map $(t, x) \rightsquigarrow M\big((t, x), \cdot\big)$ taking $[0, \infty) \times \mathbb{Z}_n$ into Lévy measures. Assume that $(t, x) \in [0, \infty) \times \mathbb{Z}_n \longmapsto M\big((t, x), \{y\}\big) \in [0, \infty)$ is continuous for all $y \in \mathbb{Z}_n$.

(i) Show that a minor modification of the arguments given in §§1.3.2–1.3.3 lead to the statement that, for each $(s, \mu) \in [0, \infty) \times \mathbf{M}_1(\mathbb{Z}_n)$ there is a unique differentiable curve $t \rightsquigarrow \mu_t^s$ with the properties that $\mu_0^s = \mu$ and $\dot{\mu}_t^s = \Lambda(s + t)_{\mu_t^s}$. We will call $t \rightsquigarrow \mu_t^s$ the *integral curve of $t \rightsquigarrow \Lambda(t)$ starting from μ at time s*.

(ii) For each $(s, x) \in [0, \infty) \times \mathbb{Z}_n$, use $t \rightsquigarrow P_t\big((s, x), \cdot\big)$ to denote the integral curve starting from δ_x at time s. Show that for any integral curve $t \rightsquigarrow \mu_t^s$ which starts at time s,

$$\mu_t^s = \int_{\mathbb{Z}_n} P_t\big((s, x), \cdot\big) \mu_0^s(dx).$$

In particular, one has the *time inhomogeneous Chapman–Kolmogorov equation*

$$P_{t_1 + t_2}\big((s, x), \cdot\big) = \int_{\mathbb{Z}_n} P_{t_2}\big((s + t_1, y), \cdot\big) P_{t_1}\big((s, x), dy\big).$$

1.4 Pathspace Realization

In the preceding section we produced paths $t \rightsquigarrow \mu_t$ in $\mathbf{M}_1(\mathbb{Z}_n)$ by integrating vector fields there. In this section we will show that these $\mathbf{M}_1(\mathbb{Z}_n)$-valued integral curves correspond, in a reasonably canonical way, to probability measures $\mathbb{P}$ on the space of $\mathbb{Z}_n$-valued paths in such a way that μ_t is the $\mathbb{P}$-distribution $\mathbb{Z}_n$-valued path at time t.

Actually, if ones only goal is to construct such a measure $\mathbb{P}$, one can do so easily. Indeed, given any measurable space $(E, \mathcal{B})$ and any mapping taking $t \in [0, \infty)$ into a probability measure μ_t on $(E, \mathcal{B})$, one can take (cf. Exercise 1.1.14 in [36]) $\mathbb{P} = \prod_{t \in [0, \infty)} \mu_t$ on the product space $E^{[0, \infty)}$ with the corresponding product σ-algebra. Since one can identify $E^{[0, \infty)}$ with the space of *all* paths $p : [0, \infty) \longrightarrow E$, and, for each t, μ_t is the $\mathbb{P}$-distribution of $p \in E^{[0,\infty)} \longmapsto p(t) \in E$, there is nothing more to do if there is nothing more that one wants to achieve.

Of course, when applied to integral curves of vector fields, the preceding construction is totally unsatisfactory: it ignores everything which is interesting about them. In particular, integral curves evolve continuously (in fact, differentiably), and we should seek to represent them using a measure on a pathspace which reflects this continuity in some way. That is, at the very least, we should hope that $\mathbb{P}$ can be chosen so that, when s, $t \in [0,\infty)$ are close, then the $\mathbb{P}$-distribution of the pair $(p(s), p(t))$ should be very nearly concentrated on the diagonal. Obviously, since the preceding choice of $\mathbb{P}$ makes $p(s)$ independent of $p(t)$, no matter how close s is to t, there is no reason why the joint $\mathbb{P}$-distribution of $p(s)$ and $p(t)$ will possess this property.[9]

§1.4.1. Kolmogorov's Approach: Suppose that $t \rightsquigarrow \mu_t$ is an integral curve of some vector field Λ on $\mathbf{M}_1(\mathbb{Z}_n)$. In order to carry out a construction of $\mathbb{P}$ which takes into account the sort of considerations alluded to above, we must make a judicious choice of

$$\mathbb{P}\big(p(t_0) \in \Gamma_1, \dots, p(t_m) \in \Gamma_m\big)$$

for $m \geq 1$, $0 = t_0 < \cdots < t_m$, and $\Gamma_0, \dots, \Gamma_m \subseteq \mathbb{Z}_n$. The choice $\prod_0^m \mu_{t_i}(\Gamma_i)$ made above was blind, not judicious.

To see how to do better, we begin with the case in which Λ is an affine vector field, and let $(t, x) \rightsquigarrow P_t(x, \cdot)$ be the transition probability function determined by Λ. Given an integral curve $t \rightsquigarrow \mu_t$ of Λ, we now want to associate with it the probability measure $\mathbb{P}$ on $(\mathbb{Z}_n)^{[0,\infty)}$ determined by the property that

$$
(1.4.1) \quad
\begin{aligned}
&\mathbb{P}\big(p(t_0) \in \Gamma_0, p(t_1) \in \Gamma_1, \dots, p(t_m) \in \Gamma_m\big) \\
&= \int_{\Gamma_0} \int_{\Gamma_1} \cdots \int_{\Gamma_m} P_{\tau_m}(y_{m-1}, dy_m) \cdots P_{\tau_1}(y_0, dy_1) \mu_0(dy_0),
\end{aligned}
$$

where $\tau_j \equiv t_j - t_{j-1}$. Of course, before going further, we must check that such a $\mathbb{P}$ exists and that, for each t, the $\mathbb{P}$-distribution of $p(t)$ is μ_t. However, both these properties are easy consequences of (1.3.9). Indeed, by Kolmogorov's consistency theorem (cf. Exercise 3.1.18 in [36]), existence will follow as soon as we verify that the measures determined by the right hand side of (1.4.1) are consistent in the sense that the one corresponding to $\{t_j : j \in \{1, \dots, m\} \setminus \{i\}\}$ can be obtained by taking $\Gamma_i = E$ in the one corresponding

---

[9] To see just how nonsensical $\mathbb{P}$ is, see §2 of Chapter II in Doob's [6].

to $\{t_j : 1 \leq j \leq m\}$. But this is tantamount to checking that

$$P_{t_{i+1}-t_{i-1}}(x, \cdot) = \int_E P_{\tau_{i+1}}(y, \cdot) P_{\tau_i}(x, dy),$$

which is just what the Chapman–Kolmogorov equation says. As for the $\mathbb{P}$-distribution of $p(t)$, simply observe that

$$\mu_t(\Gamma) = \int_E P_t(x, \Gamma) \mu_0(dx) = \mathbb{P}(p(t) \in \Gamma).$$

The analogous construction when Λ is not affine entails the following subterfuge. Namely, let $(x, \mu) \rightsquigarrow M((x, \mu), \cdot)$ determine Λ. Next, given an integral curve $t \rightsquigarrow \mu_t$ of Λ, set $M((t, x), \cdot) = M((x, \mu_t), \cdot)$, and define $t \rightsquigarrow L(t)$ so that

$$(1.4.2) \qquad L(t)\varphi(x) \equiv \int_{\mathbb{Z}_n} \big(\varphi(x + y) - \varphi(x)\big) M((t, x), dy).$$

Finally, determine the time dependent affine vector field $t \rightsquigarrow \Lambda(t)$ so that $\Lambda(t)_\mu \varphi = \langle L(t)\varphi, \mu \rangle$.

THEOREM 1.4.3. *Let* $t \rightsquigarrow \mu_t$ *be an integral curve of the vector field* Λ *on* $\mathbf{M}_1(\mathbb{Z}_n)$, *and define the time dependent affine vector field* $t \rightsquigarrow \Lambda(t)$ *as in* (1.4.2). *For each* $(s, x) \in [0, \infty) \times \mathbb{Z}_n$, *use (cf. Exercises 1.3.16)* $t \rightsquigarrow P_t((s, x), \cdot)$ *to denote the integral curve starting from* δ_x *at time* s. *Then, for all* $s \in [0, \infty)$ *and* $t > 0$,

$$\mu_{s+t} = \int_{\mathbb{Z}_n} P_t((s, x), \cdot) \mu_s(dx).$$

Furthermore, there exists a unique probability measure $\mathbb{P}$ *on* $(\mathbb{Z}_n)^{[0,\infty)}$ *with the property that*

$$(1.4.4)
\begin{aligned}
&\mathbb{P}\Big(\{p : p(t_0) \in \Gamma_0, p(t_1) \in \Gamma_1, \ldots, p(\iota_m) \in \Gamma_m\}\Big) \\
&= \int_{\Gamma_0} \int_{\Gamma_1} \cdots \int_{\Gamma_m} P_{\tau_m}((t_{m-1}, y_{m-1}), dy_m) \cdots P_{\tau_1}((t_0, y_0), dy_1) \mu_0(dy_0),
\end{aligned}$$

where $\tau_j \equiv t_j - t_{j-1}$. *In particular, for each* t, *the* $\mathbb{P}$-*distribution of* $p(t)$ *is* μ_t. *Alternatively, an equivalent description of* $\mathbb{P}$ *is that the* $\mathbb{P}$-*distribution of* $p \rightsquigarrow p(0)$ *is* μ_0 *and, for each* $(s, t) \in [0, \infty)^2$, *and* $\varphi \in C(\mathbb{Z}_n; \mathbb{R})$,[10]

$$(1.4.5) \quad \mathbb{E}^\mathbb{P}\Big[\varphi\big(p(s + t)\big) \,\Big|\, \sigma(\{p(\tau) : \tau \in [0, s]\})\Big] = \int_{\mathbb{Z}_n} \varphi(y) P_t\big((s, p(s)), dy\big)$$

---

[10] Given a family $\mathfrak{F}$ of maps from a set into a measurable space, we use $\sigma(\mathfrak{F})$ to denote the σ-algebra generated by $\mathfrak{F}$. That is, the smallest σ-algebra with respect to which each $f \in \mathfrak{F}$ is measurable. In this connection, when $\mathfrak{A}$ is a collection of subsets, $\sigma(\mathfrak{A}) = \sigma(\mathfrak{F})$ with $\mathfrak{F} = \{\mathbf{1}_A : A \in \mathfrak{A}\}$.

P-almost surely. Finally, (1.4.5) is equivalent to saying that, for all $(s,t) \in [0,\infty)^2$, $A \in \sigma(\{p(\tau) : \tau \in [0,s]\})$, and $\varphi \in C(\mathbb{Z}_n; \mathbb{R})$,

$$(1.4.6) \quad \mathbb{E}^{\mathbb{P}}\Big[\varphi\big(p(s+t)\big) - \varphi\big(p(s)\big),\, A\Big] = \int_0^t \mathbb{E}^{\mathbb{P}}\Big[L(s+\tau)\varphi\big(p(s+\tau)\big),\, A\Big]\, d\tau.$$

PROOF: Everything except the final two assertions is derived by first repeating the reasoning which led to (1.3.9), then applying the result in Exercise 1.3.16, and finally repeating the reasoning just given in the case of affine vector fields.

To prove that $\mathbb{P}$ is characterized by its distribution of $p \leadsto p(0)$ plus (1.4.5), suppose that $\mathbb{P}$ satisfies (1.4.4) and conclude that μ_0 is the $\mathbb{P}$-distribution of $p \leadsto p(0)$. Further, by taking $t_m = s + t$ and $0 = t_0 < \cdots < t_{m-1} = s$, see that

$$\mathbb{E}^{\mathbb{P}}\Big[p(s+t) \in \Gamma,\, A\Big] = \mathbb{E}^{\mathbb{P}}\Big[P_t\big((s, p(s)), \Gamma\big),\, A\Big]$$

for any $\Gamma \subseteq \mathbb{Z}_n$ and A of the form $A = \{p : p(t_0) \in \Gamma_1, \ldots, p(t_{m-1}) \in \Gamma_{m-1}\}$. Since the collection of such A's is closed under intersection and generates $\sigma(\{p(\sigma) : \sigma \in [0,s]\})$, (1.4.5) follows. To go the other way, assume that μ_0 is the $\mathbb{P}$-distribution of $p \leadsto p(0)$ and that $\mathbb{P}$ satisfies (1.4.5), and, by using induction on $m \geq 0$, check that $\mathbb{P}$ must satisfy (1.4.4).

Turning to the final statement, first suppose that $\mathbb{P}$ satisfies (1.4.5). Then

$$\frac{d}{dt}\mathbb{E}^{\mathbb{P}}\Big[\varphi\big(p(s+t)\big),\, A\Big] = \mathbb{E}^{\mathbb{P}}\Big[\langle L(s+t)\varphi, P_t((s+\tau), \cdot)\rangle,\, A\Big]$$

$$= \mathbb{E}^{\mathbb{P}}\Big[L(s+t)\varphi\big(p(s+t)\big),\, A\Big],$$

and so (1.4.6) follows after one integrates. Conversely, if (1.4.6) holds for $\mathbb{P}$ and if $A \in \sigma(\{p(\tau) : \tau \in [0,s]\})$ with $\mathbb{P}(A) > 0$, define $t \leadsto \nu_t^s$ by

$$\nu_t^s(\Gamma) = \mathbb{P}\big(p(s+t) \in \Gamma \,\big|\, A\big),$$

and conclude that $t \leadsto \nu_t^s$ is an integral curve of $t \leadsto \Lambda(t)$ starting at time s. In particular, $\nu_t^s = \int P_t\big((s,x), \cdot\big)\, \nu_0^s(dx)$, which means that

$$\mathbb{E}^{\mathbb{P}}\Big[\varphi\big(p(s+t)\big) \,\Big|\, A\Big] = \mathbb{E}^{\mathbb{P}}\Big[\big\langle \varphi, P_t\big((s, p(s)), \cdot\big)\big\rangle \,\Big|\, A\Big],$$

from which (1.4.5) is an easy step. $\square$

REMARK 1.4.7. There are two comments which are worth making about Theorem 1.4.3. First, the second formulation is tantamount to saying that the measure $\mathbb{P}$ which we have constructed is *time inhomogeneous Markov* with transition probability function $(s, t, x) \rightsquigarrow P_t\big((s, x), \cdot\big)$. In general, this transition probability function will depend on the integral curve under consideration. However, the affine case is distinguished by the fact that the transition mechanism is independent of the particular integral curve and it is only the initial condition μ_0 which changes. Secondly, it should be noted that the measure $\mathbb{P}$ produced in Theorem 1.4.3 achieves the goal at which we were aiming. Namely, because $P_t\big((s, x), \cdot\big)$ tends to δ_x as $t \searrow 0$, if Δ denotes the diagonal in $\mathbb{Z}_n \times \mathbb{Z}_n$, then

$$\mathbb{P}\Big(\big(p(s), p(s+t)\big) \notin \Delta\Big) = \int_{\mathbb{Z}_n} P_t\big((s, x), \mathbb{Z}_n \setminus \{x\}\big) \mu_s(dx) \longrightarrow 0 \quad \text{as } t \searrow 0.$$

In the subsections which follow, we will say more.

§1.4.2. Lévy Processes on $\mathbb{Z}_n$: In preparation for what follows, we will devote this subsection to a detailed examination of the pathspace measure corresponding to a translation invariant vector field Λ. Thus, let M be the Lévy measure such that $\Lambda_{\delta_0} \varphi = \int \big(\varphi(y) - \varphi(0)\big) M(dy)$, and recall that the transition probability function corresponding to Λ is given by $(t, x) \rightsquigarrow (\tau_x)_* \lambda_t$, where $t \rightsquigarrow \lambda_t$ is the infinitely divisible flow given by (1.2.2). From these facts it is easy to see that if $\mathbb{P}^M$ is the measure $\mathbb{P}$ in Theorem 1.4.3 corresponding to $t \rightsquigarrow \mu_t$, then $\mathbb{P}^M$ is characterized by the properties that $\mathbb{P}^M\big(p(0) = 0\big) = 1$ and, for any $m \geq 1$ and $0 \leq t_0 < \cdots < t_m$, the $\mathbb{P}^M$-distribution of $p \rightsquigarrow \big(p(t_1) - p(t_0), \ldots, p(t_m) - p(t_{m-1})\big)$ is $\prod_{\ell=1}^{m} \lambda_{\tau_\ell}$, where $\tau_\ell = t_\ell - t_{\ell-1}$. Equivalently, under $\mathbb{P}^M$, $p(0) = 0$ almost surely, increments of p over disjoint time intervals are mutually independent, and the distribution of the increment $p \rightsquigarrow p(s+t) - p(s)$ is λ_t. In the jargon of probability theory, $\mathbb{P}^M$ is the distribution of the process starting at 0 with *homogeneous, independent increments* for which λ_t is the distribution of an increment over a time interval of length t. In accordance with more recent terminology, we will call such a process a *Lévy process*.

In order to develop a better understanding of $\mathbb{P}^M$, we will now give a hands-on construction. Let $(\Omega, \mathcal{F}, \mathbb{Q})$ be a probability space on which there exists a family $\big\{\tau_{\ell,y}, \ (\ell, y) \in \mathbb{Z}^+ \times (\mathbb{Z}_n \setminus \{0\})\big\}$ of mutually independent, unit exponential random variables. That is, the $\tau_{\ell,y}$'s are mutually independent, and $\mathbb{Q}\big(\tau_{\ell,y} > t\big) = e^{-t}$ for all $t \geq 0$. Without loss in generality, we will assume that enough sets of $\mathbb{Q}$-measure 0 have been eliminated from Ω so that, for *all* ω,

$$\tau_{\ell,y}(\omega) > 0, \quad \sum_{\ell=1}^{\infty} \tau_{\ell,y}(\omega) = \infty,$$

and $(m, y) \neq (m', y') \implies \sum_{\ell=1}^{m} \tau_{\ell,y}(\omega) \neq \sum_{\ell=1}^{m'} \tau_{\ell',y'}(\omega).$

Next, define the random variables

$$\mathbf{T}_{m,y}(\omega) = \begin{cases} 0 & \text{if } m = 0 \\ \sum_{\ell=1}^{m} \tau_{\ell,y}(\omega) & \text{if } m \in \mathbb{Z}^+, \end{cases}$$

and

$$\mathbf{N}_y(t, \omega) = \max\{m : \mathbf{T}_{m,y}(\omega) \leq tM(\{y\})\}.$$

Obviously, $\sigma(\{\mathbf{N}_y(t) : t \geq 0\})$ is independent of $\sigma(\{\mathbf{N}_{y'}(t) : t \geq 0\})$ when $y' \neq y$. In addition, for each y and ω, $\mathbf{N}_y(\cdot, \omega)$ is an $\mathbb{N}$-valued, right continuous, nondecreasing path which starts at 0; and (cf. Lemma 3.3.4 in [36] or Exercise 1.4.11 below)

$$\mathbb{Q}\big(\mathbf{N}_y(t_1) = m_1, \ldots, \mathbf{N}_y(t_k) = m_k\big)$$

(1.4.8)

$$= e^{-t_k M(\{y\})} \big(M(\{y\})\big)^{m_k} \prod_{j=1}^{k} \frac{(t_j - t_{j-1})^{m_j - m_{j-1}}}{(m_j - m_{j-1})!}$$

for all $k \geq 1$, $0 = t_0 < t_1 < \cdots t_k$, and $0 = m_0 \leq m_1 \leq \cdots \leq m_k$. Equivalently, $\mathbf{N}_y(s + t) - \mathbf{N}_y(s)$ is $\mathbb{Q}$-independent of $\sigma(\{\mathbf{N}(\sigma) : \sigma \in [0, s]\})$ and has $\mathbb{Q}$-distribution given by $\mathbb{Q}(\mathbf{N}_y(t) = m) = e^{-tM(\{y\})} \frac{(tM(\{y\}))^m}{m!}$. That is, however one states it, $\{\mathbf{N}_y(t) : t \geq 0\}$ is a *simple Poisson process* with intensity $M(\{y\})$.

Now set $\mathbf{N}(t, \omega) = \sum_{y \in \mathbb{Z}_n \setminus \{0\}} \mathbf{N}_y(t, \omega)$. Then $\{\mathbf{N}(t) : t \geq 0\}$ is again a simple Poisson process, this time with intensity $M(\mathbb{Z}_n)$. In particular, if $\mathbf{T}_0(\omega) \equiv 0$ and

$$\mathbf{T}_m(\omega) = \min\{t \geq \mathbf{T}_{m-1}(\omega) : \mathbf{N}(t, \omega) > \mathbf{N}(\mathbf{T}_{m-1}(\omega), \omega)\},$$

then the $\mathbf{T}_m - \mathbf{T}_{m-1}$'s are mutually independent and $\mathbb{Q}(\mathbf{T}_m - \mathbf{T}_{m-1} > t) = e^{-tM(\mathbb{Z}_n)}$ for $m \geq 1$ and $t \geq 0$. Furthermore, for each $m \geq 1$ and $x \in \mathbb{Z}_n$,

(1.4.9)

$$\mathbb{Q}\big(\mathbf{N}_x(\mathbf{T}_m) - \mathbf{N}_x(\mathbf{T}_{m-1}) = 1 \,\big|\, \mathcal{F}_m\big) = \frac{M(\{x\})}{M(\mathbb{Z}_n)}$$

where $\mathcal{F}_m = \sigma\big(\{\mathbf{T}_m\} \cup \{\mathbf{N}_y(t) : t \in [0, \mathbf{T}_m) \,\&\, y \in \mathbb{Z}_n \setminus \{0\}\}\big)$.

To check (1.4.9), let $0 < h < s$ and $A \in \sigma(\{\mathbf{N}_y(\sigma) : \sigma \in [0, s - h]$ and $y \in \mathbb{Z}_n \setminus \{0\}\})$ with $\mathbb{Q}(A) > 0$ be given, and observe that, because $\mathbb{Q}(\mathbf{N}(s) - \mathbf{N}(s - h) \geq 2) = \mathcal{O}(h^2)$ as $h \searrow 0$,

$$\mathbb{Q}\big(\mathbf{N}_x(\mathbf{T}_m) - \mathbf{N}_x(\mathbf{T}_{m-1}) = 1 \,\&\, s - h < \mathbf{T}_m \leq s \,\&\, A\big)$$
$$= \mathbb{Q}\big(\mathbf{N}_x(s) - \mathbf{N}_x(s - h) = 1 \,\&\, \mathbf{T}_m > s - h \,\&\, A\big) + \mathcal{O}(h^2)$$
$$= hM(\{x\})\mathbb{Q}\big(\mathbf{T}_m > s - h \,\&\, A\big) + \mathcal{O}(h^2),$$

and, similarly,

$$\mathbb{Q}(s - h < \mathbf{T}_m \le s \ \& \ A) = hM(\mathbb{Z}_n)\mathbb{Q}(\mathbf{T}_m > s - h \ \& \ A) + \mathcal{O}(h^2).$$

We are, at last, ready to introduce the advertised understanding of $\mathbb{P}^M$. Namely, set

$$X(t, \omega) = \sum_{y \in \mathbb{Z}_n \setminus \{0\}} y \mathbf{N}_y(t, \omega),$$

where here the sum is taken in $\mathbb{Z}_n$ (i.e., modulo n). Then, $\mathbb{Q}(X(0) = 0) = 1$ and

$$\mathbb{E}^{\mathbb{Q}}\left[\mathbf{e}_{\xi,n}(X(s + t) - X(s)) \,\middle|\, \sigma(\{X(\sigma) : \sigma \in [0, s]\})\right]$$

$$= \prod_y \mathbb{E}^{\mathbb{Q}}\left[\mathbf{e}_{\xi,n}(y\mathbf{N}_y(t))\right] = \prod_y \left(e^{-tM(\{y\})} \sum_{m=0}^{\infty} \frac{(tM(\{y\})\mathbf{e}_{\xi,n}(y))^m}{m!} \right)$$

$$= \exp\left(\int_{\mathbb{Z}_n} (\mathbf{e}_{\xi,n}(y) - 1) \right) M(dy) = \hat{\lambda}_t(\xi).$$

Hence, the $\mathbb{Q}$-distribution of $\omega \in \Omega \longmapsto X(\cdot, \omega) \in (\mathbb{Z}_n)^{[0,\infty)}$ is $\mathbb{P}^M$, and this representation of $\mathbb{P}^M$ provides us with lots of information about $\mathbb{P}^M$. For one thing, the paths $X(\cdot, \omega)$ are much better than generic paths in $(\mathbb{Z}_n)^{[0,\infty)}$. In fact, for each ω, $X(\cdot, \omega)$ is an element of the space $D([0,\infty); \mathbb{Z}_n)$ of piecewise constant paths $p : [0, \infty) \longrightarrow \mathbb{Z}_n$ which are right continuous.[11] Hence, $\mathbb{P}^M$ *lives* on the space $D([0, \infty); \mathbb{Z}_n)$,[12] and from now on we will always think of it living there. Something else which our construction reveals is the mechanism underlying the evolution of paths under $\mathbb{P}^M$. Namely, the paths are piecewise constant, the times between jumps are independently and exponentially distributed with intensity $M(\mathbb{Z}_n)$, and (cf. (1.4.9)) when a jump occurs, the size of the jump is independent of everything that happened before and has distribution $\frac{M}{M(\mathbb{Z}_n)}$. All this and more is summarized in the following.

THEOREM 1.4.10. *Given $p \in D([0, \infty); \mathbb{Z}_n)$, define $\eta(t, \Gamma, p)$ for $t \ge 0$ and $\Gamma \subseteq \mathbb{Z}_n$ so that $\eta(0, \Gamma, p) = 0$ and*[13]

$$\eta(t, \Gamma, p) \equiv \sum_{0 < \tau \le t} \mathbf{1}_{\Gamma \cap (\mathbb{Z}_n \setminus \{0\})}(p(\tau) - p(\tau-)) \quad if \ t \in (0, \infty).$$

---

[11] Equivalently, the paths in $D([0, \infty); \mathbb{Z}_n)$ are right continuous and possess a left-limit at each $t \in (0, \infty)$.

[12] The term *lives* is somewhat misleading here because $D([0, \infty); \mathbb{Z}_n)$ is not a measurable subset of $\mathbb{Z}_n^{[0,\infty)}$ The proper statement is that the $\mathbb{P}$ in Theorem 1.4.3 gives $D([0, \infty); \mathbb{Z}_n)$ outer measure one.

[13] Given a path $p \in D([0, \infty); \mathbb{Z}_n)$ and a $t \in (0, \infty)$, $p(t-) \equiv \lim_{s \nearrow t} p(s)$ is the left limit of p at t.

Then, for each (t,p), $\eta(t,\,\cdot\,,p)$ is an $\mathbb{N}$-valued measure on $\mathbb{Z}_n$. Moreover, under $\mathbb{P}^M$, for each $y \in \mathbb{Z}_n$, $p \rightsquigarrow \eta(\,\cdot\,,\{y\},p)$ has the distribution of a simple Poisson process with intensity $M(\{y\})$ and is independent of $\sigma(\{\eta(t,\{y'\}) : t \geq 0 \ \& \ y' \neq y\})$. Hence, for any Γ, the $\mathbb{P}^M$-distribution of $p \rightsquigarrow \eta(\,\cdot\,,\Gamma,p)$ is that of a simple Poisson process with intensity $M(\Gamma)$. Finally, assume that $M(\mathbb{Z}_n) > 0$, and define $\mathbf{T}_0(p) \equiv 0$ and $\mathbf{T}_m(p) = \inf\{t \geq \mathbf{T}_{m-1}(p) : p(t) \neq p(\mathbf{T}_{m-1})\}$. Then, for all $m \in \mathbb{Z}^+$, $\mathbb{P}^M(\mathbf{T}_m < \infty) = 1$,

$$\mathbb{P}^M\big(\eta(t + \mathbf{T}_{m-1},\Gamma) - \eta(\mathbf{T}_{m-1},\Gamma) = 0 \,\big|\, \overline{\mathcal{T}_{m-1}}\big) = e^{-tM(\Gamma)}, \quad \Gamma \subseteq \mathbb{Z}_n,$$

and

$$\mathbb{P}^M\big(p(\mathbf{T}_m) - p(\mathbf{T}_{m-1}) = y \,\big|\, \mathcal{T}_m\big) = \frac{M(\{y\})}{M(\mathbb{Z}_n)},$$

where $\mathcal{T}_m$ and $\overline{\mathcal{T}_{m-1}}$ are the σ-algebras generated, respectively, by $\{\mathbf{T}_m\} \cup \{p(t) : t \in [0,\mathbf{T}_m(p))\}$ and by $\{\mathbf{T}_{m-1}\} \cup \{p(t) : t \in [0,\mathbf{T}_{m-1}(p)]\}$.

§1.4.3. Exercises

EXERCISE 1.4.11. The purpose of this exercise is to verify (1.4.8). Thus, let $\{\tau_\ell : \ell \geq 1\}$ be a sequence of mutually independent, unit exponential random variables, set $\mathbf{T}_0 = 0$ and $\mathbf{T}_m = \sum_1^m \tau_\ell$ for $m \geq 1$, and take $\mathbf{N}(t) = \sup\{m : \mathbf{T}_m \leq \lambda t\}$, where $\lambda \in [0,\infty)$. If $\lambda = 0$, then it is clear that $\mathbf{N}(t) = 0$ almost surely for all $t \geq 0$. Thus, assume that $\lambda > 0$, and justify:

$$\mathbb{P}\big(\mathbf{N}(t_1) = m_1, \ldots, \mathbf{N}(t_\ell) = m_\ell\big)$$
$$= \mathbb{P}\big(\mathbf{T}_{m_1} \leq \lambda t_1 < \mathbf{T}_{m_1+1}, \ldots, \mathbf{T}_{m_\ell} \leq \lambda t_\ell < \mathbf{T}_{m_\ell+1}\big)$$
$$= \lambda^{m_\ell+1} \int \cdots \int_A \exp\left(-\lambda \sum_{j=1}^{m_\ell+1} \xi_j\right) d\xi_1 \cdots d\xi_{m_\ell+1}$$
$$= \lambda^{m_\ell+1} \int \cdots \int_B e^{-\lambda \eta_{m_\ell+1}} \, d\eta_1 \cdots d\eta_{m_\ell+1}$$
$$= \lambda^{m_\ell} e^{-\lambda t_\ell} \prod_{j=1}^{\ell} \mathrm{vol}(\Delta_j) = \lambda^{m_\ell} e^{-\lambda t_\ell} \prod_{j=1}^{\ell} \frac{(t_j - t_{j-1})^{m_j - m_{j-1}}}{(m_j - m_{j-1})!},$$

where

$$A = \left\{\boldsymbol{\xi} \in (0,\infty)^{m_\ell+1} : \sum_1^{m_j} \xi_k \leq t_j < \sum_1^{m_j+1} \xi_k \text{ for } 1 \leq j \leq \ell\right\},$$
$$B = \left\{\boldsymbol{\eta} \in (0,\infty)^{m_\ell+1} : \eta_j < \eta_{j+1} \ \& \ \eta_{m_j} \leq t_j < \eta_{m_j+1} \text{ for } 1 \leq j \leq \ell\right\},$$
$$\Delta_j = \left\{\mathbf{u} \in \mathbb{R}^{m_j - m_{j-1}} : t_{j-1} \leq u_1 < \cdots < u_{m_j - m_{j-1}} \leq t_j\right\},$$

and $t_0 \equiv 0$.

Exercise 1.4.12. A probability measure $\mathbb{P}$ on $(\mathbb{Z}_n)^{[0,\infty)}$ is said to be *time homogeneous Markov with transition probability* $(t, x) \rightsquigarrow P_t(x, \cdot)$ if (1.4.1) holds for all $m \geq 0$, $0 = t_0 < t_1 \cdots < t_m$, and $\{\Gamma_\ell\}_0^m \subseteq \mathcal{B}_E$.

(i) Show that $\mathbb{P}$ satisfies (1.4.1) if and only if $\mathbb{P}\big(p(0) \in \Gamma\big) = \mu_0(\Gamma)$ and, for each $s \in [0,\infty)$ and $t > 0$:

$$(1.4.13) \qquad \mathbb{P}\Big(p(s+t) \in \Gamma \,\Big|\, \sigma\big(\{p(\sigma) : \sigma \in [0,s]\}\big)\Big) = P_t\big(p(s),\Gamma\big).$$

(ii) The goal here[14] is to show that (1.4.1) is not implied if (1.4.13) is replaced by the weaker condition

$$(*) \qquad \mathbb{P}\Big(p(s+t) \in \Gamma \,\Big|\, \sigma\big(p(s)\big)\Big) = P_t\big(p(s);\Gamma\big).$$

To this end, take $E = \mathbb{Z}_2$, and let $(\Omega, \mathcal{F}, \mathbb{Q})$ be a probability space on which there exists a sequence $\{X_\ell\}_1^\infty$ of mutually independent, $\mathbb{Z}_2$-valued random variables with $\mathbb{Q}(X_\ell = 0) = \frac{1}{2} = \mathbb{Q}(X_\ell = 1)$. Further, assume that $\{\mathbf{N}(t) : t \geq 0\}$ is a Poisson process with intensity 1 on $(\Omega, \mathcal{F}, \mathbb{Q})$ and that $\sigma(\{X_\ell : \ell \geq 1\})$ is $\mathbb{Q}$-independent of $\sigma(\{\mathbf{N}(t) : t \geq 0\})$. Finally, set $X_0 = X_1 + X_2$ (addition in the sense of $\mathbb{Z}_2$) and $Y(t) = \sum_{\ell=0}^{\mathbf{N}(t)} X_\ell$. Show that $(*)$ holds with $P_t(x, \cdot) = e^{-t}\delta_x + (1-e^{-t})\lambda_{\mathbb{Z}_2}$, but that (1.4.13) is false.

Exercise 1.4.14. It will be useful to know that Theorem 1.4.10 readily generalizes to time dependent, translation invariant vector fields. That is, suppose that $t \rightsquigarrow M(t, \cdot)$ is a continuous (i.e., $t \rightsquigarrow M(t, \{y\})$ is continuous for each y) map from $[0,\infty)$ into Lévy measures, and determine the time dependent, translation invariant vector field $t \rightsquigarrow \Lambda(t)$ by

$$\Lambda(t)\delta_x \varphi = \int_{\mathbb{Z}_n} \big(\varphi(x+y) - \varphi(x)\big) M(t, dy)$$

for $x \in \mathbb{Z}_n$ and $\varphi \in C(\mathbb{Z}_n; \mathbb{R})$. Finally, for each $s \in [0,\infty)$, let (cf. Exercise 1.3.16) $t \rightsquigarrow \lambda_{s,t}$ be the integral curve of $t \rightsquigarrow \Lambda(t)$ starting from δ_0 at time s.

(i) Proceeding in exactly the same way as we did above and using the notation introduced in Theorem 1.4.10, show that there is a unique probability measure $\mathbb{P}^{M \cdot}$ on $D\big([0,\infty); \mathbb{Z}_n\big)$ such that $\mathbb{P}^{M \cdot}\big(p(0) = 0\big) = 1$, the processes $p \rightsquigarrow \eta(\cdot, \{y\}, p)$ for $y \in \mathbb{Z}_n \setminus \{0\}$ are mutually $\mathbb{P}^{M \cdot}$-independent and, for any fixed $y \in \mathbb{Z}_n \setminus \{0\}$,

$$\mathbb{P}^{M \cdot}\Big(\eta(s+t, \{y\}) - \eta(s, \{y\}) = m \,\Big|\, \sigma\big(p(\sigma) : \sigma \in [0,s]\}\big)\Big)$$

$$= \exp\left(-\int_0^t M(s+\tau, \{y\})\, d\tau\right) \frac{\left(\int_0^t M(s+\tau, \{y\})\, d\tau\right)^m}{m!}.$$

---

[14] I got the key idea for this exercise from Richard Dudley.

(ii) Show that the measure $\mathbb{P}^{M \cdot}$ described in (i) satisfies (1.4.4) when $\mu_0 = \delta_0$ and $P_t\big((s, x), \cdot\big) = (\tau_x)_* \lambda_{s,s+t}$. In addition, show that $\mathbb{P}^{M \cdot}$-almost surely

$$\mathbb{P}_0^M \cdot \left(\eta(\mathbf{T}_{m-1}+t, \Gamma) - \eta(\mathbf{T}_{m-1}, \Gamma) = 0 \,\big|\, \overline{\mathcal{I}_{m-1}}\right) = \exp\left(-\int_{\mathbf{T}_{m-1}}^{\mathbf{T}_{m-1}+t} M(\tau, \Gamma)\, d\tau\right)$$

on $\{\mathbf{T}_{m-1} < \infty\}$ and that

$$\mathbb{P}_0^M \cdot \big(p(\mathbf{T}_m) - p(\mathbf{T}_{m-1}) = y \,\big|\, \mathcal{I}_m\big) = \frac{M(\mathbf{T}_m, \{y\})}{M(\mathbf{T}_m, \mathbb{Z}_n)} \quad \text{on } \{\mathbf{T}_m < \infty\}.$$

1.5 Itô's Idea

Theorem 1.4.10 provides us with a rather complete description of the way in which the paths of an independent increment process evolve on $\mathbb{Z}_n$. In particular, it gives us the path decomposition

$$(1.5.1) \qquad\qquad p(t) = \int_{\mathbb{Z}_n} y\, \eta(t, dy, p);$$

of p in terms of the family $\{\eta(\cdot, \{y\}) : y \in \mathbb{Z}_n\}$ of mutually independent, simple Poisson processes. Itô's method allows us to extend this type of description to the sort of process described in Theorem 1.4.3 corresponding to a general affine vector field.

§**1.5.1. Itô's Construction:** In order to explain Itô's idea in the present context, we must begin by examining what happens when we replace the integrand "y" in (1.5.1) by "$F(y)$", where $F : \mathbb{Z}_{n'} \longrightarrow \mathbb{Z}_n$ maps $0 \in \mathbb{Z}_{n'}$ to $0 \in \mathbb{Z}_n$. That is, for $p' \in D\big([0, \infty); \mathbb{Z}_{n'}\big)$, set

$$(1.5.2) \qquad\qquad X_F(t, p') = \int_{\mathbb{Z}_{n'}} F(y')\, \eta(t, dy', p').$$

Clearly, $X_F(\cdot, p') \in D\big([0, \infty); \mathbb{Z}_n\big)$. In addition, for each $s \in [0, \infty)$, $p' \rightsquigarrow X_F(s, p')$ is $\sigma(\{p'(\sigma) : \sigma \in [0, s]\})$-measurable and, for every $t \geq 0$ and $M' \in \mathfrak{M}(\mathbb{Z}_{n'})$, the increment $X_F(s + t) - X_F(s)$ is $\mathbb{P}^{M'}$-independent of $\sigma(\{p'(\sigma) : \sigma \in [0, s]\})$ and has the same distribution as $p' \rightsquigarrow X_F(t, p')$. Finally, for each $\xi \in \mathbb{Z}_n$,

$$\mathbb{E}^{\mathbb{P}^{M'}}\left[\mathbf{e}_{\xi,n}\big(X_F(t)\big)\right] = \prod_{y' \in \mathbb{Z}_{n'}} E^{\mathbb{P}^{M'}}\left[\mathbf{e}_{\xi,n}\big(\eta(t, \{y'\}) F(y')\big)\right]$$

$$= \prod_{y' \in \mathbb{Z}_{n'}} \exp\left(tM'(\{y'\})\big(\mathbf{e}_{\xi,n}(F(y')) - 1\big)\right)$$

$$= \exp\left(t \int_{\mathbb{Z}_{n'}} \big(\mathbf{e}_{\xi,n}(F(y')) - 1\big) M'(dy')\right).$$

Hence, we have now shown that the $\mathbb{P}^{M'}$-distribution of $p' \rightsquigarrow X_F(\,\cdot\,,p')$ is that of the independent increment process on $\mathbb{Z}_n$ determined by the Lévy measure F_*M' given by

$$(1.5.3) \quad F_*M'(\{y\}) = \begin{cases} 0 & \text{if } y = 0 \\ M'(\{y' \in \mathbb{Z}_{n'} : F(y') = y\}) & \text{if } y \in \mathbb{Z}_n \setminus \{0\}. \end{cases}$$

Next, suppose that we are given a map $F : \mathbb{Z}_n \times \mathbb{Z}_{n'} \longrightarrow \mathbb{Z}_n$ which takes $\mathbb{Z}_{n'} \times \{0\}$ to $0 \in \mathbb{Z}_n$, define $x \rightsquigarrow M(x, \cdot)$ by (cf. (1.5.3)) $M(x, \cdot) = (F(x, \cdot))_*M'$, and let Λ be the affine vector field on $\mathbf{M}_1(\mathbb{Z}_n)$ determined by $x \rightsquigarrow M(x, \cdot)$. Itô's idea is to construct the process corresponding to Λ by modifying the construction in (1.5.2) to reflect the dependence of $F(x, y)$ on x. To be precise, refer to the notation in Theorem 1.4.10 and define

$$(x, p') \in \mathbb{Z}_n \times D([0, \infty); \mathbb{Z}_{n'}) \longmapsto X(\,\cdot\,, x, p') \in D([0, \infty); \mathbb{Z}_n)$$

so that

$$(1.5.4) \quad \begin{aligned} & X(0, x, p') = x \\ & X(\mathbf{T}_\ell(p'), x, p') - X(\mathbf{T}_{\ell-1}(p'), x, p') \\ & \qquad = F(X(\mathbf{T}_{\ell-1}(p'), x, p'), p'(\mathbf{T}_\ell) - p'(\mathbf{T}_{\ell-1})) \\ & X(t, x, p') = X(\mathbf{T}_{\ell-1}(p'), x, p') \quad \text{if } \mathbf{T}_{\ell-1}(p') \le t < \mathbf{T}_\ell(p'). \end{aligned}$$

It should be evident that, for each $(t, x) \in [0, \infty) \times \mathbb{Z}_n$, the map $p' \rightsquigarrow X(t, x, p')$ is $\sigma(\{p'(\tau)) : \tau \in [0, t]\})$-measurable.

THEOREM 1.5.5. *Referring to the preceding, let $t \rightsquigarrow \mu_t$ be the integral curve of Λ starting from μ, and let $\mathbb{P}$ be the associated measure described in Theorem (1.3.9). If $(t, x, p') \rightsquigarrow X(t, x, p')$ is defined as in (1.5.4), then $\mathbb{P}$ is the $\mu \times \mathbb{P}^{M'}$ distribution of $(x, p') \rightsquigarrow X(\,\cdot\,, x, p')$. In particular, $\mathbb{P}$ lives on $D([0, \infty); \mathbb{Z}_n)$.*

PROOF: The first step is to observe that $\mathbb{P}$ is characterized by the facts that μ is the $\mathbb{P}$-distribution of $p \rightsquigarrow p(0)$ and that, for any $s \ge 0$, $A \in \sigma(\{p(\sigma) : \sigma \in [0, s]\})$, and $\varphi \in C(\mathbb{Z}_n; \mathbb{R})$,

$$(1.5.6) \quad \mathbb{E}^{\mathbb{P}}\Big[\varphi(p(s+t)) - \varphi(p(s)), A\Big] = \int_0^t \mathbb{E}^{\mathbb{P}}\Big[L\varphi(p(s+\tau)), A\Big]\, d\tau,$$

where

$$L\varphi(x) = \Lambda_{\delta_x}\varphi = \int_{\mathbb{Z}_n} (\varphi(x+y) - \varphi(x))\, M(x, dy).$$

Indeed, because Λ is affine, this is precisely the characterization given in the final part of Theorem 1.4.3.

In view of the preceding paragraph and the remark following (1.5.4), it suffices for us to check that for $(s, x) \in [0, \infty) \times \mathbb{Z}_n$ and $A' \in \sigma(\{p'(\sigma) : \sigma \in [0, s]\})$,

(*)
$$\mathbb{E}^{\mathbb{P}^{M'}} \left[\varphi(X(s+t, x)) - \varphi(X(s, x)), \ A'\right]$$
$$= \int_0^t \mathbb{E}^{\mathbb{P}^{M'}} \left[L\varphi(X(s+\tau, x)), \ A'\right] d\tau.$$

To this end, note that

$$\mathbb{E}^{\mathbb{P}^{M'}} \left[\varphi(X(s+t+h, x)) - \varphi(X(s+t, x)), \ A'\right]$$
$$= \sum_{y' \in \mathbb{Z}_{n'} \setminus \{0\}} \mathbb{E}^{\mathbb{P}^{M'}} \left[\varphi(X(s+t+h, x)) - \varphi(X(s+t, x)), \ A' \cap J'(h, y')\right]$$
$$+ \mathbb{E}^{\mathbb{P}^{M'}} \left[\varphi(X(s+t+h, x)) - \varphi(X(s+t, x)), \ A' \cap B'(h)\right]$$

where

$$J'(h, y') \equiv \{p' : \eta(s+t+h, \mathbb{Z}_{n'}, p') - \eta(s+t, \mathbb{Z}_{n'}, p') = 1$$
$$= \eta(s+t+h, \{y'\}, p') - \eta(s+t, \{y'\}, p')\}$$
$$B'(h) \equiv \{p' : \eta(s+t+h, \mathbb{Z}_{n'}, p') - \eta(s+t, \mathbb{Z}_{n'}, p') \geq 2\}].$$

Because $\mathbb{P}^{M'}(B'(h))$ tends to 0 like h^2 as $h \searrow 0$, while

$$\mathbb{E}^{\mathbb{P}^{M'}} \left[\varphi(X(s+t+h, x)) - \varphi(X(s+t, x)), \ A' \cap J'(y')\right]$$
$$= \mathbb{E}^{\mathbb{P}^{M'}} \left[\varphi(X(s+t, x)) + F(X(s+t, x), y') - \varphi(X(s+t, x)), \ A'\right]$$
$$\times \mathbb{P}_0^{M'}(J'(h, y'))$$

and

$$\mathbb{P}^{M'}(J'(h, y')) = \mathbb{P}^{M'}(\eta(h, \{y'\}) = 1 = \eta(h, \mathbb{Z}_{n'}))$$
$$= e^{-hM'(\{y'\})} hM'(\{y'\}) - \mathbb{P}^{M'}(\eta(h, \{y'\}) = 1 < \eta(h, \mathbb{Z}_{n'})),$$

we conclude that

$$\lim_{h \searrow 0} \frac{\mathbb{E}^{\mathbb{P}^{M'}} \left[\varphi(X(s+t+h, x)) - \varphi(X(s+t, x)), \ A'\right]}{h}$$
$$= \mathbb{E}^{\mathbb{P}^{M'}} \left[\int \left(\varphi(X(s+t, x) + y) - \varphi(X(s+t, x))\right) M(X(s+t, x), dy), \ A'\right],$$

where the convergence is uniform with respect to t. Thus, (*) follows by integration. $\square$

We now want to show that Theorem 1.5.5 applies to any affine vector field. That is, we want to show that every map $x \rightsquigarrow M(x, \cdot)$ from $\mathbb{Z}_n$ into Lévy measures can be realized as $F(x, \cdot)_* M'$ for some choice of n', Lévy measure M' on $\mathbb{Z}_{n'}$, and $F : \mathbb{Z}_n \times \mathbb{Z}_{n'} \longrightarrow \mathbb{Z}_n$ which takes $\mathbb{Z}_n \times \{0\}$ into 0. For this purpose, take $n' = n(n-1) + 1$, and let $f : \mathbb{Z}_n \times (\mathbb{Z}_n \setminus \{0\}) \longrightarrow \mathbb{Z}_{n'} \setminus \{0\}$ be one-to-one and onto. Next, define $F : \mathbb{Z}_n \times \mathbb{Z}_{n'} \longrightarrow \mathbb{Z}_n$ so that

$$(1.5.7) \quad F(x, y') = \begin{cases} 0 & \text{if } y' = 0 \text{ or } y' \neq f(x, y) \text{ for any } y \in \mathbb{Z}_n \setminus \{0\} \\ y & \text{if } y' = f(x, y). \end{cases}$$

Finally, take M' to be the Lévy measure on $\mathbb{Z}_{n'}$ given by $M'(\{y'\}) = M(x, \{y\})$ when $y' = f(x, y)$.

By combining the preceding with Theorems 1.5.5 and 1.4.10, we arrive at the following dividend of Itô's construction.

COROLLARY 1.5.8. *Let Λ be the affine vector field determined by $x \rightsquigarrow M(x, \cdot)$, and let $(t, x) \rightsquigarrow P_t(x, \cdot)$ be the associated transition probability function. Then, for each $\mu \in \mathbf{M}_1(\mathbb{Z}_n)$ there is a unique probability measure $\mathbb{P}$ on $D([0, \infty); \mathbb{Z}_n)$ with the property that, for all $m \geq 0$, $0 = t_0 < \cdots < t_m$, and $\{\Gamma_\ell\}_0^m$, (1.4.1) holds with $\mu_0 = \mu$. Further, if $p \in D([0, \infty); \mathbb{Z}_n) \longmapsto \mathbf{T}_m(p) \in \mathbb{N}$ and the σ-algebras $\mathcal{T}_m$ and $\overline{\mathcal{T}_{m-1}}$ are defined as in Theorem 1.4.10, then, for each $m \geq 1$, $\mathbb{P}$-almost surely:*

$$\mathbb{P}\left(\mathbf{T}_m - \mathbf{T}_{m-1} > t \,\middle|\, \overline{\mathcal{T}_{m-1}}\right) = \exp\left(-t M\left(p(\mathbf{T}_{m-1}), \mathbb{Z}_n\right)\right)$$

on $\{\mathbf{T}_{m-1} < \infty\}$ and

$$\mathbb{P}\left(p(\mathbf{T}_m) - p(\mathbf{T}_{m-1}) = y \,\middle|\, \mathcal{T}_m\right) = \frac{M\left(p(\mathbf{T}_{m-1}), \{y\}\right)}{M\left(p(\mathbf{T}_{m-1}), \mathbb{Z}_n\right)}$$

on $\{\mathbf{T}_m < \infty\}$.

PROOF: Only the final assertion requires comment. To verify the first computation, it suffices to treat the case when $\{\mathbf{T}_{m-1} < \infty\} \cap \{p(\mathbf{T}_{m-1}) = \xi\} \supseteq A \ni \overline{\mathcal{T}_{m-1}}$ for some $\xi \in \mathbb{Z}_n$ and to show that

$$\mathbb{P}\left(\{\mathbf{T}_m - \mathbf{T}_{m-1} > t\} \cap A\right) = e^{-t M(\xi, \mathbb{Z}_n)} \mathbb{P}(A).$$

To this end, set $\Gamma' = \{y' : F(\xi, y') \neq 0\}$ and

$$A'(k) = \left\{p' : X(\cdot, x, p') \in A \ \& \ \mathbf{T}_{m-1}\left(X(\cdot, x, p')\right) = \mathbf{T}_{k-1}(p')\right\} \quad \text{for } k \geq 1.$$

Then, because $A'(k) \in \overline{T'_{k-1}}$, Theorem 1.4.10 applied to $\mathbb{P}^{M'}$ implies

$$\mathbb{P}\big(\{\mathbf{T}_m - \mathbf{T}_{m-1} > t\} \cap A\big)$$

$$= \sum_{k=1}^{\infty} \mathbb{P}^{M'}\Big(\big\{\eta\big(\mathbf{T}_{k-1} + t, \Gamma'\big) - \eta\big(\mathbf{T}_{k-1}, \Gamma'\big) = 0\big\} \cap A'(k)\Big)$$

$$= e^{-tM'(\Gamma)} \sum_{k=1}^{\infty} \mathbb{P}^{M'}\big(A'(k)\big) = e^{-tM(\xi, \mathbb{Z}_n)} \mathbb{P}(A).$$

Turning to the second computation, let $\{\mathbf{T}_m < \infty\} \cap \{p(\mathbf{T}_{m-1}) = \xi\} \supseteq A \ni \mathcal{T}_m$ be given, set $\Gamma'(y) = \{y' : F(\xi, y') = y\}$, and define

$$A'(\ell) = \big\{p' : X(\,\cdot\,, x, p') \in A \text{ and } \mathbf{T}_m\big(X(\,\cdot\,, x, p')\big) = \mathbf{T}_\ell(p')\big\}$$

$$B'(\ell) = \big\{p' : X(\,\cdot\,, x, p') \in A$$

$$\text{and } \mathbf{T}_{m-1}\big(X(\,\cdot\,, x, p')\big) \leq \mathbf{T}_{\ell-1}(p') < \mathbf{T}_m\big(X(\,\cdot\,, x, p')\big)\big\}$$

$$C'(\ell) = \big\{p' : F\big(\xi, p'(\mathbf{T}_\ell) - p'(\mathbf{T}_{\ell-1})\big) \neq 0\big\}$$

for $\ell \geq 1$. Then $B'(\ell)$ is $\mathbb{P}^{M'}$-independent of $p' \rightsquigarrow p'(\mathbf{T}_\ell) - p'(\mathbf{T}_{\ell-1})$, $A'(\ell) = B'(\ell) \cap C'(\ell)$, and therefore

$$\mathbb{P}\big(\{p(\mathbf{T}_m) - p(\mathbf{T}_{m-1}) = y\} \cap A\big)$$

$$= \sum_{\ell=1}^{\infty} \mathbb{P}^{M'}\big(\{p'(\mathbf{T}_\ell) - p'(\mathbf{T}_{\ell-1}) \in \Gamma'(y)\} \cap A'(\ell)\big)$$

$$= \sum_{\ell=1}^{\infty} \mathbb{P}^{M'}\big(p'(\mathbf{T}_\ell) - p'(\mathbf{T}_{\ell-1}) \in \Gamma'(y) \,\big|\, C'(\ell)\big) \mathbb{P}\big(A'(\ell)\big) = \frac{M(\xi, \{y\})}{M(\xi, \mathbb{Z}_n)} \mathbb{P}(A).$$

$$\square$$

Now that we know that the measure $\mathbb{P}$ satisfying (1.4.1) lives on the space $D\big([0, \infty); \mathbb{Z}_n\big)$, we will always think of it as a probability measure on that space. Among the many advantages to doing so is that it allows us to give the following characterization of $\mathbb{P}$. The importance of this sort of characterization will not be become clear until we move our considerations from $\mathbb{Z}_n$ to $\mathbb{R}^n$.

COROLLARY 1.5.9. For each $t \in [0, \infty)$, let $\mathcal{B}_t$ be the σ-algebra generated over $D\big([0, \infty); \mathbb{Z}_n\big)$ by the maps $p \rightsquigarrow p(\tau)$ as τ runs over $[0, t]$, and let $\mathcal{B}$ be $\bigvee_{t \geq 0} \mathcal{B}_t$ be the smallest σ-algebra containing the algebra $\bigcup_{t \geq 0} \mathcal{B}_t$. Given an affine vector field Λ determined by $x \rightsquigarrow M(x, \cdot)$, define the operator L on $C(\mathbb{Z}_n; \mathbb{R})$ by

$$L\varphi(x) = \Lambda_{\delta_x} \varphi = \int_{\mathbb{Z}_n} \big(\varphi(x + y) - \varphi(x)\big) M(x, dy).$$

Then, for each $x \in \mathbb{Z}_n$ there is precisely one probability measure $\mathbb{P}_x^L$ on $\big(D([0,\infty);\mathbb{Z}_n),\mathcal{B}\big)$ with the properties that $\mathbb{P}_x^L\big(p(0) = x\big) = 1$ and

$$\left(\varphi(p(t)) - \int_0^t L\varphi(p(\tau))\,d\tau, \mathcal{F}_t, \mathbb{P}_x^L\right)$$

is a martingale for all $\varphi \in C(\mathbb{Z}_n; \mathbb{R})$. In fact, if $(t, x) \rightsquigarrow P_t(x, \cdot)$ is the transition probability function associated with Λ, then $\mathbb{P}_x^L$ is the restriction to $D([0,\infty);\mathbb{Z}_n)$ of the measure $\mathbb{P}$ satisfying (1.4.1) with $\mu = \delta_x$.

PROOF: The first part of the proof of Theorem 1.5.5 was based on the observation that the $\mathbb{P}$ satisfying (1.4.1) is characterized by the way it distributes $p \rightsquigarrow p(0)$ and (1.5.6). Hence, all that we have to do is remark that, because we now know that $(t, p) \rightsquigarrow p(t)$ is measurable and therefore

$$\int_0^t \mathbb{E}^{\mathbb{P}}\Big[L\varphi(p(\tau)),\, A\Big]\,d\tau = \mathbb{E}^{\mathbb{P}}\left[\int_0^t L\varphi(p(\tau))\,d\tau,\, A\right],$$

(1.5.6) is equivalent to the martingale property in the statement here. □

It is convenient to restate the preceding result in slightly different language. Namely, we say that a probability measure $\mathbb{P}$ on $D([0,\infty);\mathbb{Z}_n)$ solves the *martingale problem* for an operator L on $C(\mathbb{Z}_n; \mathbb{R})$ if

$$\left(\varphi(p(t)) - \int_0^t L\varphi(p(\tau))\,d\tau, \mathcal{B}_t, \mathbb{P}\right)$$

is a martingale for all $\varphi \in C(\mathbb{Z}_n; \mathbb{R})$. In this terminology, Corollary 1.5.9 is the statement that if L has the form it has there, then, for each $x \in \mathbb{Z}_n$, there is precisely one solution $\mathbb{P}$ to the martingale problem for L satisfying the initial condition $\mathbb{P}\big(p(0) = x\big) = 1$ and that solution satisfies (1.4.1) when $\mu = \delta_x$.

§1.5.2. Exercises

EXERCISE 1.5.10. In this exercise we will show that Itô's analysis admits an easy extension to the $\mathbb{P}$ in Theorem 1.4.3 corresponding to the integral curve of any vector field on $\mathbf{M}_1(\mathbb{Z}_n)$. Thus, let $(x, \mu) \rightsquigarrow M\big((x, \mu), \cdot\big)$ determine the vector field Λ, suppose that $t \rightsquigarrow \mu_t$ is an integral curve of Λ, and take $\mathbb{P}$ to be the measure in Theorem 1.4.3 corresponding to $t \rightsquigarrow \mu_t$.

(i) Take $n' = n(n-1) + 1$, and define $F : \mathbb{Z}_n \times \mathbb{Z}_{n'} \longrightarrow \mathbb{Z}_n$ as in (1.5.7). Next, define the Lévy measure valued map $t \rightsquigarrow M'(t)$ so that $M'(t, \{y'\}) = M\big((x, \mu_t), \{y\}\big)$ if $y' = f(x, y)$, and let $\mathbb{P}^{M'(\cdot)}$ be the probability measure on $D([0,\infty);\mathbb{Z}_{n'})$ associated with $t \rightsquigarrow M'(t)$ by the prescription in Exercise 1.4.14. Finally, define $p' \rightsquigarrow X(\cdot, x, p')$ as in (1.5.4), and show that $\mathbb{P}$ is the distribution of $(x, p') \rightsquigarrow X(\cdot, x, p')$ under $\mu_0 \times \mathbb{P}^{M'(\cdot)}$. In particular, conclude that $\mathbb{P}$ lives on $D([0,\infty;\mathbb{Z}_n)$.

(ii) Thinking of $\mathbb{P}$ as a measure on $D([0,\infty);\mathbb{Z}_n)$, show that (cf. Theorem 1.4.10 for the notation)

$$\mathbb{P}\big(\mathbf{T}_m - \mathbf{T}_{m-1} > t \,\big|\, \overline{\mathcal{T}_{m-1}}\big) = \exp\left(-\int_{\mathbf{T}_{m-1}}^{\mathbf{T}_{m-1}+t} M\big((p(\tau),\mu_\tau),\mathbb{Z}_n\big)\,d\tau\right)$$

and

$$\mathbb{P}\big(p(\mathbf{T}_m) - p(\mathbf{T}_{m-1}) = y \,\big|\, \mathcal{T}_m\big) = \frac{M\big((p(\mathbf{T}_{m-1}),\mu_{\mathbf{T}_m}),\{y\}\big)}{M\big((p(\mathbf{T}_m),\mu_{\mathbf{T}_{m-1}}),\mathbb{Z}_n\big)}$$

$\mathbb{P}$-almost surely on $\{\mathbf{T}_{m-1} < \infty\}$ and $\{\mathbf{T}_m < \infty\}$, respectively.

1.6 Another Approach

Although the prescription given by (1.5.4) has served us well, it will not work when we move our considerations to $\mathbb{R}^n$. On the other hand, there is another way to formulate the same prescription, namely, as an integral equation, and this formulation lends itself better to generalization. Namely, (cf. Theorem 1.4.10 for the notation here) one can express the relationship between p' and $X(\cdot, x, p')$ as the integral equation

$$(1.6.1) \quad X(t,x,p') = x + \int_{[0,t]} F\big(X(\tau-,x,p'),p'(\tau) - p'(\tau-)\big)\,\eta(d\tau,p'),$$

where $\eta(\cdot, p')$ is the $\mathbb{N}$-valued measure on $[0,\infty)$ determined by the non-decreasing function $t \rightsquigarrow \eta(t,\mathbb{Z}_n,p')$. However, it is essential that one be extremely careful about one's interpretation of (1.6.1)! In particular, as the notation is intended to indicate, the integral on the right hand side is a Lebesgue and definitely *not* a Riemann integral. Indeed, because they can jump simultaneously, $t \rightsquigarrow F\big(X(t-,x,p')\big)$ cannot be Riemann integrable with respect to $t \rightsquigarrow \eta(t,\mathbb{Z}_n,p')$. Thus, the sense in which this integral must be taken is the one given by Lebesgue. Equivalently, this integral is equal to

$$\sum_{y\in\mathbb{Z}_n}\sum_{y'\in\mathbb{Z}'_n\setminus\{0\}} F(y,y')\eta\big(T(t,y,y',p'),p'\big),$$

where $T(t,y,y',p') = \{\tau \in [0,t] \,:\, X(\tau-,x,p') = y \ \& \ p(\tau) - p(\tau-) = y'\}$ and

$$\eta\big(T(t,y,y',p'),p'\big) = \sum_{\tau\geq 0} \mathbf{1}_{T(t,y,y',p')}(\tau)$$

is the measure which the measure $\eta(\cdot, p')$ assigns to the set $T(t,y,y',p') \subseteq [0,\infty)$. Equally important is the observation that the decision to evaluate the integrand at $X(\tau-, x, p')$, and not $X(\tau, x, p')$, is forced on us by (1.5.4). We have no choice in the matter.

Our purpose in this section is to show that the integral equation (1.6.1) can be solved by an Euler approximation scheme and that the approximants are the pathspace counterpart of the measures $\{\mu_t^N : t \geq 0\}$ given by (1.3.4) in the same way as $p' \rightsquigarrow X(\cdot, x, p')$ is the pathspace counterpart of $\{\mu_t : t \geq 0\}$.

§1.6.1. Itô's Approximation Scheme: Let $x \rightsquigarrow M(x, \cdot)$ determine the affine vector field Λ on $\mathbf{M}_1(\mathbb{Z}_n)$, and choose n', a Lévy measure M' on $\mathbb{Z}_{n'}$, and $F : \mathbb{Z}_n \times \mathbb{Z}_{n'} \longrightarrow \mathbb{Z}_n$ (cf. (1.5.3) and the discussion preceding Corollary 1.5.8) so that $M(x, \cdot) = F(x, \cdot)_* M'$ for each $x \in \mathbb{Z}_n$. Given $x \in \mathbb{Z}_n$ and $N \geq 0$, define, as in (1.3.4), the Euler approximation $t \rightsquigarrow \mu_t^N$ to the integral curve of Λ starting at δ_x. Itô's idea is to mimic Euler's scheme path-by-path. To this end, for each $p' \in D([0, \infty); \mathbb{Z}_{n'})$, and $N \geq 0$, define $X^N(\cdot, x, p') \in D([0, \infty); \mathbb{Z}_n)$ so that

$$X^N(0, x, p') = x$$

$$(1.6.2) \quad \begin{aligned} X^N(t, x, p') &- X^N(m2^{-N}, x, p') \\ &= \int_{\mathbb{Z}_{n'}} F(X^N(m2^{-N}, x, p'), y')(\eta(t, dy', p') - \eta(m2^{-N}, dy', p')) \end{aligned}$$

when $m2^{-N} < t \leq (m+1)2^{-N}$. Again it is clear that $p' \rightsquigarrow X^N(s, x, p)$ is $\sigma(\{p'(\sigma) : \sigma \in [0, s]\})$-measurable. Further, given $\sigma(\{p'(\sigma) : \sigma \in [0, k2^{-m}]\})$, the conditional distribution under $\mathbb{P}^{M'}$ of the restriction of $p' \rightsquigarrow X_F(\cdot, p')$ to $[m2^{-N}, (m+1)2^{-N}]$ is exactly the same as the $\mathbb{P}^{M'}$-distribution of the restriction of (cf. (1.5.2))

$$q' \rightsquigarrow X^N(m2^{-N}, p') + X_{F(X^N(m2^{-N}, x, p'), \cdot)}(\cdot - m2^{-N}, q')$$

to the same interval. In particular, this proves that the $\mathbb{P}^{M'}$-distribution of $p' \rightsquigarrow X^N(t, x, p')$ is exactly μ_t^N, thereby supporting our claim that the procedure here is mimicking Euler's scheme path-by-path. What we want to show now is that, for $\mathbb{P}^{M'}$-almost every p', the sequence $\{X^N(\cdot, x, p') : N \geq 0\}$ converges, uniformly on finite intervals, to the process $p' \rightsquigarrow X(\cdot, x, p')$ given by (1.5.4).

THEOREM 1.6.3. *Referring to the preceding,*

$$(1.6.4) \quad \begin{aligned} \mathbb{P}^{M'}(\{p' : \exists N' \geq N \ X(\cdot, x, p') \upharpoonright [0, T] &\neq X^{N'}(\cdot, x, p') \upharpoonright [0, T]\}) \\ &\leq (T+1)M'(\mathbb{Z}_{n'})^2 2^{-N}. \end{aligned}$$

In particular, $X^N(\cdot, x, p') \longrightarrow X(\cdot, x, p')$ uniformly on finite intervals for $\mathbb{P}^{M'}$-almost every p'.

PROOF: Observe that as long as at most one jump of p' occurs in each of the intervals $(m2^{-N}, (m+1)2^{-N}]$ for $0 \le m < 2^N T$, $X(\cdot, x, p')$ coincides with $X^N(\cdot, x, p')$ on the interval $[0, T]$. Thus,

$$
\mathbb{P}^{M'}\big(\{p' : \exists N' \ge N \ X(\cdot, x, p') \restriction [0,T] \ne X^{N'}(\cdot, x, p') \restriction [0,T]\}\big)
$$

$$
\le \mathbb{P}^{M'}\big(\{p' : \exists m < 2^N T \ \eta((m+1)2^{-N}, \mathbb{Z}_{n'}) - \eta(m2^{-N}, \mathbb{Z}_{n'}) \ge 2\}\big)
$$

$$
\le 2^N(T+1)\mathbb{P}^{M'}\big(\eta(2^{-N}, \mathbb{Z}_{n'}) \ge 2\big)
$$

$$
= 2^N(T+1)e^{-2^{-N}M'(\mathbb{Z}_{n'})} \sum_{\ell=2}^{\infty} \frac{\big(2^{-N}M'(\mathbb{Z}_{n'})\big)^{\ell}}{\ell!} \le \frac{(T+1)M'(\mathbb{Z}_{n'})^2}{2^N}. \quad \square
$$

§1.6.2. Exercises

EXERCISE 1.6.5. Refer to (1.5.2) and (1.6.2).

(i) Set

$$
X_m^N(p') = X^N(m2^{-N}, x, p') \quad \text{and} \quad \tilde{X}_m^N(t, p', q') = X_{F(X_m^N(p'), \cdot)}(t, q').
$$

Assuming that $m2^{-N} \le s < (m+1)2^{-N}$, show that the $\mathbb{P}^{M'}$-distribution of $p' \rightsquigarrow X^N(s + \cdot, x, p') \restriction [0, (m+1)2^{-N} - s]$ coincides with the $\mathbb{P}^{M'} \times \mathbb{P}^{M'}$-distribution of

$$
(p', q') \rightsquigarrow X^N(s, x, p') + \tilde{X}^N(\cdot, p', q') \restriction [0, (m+1)2^{-N} - s].
$$

(ii) Given $t \in [0, \infty)$, use $[t]$ to denote the integer part (i.e., the largest integer dominated by) t, and, for $N \in \mathbb{N}$, set $[t]_N = [2^N t]2^{-N}$. Show that for any $s \ge 0$ and $A \in \sigma(\{p'(\sigma) : \sigma \in [0, s]\})$,

$$
\mathbb{E}^{\mathbb{P}^{M'}}\big[\varphi(X^N(s+t, x)) - \varphi(X^N(s, x)), A\big]
$$

$$
= \mathbb{E}^{\mathbb{P}^{M'}}\bigg[\int_0^t \bigg(\int \big(\varphi(X^N(s+\tau, x) + y)
$$

$$
- \varphi(X^N(s+\tau, x))\big)\, M\big(X^N([s+\tau]_N, x), dy\big)\bigg)\, d\tau, A\bigg].
$$

In fact, show that this property, together with the initial condition

$$
\mathbb{P}^{M'}\big(X^N(0, x) = x\big) = 1
$$

completely characterizes the $\mathbb{P}^{M'}$-distribution of $p' \rightsquigarrow X^N(\cdot, x, p')$.

(iii) Starting from the result in (ii) and using the estimates (1.6.4) and

$$
\mathbb{P}^{M'}\big(\eta((m+1)2^{-N}, \mathbb{Z}_{n'}) - \eta(m2^{-N}, \mathbb{Z}_{n'}) \ge 1\big) \le M(\mathbb{Z}_{n'})2^{-N},
$$

give another derivation of Theorem 1.5.5.

CHAPTER 2

Moving to Euclidean Space, the Real Thing

The preceding chapter was an extended introduction which was offered in the hope that it would help to elucidate the ideas on which this book rests. Whether that hope was realistic will be apparent soon. In any case, in this and the next chapters we will be attempting to mimic for $\mathbb{R}^n$ what we did for $\mathbb{Z}_n$ in Chapter 1.

2.1 Tangent Vectors to $\mathbf{M}_1(\mathbb{R}^n)$

Throughout, the topology on $\mathbf{M}_1(\mathbb{R}^n)$ will be the weak topology: the topology corresponding to convergence of measures when tested against bounded continuous functions. As is well known (cf. §3.1 [36]), the weak topology makes $\mathbf{M}_1(\mathbb{R}^n)$ into a Polish space. That is, $\mathbf{M}_1(\mathbb{R}^n)$ with the weak topology is separable and admits a complete metric.

§**2.1.1. Differentiable Curves on $\mathbf{M}_1(\mathbb{R}^n)$:** We now want to put a differentiable structure on $\mathbf{M}_1(\mathbb{R}^n)$, and for that purpose we begin by describing what we will mean when we talk about differentiable curves there. Namely, we will say that the curve $t \in [0, \infty) \longmapsto \mu_t \in \mathbf{M}_1(\mathbb{R}^n)$ is *differentiable* if there exists for each $t \in [0, \infty)$ a linear functional $\dot{\mu}_t$ on $C_b^\infty(\mathbb{R}^n; \mathbb{C})$ such that

(1) $t \rightsquigarrow \dot{\mu}_t$ is *continuous* in the sense that

$$(t, \varphi) \in [0, \infty) \times C_b^\infty(\mathbb{R}^n; \mathbb{C}) \longmapsto \dot{\mu}_t \varphi \in \mathbb{C}$$

is continuous.[1]

(2) $t \rightsquigarrow \dot{\mu}_t$ is the *derivative* of $t \rightsquigarrow \mu_t$ in the sense that, for each $(t, \varphi) \in [0, \infty) \times C_b^\infty(\mathbb{R}^n; \mathbb{C})$,

$$\langle \varphi, \mu_t \rangle - \langle \varphi, \mu_0 \rangle = \int_0^t \dot{\mu}_\tau \varphi \, d\tau.$$

---

[1] For each $k \in \mathbb{N}$, the topology which we put on $C_b^k(\mathbb{R}^n; \mathbb{C})$ is the one for which $\varphi_m \longrightarrow \varphi$ if and only if $\sup_m \|\varphi_m\|_{C_b^k(\mathbb{R}^n; \mathbb{C})} < \infty$ and $\partial^\alpha \varphi_m \longrightarrow \partial^\alpha \varphi$ uniformly on compacts for each $\alpha \in \mathbb{N}^n$ with $\|\alpha\| \equiv \sum \alpha_i \leq k$. The topology on $C_b^\infty(\mathbb{R}^n; \mathbb{C})$ is the one for which $\varphi_m \longrightarrow \varphi$ means that $\varphi_m \longrightarrow \varphi$ in $C_b^k(\mathbb{R}^n; \mathbb{C})$ for every $k \geq 0$.

At this point, the reader should be expecting me to say that the tangent space to $\mathbf{M}_1(\mathbb{R}^n)$ at $\mu \in \mathbf{M}_1(\mathbb{R}^n)$ is the set of linear functionals on $C_b^\infty(\mathbb{R}^n; \mathbb{C})$ which can be represented as $\dot{\mu}_0$ for some differentiable $t \rightsquigarrow \mu_t$ starting from μ. Indeed, this is the definition which I would like, and it is the one which I will adopt when μ is a point mass δ_x. However, for technical reasons, it is not the definition which I will take at μ's other than a point mass.

§2.1.2. **Infinitely Divisible Flows on $\mathbf{M}_1(\mathbb{R}^n)$:** In preparation for our analysis of the tangent space to $\mathbf{M}_1(\mathbb{R}^n)$ at δ_0, we will need to review some classical facts about the semigroup structure of $\mathbf{M}_1(\mathbb{R}^n)$. In particular, because they play the role of rays in $\mathbf{M}_1(\mathbb{R}^n)$, we will need to recall the notion of an infinitely divisible flow on $\mathbf{M}_1(\mathbb{R}^n)$.

It is probably unnecessary to mention, but $\mathbb{R}^n$ is an Abelian group under addition, and the associated convolution product $M_1 \star M_2$ of finite, Borel measures M_1 and M_2 is determined by[2]

$$M_1 \star M_2(\Gamma) = \iint_{\mathbb{R}^n \times \mathbb{R}^n} \mathbf{1}_\Gamma(x + y)\, M_1(dx) M_2(dy), \quad \text{for } \Gamma \in \mathcal{B}_{\mathbb{R}^n}.$$

In particular, $\mathbf{M}_1(\mathbb{R}^n)$ becomes an Abelian semigroup under convolution. In this connection, the Fourier transform $\hat{\mu} \in C_b(\mathbb{R}^n; \mathbb{C})$ of $\mu \in \mathbf{M}_1(\mathbb{R}^n)$ is given by

$$(2.1.1) \qquad \hat{\mu}(\xi) = \int_{\mathbb{R}^n} \mathbf{e}_\xi(y)\, \mu(dy), \quad \text{where } \mathbf{e}_\xi(y) \equiv \exp\big(\sqrt{-1}\,(\xi, y)_{\mathbb{R}^n}\big),$$

and

$$\widehat{\mu \star \nu} = \hat{\mu}\hat{\nu} \quad \text{for all } \mu, \nu \in \mathbf{M}_1(\mathbb{R}^n).$$

Finally, for each $\varphi \in L^1(\mathbb{R}^n; \mathbb{C}) \cap C_b(\mathbb{R}^n; \mathbb{C})$ whose Fourier transform

$$\hat{\varphi}(\xi) \equiv \int_{\mathbb{R}^n} \mathbf{e}_\xi(y)\, \varphi(y)\, dy$$

is in $L^1(\mathbb{R}^n; \mathbb{C})$: (cf. Lemma 2.2.8 in [36])

$$(2.1.2) \qquad\qquad \langle \varphi, \mu \rangle = \frac{1}{(2\pi)^n} \int_{\mathbb{R}^n} \hat{\varphi}(\xi)\hat{\mu}(-\xi)\, d\xi.$$

A curve $t \in [0, \infty) \longmapsto \lambda_t \in \mathbf{M}_1(\mathbb{R}^n)$ is said to be an *infinitely divisible flow* if $t \rightsquigarrow \lambda_t$ is continuous and $\lambda_{s+t} = \lambda_s \star \lambda_t$ for all $s, t \geq 0$. Notice that, as distinguished from the situation on $\mathbb{Z}_n$, all infinitely divisible flows on $\mathbb{R}^n$

---

[2] When E is a topological space, $\mathcal{B}_E$ is the Borel field over E.

start at 0. Indeed, because (cf. part (ii) in Exercise 1.2.8) $\lambda_0 = \lambda_0 \star \lambda_0$, λ_0 must be the normalized Haar measure for some subgroup of $\mathbb{R}^n$, and $\{0\}$ is the only subgroup of $\mathbb{R}^n$ which admits a normalized Haar measure. We now want to show that each infinitely divisible flow $t \rightsquigarrow \lambda_t$ is differentiable and that the form of the tangent $\dot{\lambda}_t$ is well understood. For this purpose, recall that a *Lévy system* is a triple (a, b, M) consisting of a non-negative definite, $n \times n$ symmetric matrix a, a vector $b \in \mathbb{R}^n$, and a *Lévy measure* M: a (non-negative) Borel measure M on $\mathbb{R}^n$ with the properties that $M(\{0\}) = 0$ and that $y \rightsquigarrow \frac{|y|^2}{1+|y|^2}$ is M-integrable. We associate with a Lévy system (a, b, M) the (pseudodifferential) operator $L^{(a,b,M)}$ given by

$$L^{(a,b,M)}\varphi(x) = \frac{1}{2}\sum_{i,j=1}^{n} a_{i,j}\partial_i\partial_j\varphi(x) + \sum_{i=1}^{n} b_i\partial_i\varphi(x)$$

$$(2.1.3) \qquad + \int_{\mathbb{R}^n}\left(\varphi(y+x) - \varphi(x) - \frac{(y, \operatorname{grad}\varphi(x))_{\mathbb{R}^n}}{1+|y|^2}\right) M(dy).$$

Our main goal in this subsection is to make the preparations which will allow us to show that each infinitely divisible flow $t \rightsquigarrow \lambda_t$ is differentiable and is associated with a unique Lévy system (a, b, M) in such a way that

$$(2.1.4) \qquad \dot{\lambda}_t\varphi = \langle L^{(a,b,M)}\varphi, \lambda_t\rangle \quad \text{for} \quad t \geq 0 \text{ and } \varphi \in C_b^2(\mathbb{R}^n; \mathbb{C}).$$

As the cognoscenti already realize, (2.1.4) is really just a restatement of the famous *Lévy–Khinchine formula* for infinitely divisible laws. Thus, it will come as no surprise to them that what we are about to do yields a proof of that formula.

LEMMA 2.1.5. *Given a Lévy system (a, b, M), the operator L in (2.1.3) is well defined on $C^2(\mathbb{R}^n; \mathbb{C})$ and, for each $R \in (0, \infty)$, satisfies the estimate*[3]

$$(2.1.6) \qquad \left|L^{(a,b,M)}\varphi(x)\right| \leq C\|\varphi\|_{C_b^2(B(x,R);\mathbb{C})} + M\big(B(x,R)\complement\big)\|\varphi\|_{\mathrm{u}}.$$

In particular, $\varphi \in C_b^2(\mathbb{R}^n; \mathbb{C}) \longmapsto L^{(a,b,M)}\varphi \in C_b(\mathbb{R}^n; \mathbb{C})$ is continuous. Finally, if φ is in the Schwartz test function class[4] $\mathcal{S}(\mathbb{R}^n; \mathbb{C})$, *then*[5]

$$\widehat{L^{(a,b,M)}\varphi}(\xi) = \ell^{(a,b,M)}(-\xi)\hat{\varphi}(\xi) \quad \text{where}$$

$$(2.1.7) \qquad \ell^{(a,b,M)}(\xi) = -\frac{(\xi, a\xi)_{\mathbb{R}^n}}{2} + \sqrt{-1}(b, \xi)_{\mathbb{R}^n}$$

$$+ \int_{\mathbb{R}^n}\left(\mathfrak{e}_\xi(y) - 1 - \frac{\sqrt{-1}(\xi, y)_{\mathbb{R}^n}}{1+|y|^2}\right) M(dy).$$

---

[3] We use $\|\cdot\|_{\mathrm{u}}$ to denote the *uniform* or supremum norm of a function.
[4] The class of smooth functions φ with the property that $x \rightsquigarrow (1 + |x|^2)^m \partial^\alpha \varphi(x)$ is bounded for each $m \in \mathbb{N}$ and all $\alpha \in \mathbb{N}^n$.
[5] In the jargon of pseudodifferential operator theory, the statement which follows identifies $\xi \rightsquigarrow \ell^{(a,b,M)}(-\xi)$ as the *symbol* of $L^{(a,b,M)}$.

PROOF: As an application of Taylor's theorem, one knows that there exists a $\kappa < \infty$, depending only on n, such that

$$\left| \varphi(y+x) - \varphi(x) - \frac{(y, \mathrm{grad}\gamma(x))_{\mathbb{R}^n}}{1 + |y|^2} \right| \leq \frac{\kappa|y|^2}{1 + |y|^2} \|\varphi\|_{C_b^2(\mathbb{R}^n;\mathbb{C})}.$$

Thus, both the definition of $L^{(a,b,M)}$ on $C_b^2(\mathbb{R}^n; \mathbb{C})$ is justified and the bound (2.1.6) is clear. Furthermore, if, for each $\epsilon > 0$, $M^\epsilon(dy) \equiv \mathbf{1}_{[\epsilon,\infty)}(|y|)M(dy)$, then, for each $\epsilon > 0$, there is no question that $L^{(a,b,M^\epsilon)}\varphi$ determines a continuous map from $C_b^2(\mathbb{R}^n; \mathbb{C})$ into $C_b(\mathbb{R}^n; \mathbb{C})$. Hence, because

$$\left| L^{(a,b,M)}\varphi(x) - L^{(a,b,M^\epsilon)}\varphi(x) \right| \leq \kappa\|\varphi\|_{C_b^2(\mathbb{R}^n;\mathbb{C})} \int_{|y|<\epsilon} \frac{|y|^2}{1 + |y|^2} \, M(dy),$$

it is clear that the same continuity result extends to $\varphi \rightsquigarrow L^{(a,b,M)}\varphi$. Finally, (2.1.7) is an elementary application of the basic theory of the Fourier transform and is therefore left to the reader. $\quad\square$

We next give an operator theoretic characterization of Lévy systems.

LEMMA 2.1.8. *Use* $D(\mathbb{R}^n; \mathbb{R})$ *to denote the space of functions* $\varphi \in C^\infty(\mathbb{R}^n; \mathbb{R})$ *for which there exists a* $\varphi(\infty) \in \mathbb{R}$ *with the property that* $\varphi - \varphi(\infty)$ *has compact support. Then a linear functional* $A : D(\mathbb{R}^n; \mathbb{R}) \longrightarrow \mathbb{R}$ *is given by (cf.* (2.1.3)) $A\varphi = L^{(a,b,M)}\varphi(0)$ *for some Lévy system* (a, b, M) *if and only if* A *satisfies*

 (1) *The minimum principle:* $A\varphi \geq 0$ *whenever 0 is a global minimum of* $\varphi \in D(\mathbb{R}^n; \mathbb{R})$,
 (2) *Tightness: for any* $\varphi \in D(\mathbb{R}^n; \mathbb{R})$, $\lim_{R\to\infty} A\varphi_R = 0$ *where* $\varphi_R(y) \equiv \varphi(R^{-1}y)$.

PROOF: First observe that, for any Lévy system (a, b, M), $\varphi \rightsquigarrow L^{(a,b,M)}\varphi(0)$ satisfies both the minimum principle and the tightness property. Indeed, the tightness property is an easy variant on the argument to prove (2.1.6): by Taylor's theorem, for any $R \geq 1$:

$$\left| L^{(a,b,M)}\varphi_R(0) \right| \leq \|\varphi\|_{C_b^2(\mathbb{R}^n;\mathbb{R})} \left(R^{-2}\|a\|_{\mathrm{op}} + R^{-1}|b| \right) + 3M\big(B_{\mathbb{R}^n}(0, \sqrt{R})\complement\big)$$

$$+ \int_{|y|\leq\sqrt{R}} \left(R^{-2}|y|^2 + R^{-1}\frac{|y|^3}{1 + |y|^2} \right) M(dy) \bigg).$$

As for the minimum principle, note that if 0 is a global minimum of $\varphi \in D(\mathbb{R}^n; \mathbb{R})$, then, by the first and second order derivative tests, its gradient vanishes at 0 and its Hessian is non-negative definite there. Hence, because a is non-negative definite,

$$L^{(a,b,M)}\varphi(0) \geq \int_{\mathbb{R}^n} \Big(\varphi(y) - \varphi(0)\Big)\, M(dy) \geq 0.$$

Now suppose that A is a linear functional on $D(\mathbb{R}^n; \mathbb{R})$ which satisfies the minimum principle and the tightness condition. Let $\psi \in C^\infty(\mathbb{R}; [0,1])$ be a function with the properties that $\psi \upharpoonright (-\infty, 1] \equiv 1$, $\psi \upharpoonright [1,2]$ is nonincreasing, and $\psi \upharpoonright [2, \infty) \equiv 0$. For $R > 0$, define $\eta_R \in C_c^\infty(\mathbb{R}^n; [0,1])$ by $\eta_R(x) = \psi(R^{-1}|x|)$ and $\varphi \rightsquigarrow A^R\varphi$ so that $A^R\varphi = A\big((1 - \eta_R)\varphi\big)$. By the minimum principle, we know that $\varphi \geq 0 \implies A^R\varphi \geq 0$, and so $\varphi_1 \leq \varphi_2 \implies A^R\varphi_1 \leq A^R\varphi_2$. In particular, for $r \geq 2R$,

$$\big|A^R\varphi\big| \leq \|\varphi\|_{C_{\mathrm{b}}(B_{\mathbb{R}^n}(0,2r); \mathbb{R})} A\big((1 - \eta_R)\eta_r\big) + \|\varphi\|_{C_{\mathrm{b}}(\mathbb{R}^n \setminus B_{\mathbb{R}^n}(0,2r); \mathbb{R})} A(1 - \eta_r),$$

and therefore, by the tightness property, for each $\epsilon > 0$ we can choose $r_\epsilon \in (2R, \infty)$ so that

$$\big|A^R\varphi\big| \leq K_R \|\varphi\|_{C_{\mathrm{b}}(B_{\mathbb{R}^n}(0,r_\epsilon); \mathbb{C})} + \epsilon\|\varphi\|_{C_{\mathrm{b}}(\mathbb{R}^n \setminus B_{\mathbb{R}^n}(0,r_\epsilon); \mathbb{C})},$$

where $K_R = A(1 - \eta_R)$. This means, by the form of the Riesz representation theorem in Lemma 3.1.7 of [36], there exists a finite Borel measure M^R on $\mathbb{R}^n$ such that $A^R\varphi = \int_{\mathbb{R}^n} \varphi(y)\, M^R(dy)$. Moreover, because $A^{R_1}\varphi = A^{R_2}\varphi$ if $R_1 \leq R_2$ and φ vanishes on $B_{\mathbb{R}^n}(0, 2R_2)$, we have now proved that there is a unique Borel measure M on $\mathbb{R}^n$ such that $M(\{0\}) = 0$ and

$$(*)\qquad A\big((1 - \eta_R)\varphi\big) = \int_{\mathbb{R}^n} \big(1 - \eta_R(y)\big)\varphi(y)\, M(dy) \quad \text{for all } R > 0.$$

Next, given $\xi \in \mathbb{R}^n$, set $\varphi_\xi(y) = (\xi, y)_{\mathbb{R}^n}$, and apply the minimum principle to see that

$$|\xi|^2 A(|y|^2\eta_1) \geq A\big(\varphi_\xi^2\eta_{R_2}\big) \geq A\big(\varphi_\xi^2\eta_{R_1}\big) \geq 0 \quad \text{for } 0 < R_1 \leq R_2 \leq 1.$$

Hence, by polarization, we conclude that there exists a symmetric, non-negative definite $a \in \mathrm{Hom}(\mathbb{R}^n; \mathbb{R}^n)$ such that

$$(**)\qquad \lim_{R \searrow 0} A\big(\varphi_\xi\varphi_{\xi'}\eta_R\big) = (\xi, a\xi')_{\mathbb{R}^n} \quad \text{for all } \xi,\, \xi' \in \mathbb{R}^n.$$

In addition, if φ and all its derivatives of first and second order vanish at 0, then there exists a $C < \infty$ for which $|\varphi(y)|\eta_R(y) \leq CR|y|^2\eta_R(y)$ and so

$$(***)\qquad \lim_{R \searrow 0} A\big(\varphi\eta_R\big) = 0 \quad \text{when } \partial^\alpha\varphi(0) = 0 \text{ for all } |\alpha| \leq 2.$$

To complete the proof from here, let $\varphi \in D(\mathbb{R}^n; \mathbb{C})$ be given, and, for $R > 0$, write $\varphi(y)$ as

$$\varphi(0) + \eta_1(y)\big(y, \operatorname{grad}\varphi(0)\big)_{\mathbb{R}^n} + \tfrac{1}{2}\eta_R(y)\big(y, \operatorname{Hess}\varphi(0)y\big)_{\mathbb{R}^n}$$
$$+ \eta_R(y)\Big(\varphi(y) - \varphi(0) - \eta_1(y)\big(y, \operatorname{grad}\varphi(0)\big)_{\mathbb{R}^n} - \tfrac{1}{2}\big(y, \operatorname{Hess}\varphi(0)y\big)_{\mathbb{R}^n}\Big)$$
$$+ \big(1 - \eta_R(y)\big)\Big(\varphi(y) - \varphi(0) - \eta_1(y)\big(y, \operatorname{grad}\varphi(0)\big)_{\mathbb{R}^n}\Big),$$

where $\operatorname{Hess}\varphi$ denotes the Hessian of φ. By the minimum principle applied to $\mathbf{1}$ and $-\mathbf{1}$, we know that A annihilates constants. Thus, by linearity,

$$A\varphi = \frac{1}{2}\sum_{i,j=1}^{n} a_{ij}^{R}\partial_{x_i}\partial_{x_j}\varphi(0) + \sum_{i=1}^{n} \tilde{b}_i \partial_{x_i}\varphi(0) + A\big(\eta_R\tilde{\varphi}\big)$$
$$+ \int_{\mathbb{R}^n} \big(1 - \eta_R(y)\big)\Big(\varphi(y) - \varphi(0) - \eta_1(y)\big(y, \operatorname{grad}\varphi(0)\big)_{\mathbb{R}^n}\Big)M(dy),$$

where

$$\tilde{b}_i = A(\eta_1\varphi_i) \quad \text{and} \quad a_{i,j}^{R} = A\big(\eta_R\varphi_i\varphi_j\big) \quad \text{with } \varphi_i(y) = y_i$$

when $y = (y_1, \ldots, y_n)$ and

$$\tilde{\varphi}(y) = \varphi(y) - \varphi(0) - \eta_1(y)\big(y, \operatorname{grad}\varphi(0)\big)_{\mathbb{R}^n} - \tfrac{1}{2}\big(y, \operatorname{Hess}\varphi(0)y\big)_{\mathbb{R}^n}.$$

Hence, by (*), (**), and (***), we obtain

$$A\varphi = \frac{1}{2}\sum_{i,j=1}^{n} a_{i,j}\partial_{x_i}\partial_{x_j}\varphi(0) + \sum_{i=1}^{n} \tilde{b}_i \partial_{x_i}\varphi(0)$$
$$+ \int_{\mathbb{R}^n} \Big(\varphi(y) - \varphi(0) - \eta_1(y)\big(y, \operatorname{grad}\varphi(0)\big)_{\mathbb{R}^n}\Big)M(dy)$$

after letting $R \searrow 0$. Starting from here, it is easy to check that $\frac{|y|^2}{1+|y|^2}$ is M-integrable and that $A\varphi = L^{(a,b,M)}\varphi(0)$ where

$$b \equiv \tilde{b} + \int_{\mathbb{R}^n} \left(\frac{y}{1 + |y|^2} - \eta_1(y)y\right) M(dy). \qquad \square$$

With these preparations, we can now prove the *Lévy–Khinchine formula*.[6]

---

[6] There are many proofs of this renowned formula, most of which rely more heavily on Fourier analysis. For example, see §3.2 in [36] for the case when $n = 1$ and Chapter VI of [26] for the result in the setting of a general locally compact, Abelian group.

THEOREM 2.1.9. *Suppose that $\lambda \in \mathbf{M}_1(\mathbb{R}^n)$ is infinitely divisible in the sense that, for each $m \geq 1$, there is an $\lambda_{1/m} \in \mathbf{M}_1(\mathbb{R}^n)$ such that $\lambda = \lambda_{1/m}^{\star m}$. Then there exists a unique Lévy system (a, b, M) with the property that (cf. (2.1.7))*

$$(2.1.10) \qquad \hat{\lambda}(\xi) = f^{(a,b,M)} \equiv e^{\ell^{(a,b,M)}(\xi)}.$$

Conversely, if (a, b, M) is a Lévy system, then there is a unique probability measure $\lambda^{(a,b,M)}$ for which (2.1.10) holds with $\lambda = \lambda^{(a,b,M)}$, and $\lambda^{(a,b,M)}$ is infinitely divisible.

PROOF: Because it is the easier part, we begin with the converse statement. First note that it suffices to prove that, for each (a, b, M), $f^{(a,b,M)}(\xi)$ is the Fourier transform of a probability measure, since infinite divisibility will then follow when one replaces (a, b, M) by $\left(\frac{1}{m}a, \frac{1}{m}b, \frac{1}{m}M\right)$. Next, observe that it suffices to treat the cases $(a, b, 0)$ and $(0, 0, M)$ separately, since, if $\lambda^{(a,b,0)}$ and $\lambda^{(0,0,M)}$ exist, then $\lambda^{(a,b,M)}$ exists and is equal to $\lambda^{(a,b,0)} \star \lambda^{(0,0,M)}$. In addition, $f^{(a,b,0)}$ is easily recognized as the Fourier transform of the Gauss measure on $\mathbb{R}^n$ with mean b and covariance a. Thus, everything comes down to proving the existence of $\lambda^{(0,0,M)}$. To this end, define M^ϵ as in the proof of Lemma 2.1.5, set $\beta^\epsilon = -\int_{\mathbb{R}^n} \frac{y}{1+|y|^2} M^\epsilon(dy)$, and define

$$\lambda^\epsilon = \delta_{\beta^\epsilon} \star \left(e^{-M^\epsilon(\mathbb{R}^n)} \sum_{m=0}^{\infty} \frac{1}{m!} (M^\epsilon)^{\star m} \right).$$

It is then an easy matter to check that $\lambda^\epsilon \in \mathbf{M}_1(\mathbb{R}^n)$ and that $\widehat{\lambda^\epsilon}(\xi) = f^{(0,0,M^\epsilon)}(\xi)$. Finally, observe that $\ell^{(0,0,M^\epsilon)} \longrightarrow \ell^{(0,0,M)}$ uniformly on compacts, and apply Lévy's continuity theorem (cf. Exercise 3.1.19 in [36]) to conclude that $\lambda^\epsilon \longrightarrow \lambda^0$ in $\mathbf{M}_1(\mathbb{R}^n)$, where $\widehat{\lambda^0} = f^{(0,0,M)}$. That is, $\lambda^0 = \lambda^{(0,0,M)}$.

We now return to first part of the theorem. Thus, suppose that λ is infinitely divisible, set $f = \hat{\lambda}$, and let $\lambda_{1/m}$ be given accordingly. Our strategy for showing that $\lambda = \lambda^{(a,b,M)}$ for some Lévy system (a, b, M) will be to prove that

$$(*) \qquad A\varphi \equiv \lim_{m \to \infty} m\big(\langle \varphi, \lambda_{1/m} \rangle - \varphi(0)\big)$$

exists for φ from the class $D(\mathbb{R}^n; \mathbb{R})$ considered in Lemma 2.1.8 and to verify that the resulting linear functional A satisfies the minimum principle and the tightness condition described in that lemma.

The first step is to show that $\lambda_{1/m}$ tends to δ_0 in $\mathbf{M}_1(\mathbb{R}^n)$ as $m \to \infty$. For this purpose, choose $\delta > 0$ so that $|\xi| \leq \delta \implies |f(\xi) - 1| \leq \frac{1}{2}$. It is then clear that

$$\widehat{\lambda_{1/m}}(\xi) = g_{1/m}\big(f(\xi) - 1\big), \quad |\xi| \leq \delta,$$

where, for any $\alpha \in \mathbb{R}$,

$$g_\alpha(z) \equiv 1 + \sum_{j=1}^\infty \binom{\alpha}{j} z^j \quad \text{with} \quad \binom{\alpha}{j} \equiv \frac{\alpha(\alpha - 1)\cdots(\alpha - j + 1)}{j!}$$

is the branch of the $z \rightsquigarrow (1 + z)^\alpha$ on the unit disk which is 1 at the origin. In particular, this means first that

$$|\xi| \leq \delta \implies \big|1 - \widehat{\lambda_{1/m}}(\xi)\big| \leq \frac{1}{m}$$

and then that, for any $\mathbf{e} \in \mathbb{S}^{n-1}$ and $r > 0$,

$$\frac{1}{m} \geq \frac{1}{\delta} \int_0^\delta \Big(1 - \mathfrak{Re}\big(\widehat{\lambda_{1/m}}(t\mathbf{e})\big)\Big)\, dt = \int_{\mathbb{R}^n} \left(1 - \frac{\sin\big(\delta(\mathbf{e}, y)_{\mathbb{R}^n}\big)}{\delta(\mathbf{e}, y)_{\mathbb{R}^n}}\right) \lambda_{1/m}(dy)$$

$$\geq \epsilon(\delta r)\lambda_{1/m}\big(\{y : |(y, \mathbf{e})_{\mathbb{R}^n}| \geq r\}\big),$$

where $\epsilon(\beta) \equiv \inf\left\{1 - \frac{\sin t}{t} : |t| \geq \beta\right\} > 0 \quad \text{for } \beta > 0.$

As a consequence, we see that $\lambda_{1/m} \longrightarrow \delta_0$.

In view of the preceding, we know that, as $m \to \infty$, $\widehat{\lambda_{1/m}}$ tends to 1 uniformly on compacts as $m \to \infty$. Hence, since $f = (\widehat{\lambda_{1/m}})^m$ for all $m \geq 1$, f cannot vanish anywhere, and therefore there exists a unique continuous $\ell : \mathbb{R}^n \longrightarrow \mathbb{C}$ such that $\ell(0) = 0$ and $f(\xi) = e^{\ell(\xi)}$. Similarly, $\widehat{\lambda_{1/m}}$ never vanishes for any m, and so there exists a unique continuous $\ell_{1/m}$ such that $\ell_{1/m}(0) = 0$ and $\widehat{\lambda_{1/m}} = e^{\ell_{1/m}}$. But this means that and $\xi \rightsquigarrow (\sqrt{-1}\,2\pi)^{-1}\big(\ell(\xi) - m\ell_{1/m}(\xi)\big)$ is a continuous $\mathbb{Z}$-valued function, and so, since $\ell(0) = 0 = m\ell_{1/m}(0)$, we now know that $\widehat{\lambda_{1/m}}(\xi) = e^{\frac{\ell(\xi)}{m}}$ for all $m \geq 1$ and $\xi \in \mathbb{C}$.

The next step is to show that

$$f_m(\xi) \equiv \exp\left(m \int_{\mathbb{R}^n} \big(\mathbf{e}_\xi(y) - 1\big) \lambda_{1/m}(dy)\right) \longrightarrow f(\xi)$$

uniformly on compacts. For this purpose, first note that, for any $R > 0$ and $m \geq R^2$,

$$\big|f_m(\xi) - f(\xi)\big| = \left|e^{-m} \sum_{k=0}^\infty \frac{m^k}{k!}\left(\widehat{\lambda_{1/m}}(\xi)^k - \widehat{\lambda_{1/m}}(\xi)^m\right)\right|$$

$$\leq 2e^{-m} \sum_{\{k \geq 0 : |k - m| \geq R\sqrt{m}\}} \frac{m^k}{k!} + \sup_{0 \leq k \leq R\sqrt{m}} \big|\widehat{\lambda_{1/m}}(\xi)^k - 1\big|.$$

By Chebychev's inequality for Poisson random variables, one knows that

$$e^{-m} \sum_{\{k \geq 0 : |k-m| \geq R\sqrt{m}\}} \frac{m^k}{k!} \leq R^{-2} \quad \text{for all } m \geq 1.$$

Finally, by the conclusion reached at the end of the preceding paragraph, we know that, for each $R > 0$, as $m \to \infty$, $\sup_{k \leq R\sqrt{m}} \left| \widehat{\lambda_{1/m}}(\xi)^k - 1 \right| \longrightarrow 0$, uniformly on compacts. In particular, we have now shown that $f_m \longrightarrow f$ and therefore

$$\lim_{m \to \infty} m \int_{\mathbb{R}^n} \left(e_\xi(y) - 1 \right) \lambda_{1/m}(dy) = \ell(\xi)$$

uniformly on compacts.

The preceding allows us to prove that $|\ell(\xi)| \leq C(1+|\xi|^2)$ for some $C < \infty$. Indeed, by the preceding, we know that
(**)
$$\sup_{\substack{m \geq 1 \\ |\xi| \leq 1}} m \left(\int_{\mathbb{R}^n} \left(1 - \cos(\xi, y)_{\mathbb{R}^n} \right) \lambda_{1/m}(dy) \vee \left| \int_{\mathbb{R}^n} \sin(\xi, y)_{\mathbb{R}^n} \lambda_{1/m}(dy) \right| \right) < \infty.$$

Hence, after integrating the first of these along the ray from the origin to $\xi \in \mathbb{S}^{n-1}$, we see that[7]

$$\sup_{m \geq 1} m \int_{\mathbb{R}^n} \left(1 - \frac{\sin(\xi, y)_{\mathbb{R}^n}}{(\xi, y)_{\mathbb{R}^n}} \right) \lambda_{1/m}(dy) < \infty \quad \text{for all } \xi \in \mathbb{S}^{n-1},$$

from which it is clear that

$$C_1 \equiv \sup_{m \geq 1} m \int_{|y| \leq 1} |y|^2 \lambda_{1/m}(dy) < \infty \text{ and } C_2 \equiv \sup_{m \geq 1} m\lambda_{1/m}\left(B_{\mathbb{R}^n}(0,1)\complement \right) < \infty.$$

In particular, $|\ell(\xi)| \leq 2C_2 + \sup_{m \geq 1} |h_m(\xi)|$, where

$$h_m(\xi) \equiv m \int_{|y| < 1} \left(e_\xi(y) - 1 \right) \lambda_{1/m}(dy).$$

In addition, by the second estimate in (**) and the preceding, we know that

$$C_3 \equiv \sup_{m \geq 1} \sup_{|\xi| \leq 1} \left| m \int_{|y| < 1} \sin(\xi, y)_{\mathbb{R}^n} \lambda_{1/m}(dy) \right| < \infty.$$

---

[7] We take $\frac{\sin 0}{0} = 1$.

Hence, since $\partial_\xi h_m(0) = \sqrt{-1} m \int_{|y|<1} (\xi, y)_{\mathbb{R}^n} \lambda_{1/m}(dy)$,

$$\left| \partial_\xi h_m(0) \right| \le C_3 + \left| m \int_{|y|<1} \left((\xi, y)_{\mathbb{R}^n} - \sin(\xi, y)_{\mathbb{R}^n} \right) \lambda_{1/m}(dy) \right| \le C_3 + C_1$$

for all $\xi \in \mathbb{S}^{n-1}$. At the same time, $h_m(0) = 0$ and all second derivatives of h_m are uniformly bounded in absolute value by C_1. Thus, we have now shown that $|\ell(\xi)| \le 2C_2 + (C_1 + C_3)|\xi| + \frac{1}{2} C_1 |\xi|^2$.

We can at last show that the limit in (*) exists. In fact, given $\varphi \in D(\mathbb{R}^n; \mathbb{R})$, set $\tilde\varphi = \varphi - \varphi(\infty)$, and note that

$$m\left(\langle \varphi, \lambda_{1/m} \rangle - \varphi(0) \right) = \frac{1}{(2\pi)^n} \int_{\mathbb{R}^n} m \left(e^{\frac{\ell(-\xi)}{m}} - 1 \right) \widehat{\tilde\varphi}(\xi) \, d\xi,$$

and so (remember that $\mathfrak{Re}\ell \le 0$), by Lebesgue's dominated convergence theorem and the preceding estimate on $|\ell(\xi)|$, we see that the limit in (*) not only exists but also that

$$(***) \qquad\qquad A\varphi = \frac{1}{(2\pi)^n} \int_{\mathbb{R}^n} \ell(-\xi) \widehat{\tilde\varphi}(\xi) \, d\xi.$$

In particular, because $\widehat{\tilde\varphi}_R(\xi) = R^n \widehat{\tilde\varphi}(R\xi)$ and $\widehat{\tilde\varphi}$ is rapidly decreasing, another application of the estimate for $|\ell(\xi)|$ shows that A satisfies the tightness property in Lemma 2.1.8. At the same time, from its definition in (*), it is clear that A satisfies the minimum principle in that lemma. Hence, by Lemma 2.1.8, we now know that there is a unique Lévy system (a, b, M) such that $A\varphi = L^{(a,b,M)}\varphi(0)$ for all $\varphi \in D(\mathbb{R}^n; \mathbb{R})$. But, by (***) and elementary Fourier analysis, this means that, for any $\varphi \in C_c^\infty(\mathbb{R}^n; \mathbb{R})$,

$$\int_{\mathbb{R}^n} \hat\varphi(\xi) \ell^{(a,b,M)}(-\xi) \, d\xi = (2\pi)^n L^{(a,b,M)}\varphi(0)$$

$$= (2\pi)^n A\varphi = \int_{\mathbb{R}^n} \hat\varphi(\xi) \ell(-\xi) \, d\xi,$$

and from this the conclusion $\ell = \ell^{(a,b,M)}$ is an easy step. Namely, using the fact that both $|\ell(\xi)|$ and $|\ell^{(a,b,M)}(\xi)|$ have at most quadratic growth, one first extends the equality $\int_{\mathbb{R}^n} \hat\varphi(\xi) \ell^{(a,b,M)}(-\xi) \, d\xi = \int_{\mathbb{R}^n} \hat\varphi(\xi) \ell(-\xi) \, d\xi$ to all $\varphi \in \mathcal{S}(\mathbb{R}^n; \mathbb{C})$. One then uses the fact that $\mathcal{S}(\mathbb{R}^n; \mathbb{C})$ is invariant under the Fourier transform to see that this equality holds when $\hat\varphi$ is replaced by φ, at which point the asserted conclusion becomes obvious. $\square$

§2.1.3. The Tangent Space at δ_x: As expected, the results in the preceding section allow us to completely characterize the tangent space at a point mass. In particular, we have the following corollary to Theorem 2.1.9.

Theorem 2.1.11. *There is a one-to-one correspondence between Lévy systems and infinitely divisible flows. Namely, the Lévy (a, b, M) system determines, and is determined by, the infinitely divisible flow (cf. the end of Theorem 2.1.9) $t \rightsquigarrow \lambda_t^{(a,b,M)} \equiv \lambda^{(ta,tb,tM)}$. In addition, $t \rightsquigarrow \lambda_t^{(a,b,M)}$ is differentiable and (cf. (2.1.3))*

$$\dot{\lambda}_t^{(a,b,M)}\varphi = \langle L^{(a,b,M)}\varphi, \lambda_t \rangle \quad \text{for all } t \in [0, \infty) \text{ and } \varphi \in C_b^2(\mathbb{R}^n; \mathbb{R}).$$

Finally, if $t \rightsquigarrow \mu_t$ is differentiable and $\mu_0 = \delta_x$ for some $x \in \mathbb{R}^n$, then there is a unique Lévy system (a, b, M) such that $\dot{\mu}_0\varphi = L^{(a,b,M)}\varphi(x)$ for all $\varphi \in C_b^2(\mathbb{R}^n; \mathbb{C})$. Thus, $T_{\delta_x}\mathbf{M}_1(\mathbb{R}^n)$ can be identified with the cone of linear functionals of the form $\varphi \rightsquigarrow L^{(a,b,M)}\varphi(x)$, where (a, b, M) is a Lévy system.

Proof: Clearly, if (a, b, M) is a Lévy system, then so is (ta, tb, tM) for each $t \geq 0$, and $t \rightsquigarrow \lambda_t^{(a,b,M)} \equiv \lambda^{(ta,tb,tM)}$ is an infinitely divisible flow. Conversely, if $t \rightsquigarrow \lambda_t$ is an infinitely divisible flow, then, for each $t \geq 0$, λ_t is an infinitely divisible measure, and therefore $\hat{\lambda}_t(\xi) = e^{\ell_t(\xi)}$ for some function ℓ_t of the form in (2.1.7). Moreover, because $t \rightsquigarrow \lambda_t$ is continuous, $t \rightsquigarrow \ell_t(\xi)$ is continuous, and, because $\lambda_{s+t} = \lambda_s \star \lambda_t$, $\ell_{s+t}(\xi) = \ell_s(\xi) + \ell_t(\xi)$. Hence, $\ell_t(\xi) = t\ell^{(a,b,M)}(\xi)$, where (a, b, M) is the Lévy system for which $\ell_1 = \ell^{(a,b,M)}$.

Turning to the proof that $t \rightsquigarrow \lambda_t$ is differentiable, note that, by the preceding,

$$\langle \varphi, \lambda_t \rangle - \varphi(0) = \frac{1}{(2\pi)^n} \int_{\mathbb{R}^n} \left(e^{t\ell^{(a,b,M)}(-\xi)} - 1 \right) \hat{\varphi}(\xi) \, d\xi$$

$$= \frac{1}{(2\pi)^n} \int_{\mathbb{R}^n} \left(\int_0^t \ell^{(a,b,M)}(-\xi) e^{\tau\ell^{(a,b,M)}(-\xi)} \hat{\varphi}(\xi) \, d\tau \right) d\xi$$

$$= \int_0^t \langle L^{(a,b,M)}\varphi, \lambda_\tau \rangle \, d\tau$$

for all $\varphi \in \mathcal{S}(\mathbb{R}^n; \mathbb{C})$. Hence, after a trivial extention argument, we find that

$$\langle \varphi, \lambda_t \rangle - \varphi(0) = \int_0^t \langle L^{(a,b,M)}\varphi, \lambda_\tau \rangle \, d\tau$$

for all $\varphi \in C_b^2(\mathbb{R}^n; \mathbb{C})$, from which it is an easy step, via Lemma 2.1.5, to the conclusion that $t \rightsquigarrow \lambda_t$ is differentiable and that $\dot{\lambda}_t\varphi = \langle L^{(a,b,M)}\varphi, \lambda_t \rangle$.

Now suppose that $t \rightsquigarrow \mu_t$ is differentiable and that $\mu_0 = \delta_x$. Then $t \rightsquigarrow \bar{\mu}_t \equiv \delta_{-x} \star \mu_t$ is differentiable, $\bar{\mu}_0 = \delta_0$, and $\dot{\mu}_0 = (\tau_x)_* \dot{\bar{\mu}}_0$. Hence, it suffices for us to handle the case when $\mu_0 = \delta_0$. But in this case it is obvious that the restriction of $\dot{\mu}_0$ to $D(\mathbb{R}^n; \mathbb{R})$ satisfies the minimum principle in Lemma

2.1.8. In addition, because $\varphi \in C_b^2(\mathbb{R}^n; \mathbb{C}) \longmapsto \dot{\mu}_0 \varphi \in \mathbb{C}$ is continuous, one sees that $\dot{\mu}_0$ has the tightness property described in that lemma. Hence, we know that there exists a Lévy system (a, b, M) such that $\dot{\mu}_0 \varphi = L^{(a,b,M)} \varphi(0)$ for all $\varphi \in D(\mathbb{R}^n; \mathbb{R})$. Finally, by the linearity and continuity of $\varphi \rightsquigarrow \dot{\mu}_0 \varphi$ and $\varphi \rightsquigarrow L^{(a,b,M)} \varphi(0)$ on $C_b^2(\mathbb{R}^n; \mathbb{C})$, it follows that the equality extends to all $\varphi \in C_b^2(\mathbb{R}^n; \mathbb{C})$.

We now know that each tangent vector to δ_x is a linear functional of the form $\varphi \rightsquigarrow L^{(a,b,M)} \varphi(x)$ for some Lévy system (a, b, M). Conversely, if (a, b, M) is a Lévy system, then we can take $\mu_t = \delta_x \star \lambda_t^{(a,b,M)}$, in which case $t \rightsquigarrow \mu_t$ is differentiable, $\mu_0 = \delta_x$, and $\dot{\mu}_0 \varphi = (\tau_x)_* \dot{\lambda}_0 \varphi = L^{(a,b,M)} \varphi(x)$. □

§2.1.4. The Tangent Space at General $\mu \in \mathbf{M}_1(\mathbb{R}^n)$: Having described the tangent spaces at point masses, we might hope (cf. the first part of Remark 1.1.7) that the tangent space at a general μ should be amenable to the sort of analysis which we applied in §1.2.4. That is, one might expect that we can simply define the tangent space $T_\mu \mathbf{M}_1(\mathbb{R}^n)$ to be the set of linear functionals $\dot{\mu}_0$ which arise as the derivative at $t = 0$ of some differentiable curve $t \rightsquigarrow \mu_t$ which starts from μ and then prove that the representation (1.1.6) holds in the present setting.

In fact, as long as we are dealing with μ's which are finite, convex combinations of point masses, a minor variant of the argument used in Lemma 1.2.7 works again. To be precise, suppose that $\mu = \sum_{\ell=1}^m \alpha_m \delta_{x_m}$, where $\{\alpha_\ell\}_1^m \subseteq (0, 1)$, $\sum_1^m \alpha_\ell = 1$, and $x_1, \ldots, x_m$ are distinct points in $\mathbb{R}^n$. Next, choose $\{\psi_\ell\}_1^m \subseteq C_b^\infty(\mathbb{R}^n; [0, 1])$ so that: for each $1 \leq \ell \leq m$, $\psi_\ell \equiv 1$ in a neighborhood of x_ℓ while $\psi_\ell \equiv 0$ in a neighborhood of x_j for each $j \neq \ell$, and $\sum_1^m \psi_\ell \equiv 1$. If $t \rightsquigarrow \mu_t$ is a differentiable curve starting from μ, there are two cases, depending on whether $\dot{\mu}_0 \psi_\ell = 0$ for each $1 \leq \ell \leq m$ or whether $\dot{\mu}_0 \psi_\ell \neq 0$ for some ℓ. In the first case, simply write $\mu_t = \sum_{\ell=1}^m \langle \psi_\ell, \mu_t \rangle \mu_t^\ell$ where $\langle \varphi, \mu_t^\ell \rangle \equiv \frac{\langle \varphi \psi_\ell, \mu_t \rangle}{\langle \psi_\ell, \mu_t \rangle}$ for small $t \geq 0$, and observe that $\dot{\mu}_0 = \int_{\mathbb{R}^n} \Lambda_{\delta_x} \mu(dx)$ where $\Lambda_x = \dot{\mu}_0^\ell$ if $x = x_\ell$ and $\Lambda_x = 0$ if $x \notin \{x_1, \ldots, x_m\}$. In the second case, note that $\dot{\mu}_0 \psi_\ell < 0$ for at least one $1 \leq \ell \leq m$. Hence, one can repeat the argument used in the proof of Lemma 1.2.7 to peel off the component of $\dot{\mu}_0$ which is in $T_{\delta_{x_\ell}} \mathbf{M}_1(\mathbb{R}^n)$.

Unfortunately, this hands-on procedure does not appear to be up to the challenge presented by general μ's. Indeed, one should anticipate that some sort of reasonably heavy machinery from convex analysis would be required to handle general μ's. Specifically, one might hope that a clever application of something like Choquet's representation theorem (cf. Theorem 27.6 in [2]) would yield the desired conclusion. However, at least to date, I have been unable to produce anything satisfactory. For this reason, I will simply use an appropriate analog of (1.1.6) for the definition pf the tangent space at μ.

More precisely, a *tangent vector* to $\mathbf{M}_1(\mathbb{R}^n)$ at μ will be a linear functional Λ_μ on $C_b^2(\mathbb{R}^n; \mathbb{R})$ of the form

$$(2.1.12) \qquad \Lambda_\mu \varphi = \langle L\varphi, \mu \rangle,$$

where

$$
\begin{aligned}
L\varphi(x) = &\frac{1}{2} \sum_{i,j=1}^n a_{ij}(x) \partial_{x_i} \partial_{x_j} \varphi + \sum_{i=1}^n b_i(x) \partial_{x_i} \varphi \\
(2.1.13) \qquad &+ \int_{\mathbb{R}^n} \left(\varphi(x+y) - \varphi(x) - \frac{(y, \mathrm{grad}_x \varphi)_{\mathbb{R}^n}}{1 + |y|^2} \right) M(x, dy)
\end{aligned}
$$

for some map taking each $x \in \mathbb{R}^n$ into a Lévy system $\big(a(x), b(x), M(x, \cdot)\big)$ in such a way that all the functions

$$x \in \mathbb{R}^n \longmapsto a(x) \in \mathbb{R}^n \otimes \mathbb{R}^n, \quad x \in \mathbb{R}^n \longmapsto b(x) \in \mathbb{R}^n,$$

$$\text{and } x \in \mathbb{R}^n \longmapsto \int_{\mathbb{R}^n} \varphi(y) \frac{|y|^2}{1 + |y|^2} M(x, dy) \in \mathbb{R} \text{ for each } \varphi \in C_b(\mathbb{R}^n, \mathbb{R})$$

are bounded and continuous. In particular, (1.1.6) will hold when $\Lambda_{\delta_x} = (\tau_x)_* \dot{\lambda}_0^x$ and $t \rightsquigarrow \lambda_t^x$ is the infinitely divisible flow whose Lévy system is $\big(a(x), b(x), M(x, \cdot)\big)$. In addition, by taking

$$t \in [0, \infty) \longmapsto \mu_t = \int_{\mathbb{R}^n} \lambda_t^x \, \mu(dx) \in \mathbf{M}_1(\mathbb{R}^n),$$

we see that $\Lambda_\mu = \dot{\mu}_0$ for a differentiable curve starting from μ. Thus, although we cannot say that for every differentiable $t \rightsquigarrow \mu_t$ starting from μ the derivative $\dot{\mu}_0$ is a tangent vector to $\mathbf{M}_1(\mathbb{R}^n)$ at μ, we can say that every tangent vector at μ can be represented as $\dot{\mu}_0$ for some differentiable $t \rightsquigarrow \mu_t$ starting from μ. Of course, the cone $T_\mu \mathbf{M}_1(\mathbb{R}^n)$ of tangent vectors to $\mathbf{M}_1(\mathbb{R}^n)$ at μ will be called the *tangent space to $\mathbf{M}_1(\mathbb{R}^n)$ at μ.*

REMARK 2.1.14. It should be emphasized that, unless μ is a point mass, no claim of uniqueness is made about the representation of Λ_μ given in (2.1.12). Indeed, even a $\Lambda_{\delta_0} \in T_{\delta_0} \mathbf{M}_1(\mathbb{R}^n)$ determines $x \rightsquigarrow \big(a(x), b(x), M(x, \cdot)\big)$ only at $x = 0$, which is, of course, the support of δ_0. On the other hand, as part (ii) of Exercise 2.1.15 demonstrates, for μ's other than point masses, there is enormous ambiguity about the choice of $\big(a(x), b(x), M(x, \cdot)\big)$, whether or not x is in the support of μ.

In a related direction, it may be interesting to note that, as a differentiable manifold, $\mathbf{M}_1(\mathbb{R}^n)$ is seriously flawed. Indeed, as one might suspect after thinking about what is happening to $\mathbf{M}_1(\mathbb{Z}_n)$ when $n \to \infty$, $\mathbf{M}_1(\mathbb{R}^n)$, *as a differentiable manifold, has a boundary which is dense.* To be more precise, identify the boundary of $\mathbf{M}_1(\mathbb{R}^n)$ as those μ with the property that $T_\mu \mathbf{M}_1(\mathbb{R}^n)$ is not a vector space. Equivalently, we will say that μ is a boundary point of $\mathbf{M}_1(\mathbb{R}^n)$ if there is a $\Lambda_\mu \in T_\mu \mathbf{M}_1(\mathbb{R}^n)$ such that $-\Lambda_\mu \notin T_\mu \mathbf{M}_1(\mathbb{R}^n)$. As is outlined in part **(iv)** of Exercise 2.1.15 below, if Δ denotes the Euclidean Laplace operator, one can show that the linear functional $\varphi \rightsquigarrow -\langle \Delta\varphi, \mu \rangle$ is not an element of the tangent space at any μ which is a finite, convex combination of point masses. On the other, $\varphi \rightsquigarrow \langle \Delta\varphi, \mu \rangle$ is an element of the tangent space to every $\mu \in \mathbf{M}_1(\mathbb{R}^n)$. Finally, as is easily checked, the set of μ's which are finite, convex combinations of point masses is dense in $\mathbf{M}_1(\mathbb{R}^n)$.

§2.1.5. Exercises

EXERCISE 2.1.15. Here are some examples which indicate just how tricky is the relationship between an element Λ_μ of $T_\mu \mathbf{M}_1(\mathbb{R}^n)$ and the representation $\int \Lambda_{\delta_x} \mu(dx)$.

(i) Take $n = 1$, $\mu_t = \frac{1+t}{2}\delta_0 + \frac{1-t}{2}\delta_1$ for $t \in \left[0, \frac{1}{2}\right)$, and find $x \rightsquigarrow \Lambda_{\delta_x}$ so that $\dot{\mu}_0 = \int \Lambda_{\delta_x} \mu(dx)$.

(ii) In this example, we show that, when μ is not a point mass, there is enormous nonuniqueness in the relationship between $\Lambda_\mu \in T_\mu \mathbf{M}_1(\mathbb{R})$ and the representation $\Lambda_\mu = \int \Lambda_{\delta_x} \mu(dx)$. For this purpose, take $\mu(dx) = c\exp(-\sqrt{1+x^2})$, where $c \in (0, \infty)$ is chosen to make μ into a probability measure, and, for each $\alpha \geq 0$ and $\beta \in \mathbb{R}$, set

$$\Lambda_{\delta_x}^{(\alpha,\beta)} = \frac{-\beta x}{\sqrt{1+x^2}}\partial_x + \alpha \partial_x^2.$$

Show that

$$\Lambda_\mu^{(\alpha,\beta)}\varphi \equiv \int_\mathbb{R} \Lambda_{\delta_x}^{(\alpha,\beta)}\varphi\, \mu(dx) = \langle(\alpha - \beta)\varphi'', \mu\rangle, \quad \varphi \in C_b(\mathbb{R};\mathbb{R}).$$

In particular, when $\beta = \alpha$, $\Lambda_\mu^{(\alpha,\beta)} = 0$.

(iii) In a different direction, the preceding can be used to demonstrate how difficult it will be to even recognize when a linear functional is in the tangent space to $\mathbf{M}_1(\mathbb{R})$ at a μ which is not a point mass. Namely, let μ be as in **(ii)**, and use the preceding considerations to see that, in spite of the "wrong sign," $\varphi \rightsquigarrow \langle-\varphi'', \mu\rangle$ is an element of $T_\mu \mathbf{M}_1(\mathbb{R})$.

(iv) Suppose that $\mu = \sum_{m=0}^N \alpha_m \delta_{x_m}$, where $\{\alpha_m\}_0^N \subseteq (0, 1]$, $\sum_0^N \alpha_m = 1$, and $x_m \neq x_{m'}$ for $m \neq m'$. Let Δ denote the standard, Euclidean Laplacian on $C^2(\mathbb{R}^n; \mathbb{R})$, and show that the linear functional $\varphi \rightsquigarrow -\langle\Delta\varphi, \mu\rangle$ cannot be an element of $T_\mu \mathbf{M}_1(\mathbb{R}^n)$.

Hint: Choose $r > 0$ so that $\{x_1, \ldots, x_N\} \cap B_{\mathbb{R}^n}(x_0, 3r) = \emptyset$, and take φ to be a smooth, non-negative function which vanishes off $B_{\mathbb{R}^n}(x_0, 2r)$ and is given by $\varphi(x) = |x - x_0|^2$ for $x \in B_{\mathbb{R}^n}(x_0, r)$. Show that $-\langle \Delta \varphi, \mu \rangle = -2n\alpha_0 < 0$ but that $\underline{\lim}_{t \searrow 0} t^{-1} \big(\langle \varphi, \mu_t \rangle - \langle \varphi, \mu \rangle \big) \geq 0$ for any differentiable curve $t \rightsquigarrow \mu_t$ starting from μ.

EXERCISE 2.1.16. We could have adopted an approach to proving Theorem 2.1.11 along a line of reasoning which bears greater resemblance to the treatment of the analogous result in §1.2.3. Namely, suppose that $t \rightsquigarrow \mu_t$ is a differentiable curve with $\mu_0 = \delta_0$, and set

$$\lambda_t^N = e^{-Nt} \sum_{m=0}^{\infty} \frac{(NT)^m}{m!} \mu_{1/m}^{\star m}.$$

Note that, for each $N \in \mathbb{Z}^+$, $t \rightsquigarrow \lambda_t^N$ is an infinitely divisible flow, and show that, as $N \to \infty$, $\widehat{\lambda_t^N}(\xi) \longrightarrow e^{t\dot{\mu}_0 \epsilon_\xi}$ uniformly on compacts. Conclude that, for each $t \geq 0$, $\lambda_t^N \longrightarrow \lambda_t$ and that $t \rightsquigarrow \lambda_t$ is an infinitely divisible flow with the property that $\dot{\lambda}_0 = \dot{\mu}_0$.

2.2 Vector Fields and Integral Curves on $\mathbf{M}_1(\mathbb{R}^n)$

A *vector field* on $\mathbf{M}_1(\mathbb{R}^n)$ is a map $\mu \in \mathbf{M}_1(\mathbb{R}^n) \longmapsto \Lambda_\mu \in T_\mu \mathbf{M}_1(\mathbb{R}^n)$. Thus, in order to describe a vector field, one must specify a map taking $(x, \mu) \in \mathbb{R}^n \times \mathbf{M}_1(\mathbb{R}^n)$ into a Lévy system $\big(a(x,\mu), b(x,\mu), M((x,\mu), \cdot)\big)$ in such a way that, for each μ,

$$x \rightsquigarrow \Lambda_{\delta_x} \equiv (\tau_x)_* \Lambda_{\delta_0}^{(a(x,\mu), b(x,\mu), M((x,\mu), \cdot))}$$

satisfies the conditions prescribed in the definition of a tangent vector. Equivalently, specifying a vector will be tantamount to giving a map taking $\mu \in \mathbf{M}_1(\mathbb{R}^n)$ into an operator L_μ on $C_b^2(\mathbb{R}^n; \mathbb{R})$, where, for each μ, L_μ has the form in (2.1.13). That is,

$$\Lambda_\mu \varphi = \langle L_\mu \varphi, \mu \rangle, \quad \text{where}$$

$$(2.2.1) \quad L_\mu \varphi(x) = \frac{1}{2} \sum_{i,j=1}^{n} a_{ij}(x,\mu) \partial_{x_i} \partial_{x_j} \varphi + \sum_{i=1}^{n} b_i(x,\mu) \partial_{x_i} \varphi$$

$$+ \int_{\mathbb{R}^n} \left(\varphi(x+y) - \varphi(x) - \frac{(y, \operatorname{grad}\varphi(x))}{1+|y|^2} \right) M((x,\mu), dy)).$$

Just as before, we say that a vector field Λ is *affine* if it satisfies

$$\Lambda_{(1-\theta)\mu + \theta\nu} \varphi = (1-\theta)\Lambda_\mu \varphi + \theta \Lambda_\nu \varphi$$

for all $\theta \in [0,1]$, $\mu, \nu \in \mathbf{M}_1(\mathbb{R}^n)$, and $\varphi \in C_b^2(\mathbb{R}^n; \mathbb{C})$.

Equivalently, Λ is affine when the associated mapping into Lévy systems can be chosen to be independent of μ. That is, when, for all $\varphi \in C_b^2(\mathbb{R}^n; \mathbb{C})$,

$$\Lambda_\mu \varphi = \langle L\varphi, \mu \rangle, \quad \text{where}$$

$$(2.2.2) \quad L\varphi(x) = \frac{1}{2} \sum_{i,j=1}^n a_{ij}(x) \partial_{x_i} \partial_{x_j} \varphi + \sum_{i=1}^n b_i(x) \partial_{x_i} \varphi$$

$$+ \int_{\mathbb{R}^n} \left(\varphi(x+y) - \varphi(x) - \frac{(y, \mathrm{grad}\varphi(x))}{1+|y|^2} \right) M(x, dy).$$

Of course, inside the cone of affine vector fields is the cone of *translation invariant vector fields*: those Λ's for which

$$\Lambda_{\mu \star \nu} \varphi = \int_{\mathbb{R}^n} (\tau_x)_* \Lambda_\mu \varphi \, \nu(dx);$$

or, equivalently, vector fields for which the associated Lévy systems can be chosen to be independent of x as well as μ.

REMARK 2.2.3. Although it is obvious, one should nonetheless remark that the ambiguity alluded in Remark 2.1.14 disappears when talking about an affine vector field. Indeed, an affine vector field is completely determined by its values at point masses, which are exactly the places at which a vector field uniquely determines its associated Lévy systems.

§2.2.1. **Existence of Integral Curves:** Given a vector field Λ, we say that $t \rightsquigarrow \mu_t$ is an *integral curve* of Λ starting from μ if

$$(2.2.4) \qquad \langle \varphi, \mu_t \rangle - \langle \varphi, \mu \rangle = \int_0^t \Lambda_{\mu_\tau} \varphi \, d\tau, \quad t \in [0, \infty),$$

both makes sense and is true for all $\varphi \in C_b^2(\mathbb{R}^n; \mathbb{C})$.

We now give conditions on Λ which guarantee that it admits integral curves. Apart from a few embellishments necessitated by some technical difficulties, the ideas differ very little from those used to prove Lemma 1.3.5 and, as such, are essentially an natural adaptation of Euler's approximation scheme.

THEOREM 2.2.5. *Let Λ be the vector field given by (2.2.1). Under the conditions that*

(a) $\qquad \sup_{(x,\mu)} \|a(x,\mu)\|_{\mathrm{op}} \vee |b(x,\mu)| \vee \int_{\mathbb{R}^n} \frac{|y|^2}{1+|y|^2} M\big((x,\mu), dy\big) < \infty,$

(b) $\qquad (x,\mu) \rightsquigarrow L_\mu \varphi(x)$ *is continuous for each $\varphi \in C_b^2(\mathbb{R}^n; \mathbb{R})$,*

(c) $\qquad \lim_{R \to \infty} \sup_{(x,\mu)} M\big((x,\mu), B_{\mathbb{R}^n}(0, R)\big) = 0,$

there exists for each $\mu_0 \in \mathbf{M}_1(\mathbb{R}^n)$ an integral curve of Λ starting from μ_0.

PROOF: For each (x, μ), determine $t \rightsquigarrow (\lambda_\mu^x)_t$ so that $t \rightsquigarrow \delta_{-x} \star (\lambda_\mu^x)_t$ is the infinitely divisible flow whose Lévy system is $\big(a(x, \mu), b(x, \mu), M((x, \mu), \cdot)\big)$. Note that $(t, x, \mu) \rightsquigarrow (\lambda_\mu^x)_t$ is continuous. In addition, if L_μ^x is the translation invariant operator on $C_b^2(\mathbb{R}^n; \mathbb{R})$ given by $L_\mu^x \varphi(y) = \big[L_\mu(\varphi \circ \tau_{y-x})\big](x)$, then $(\dot{\lambda}_\mu^x)_t \varphi = \langle L_\mu^x \varphi, (\lambda_\mu^x)_t \rangle$.

Next, for each $N \geq 1$, determine $t \in [0, \infty) \longmapsto \mu_t^N \in \mathbf{M}_1(\mathbb{R}^n)$ so that $\mu_0^N = \mu_0$ and

$$\mu_t^N = \int_{\mathbb{R}^n} (\tau_x)_* (\lambda_{\mu_{[t]_N}^x}^x)_{]t[_N} \, \mu_{[t]_N}^N (dx),$$

where $[t]_N \equiv 2^{-N}[2^N t]$, $]t[_N \equiv t - [t]_N$, and $[t] = [t]_1$ denotes the integer part of t. Then, by the preceding, we know that, for any $\varphi \in C_b^2(\mathbb{R}^n; \mathbb{C})$,

$$(*) \qquad \langle \varphi, \mu_T^N \rangle - \langle \varphi, \mu_0 \rangle = \int_0^T \left(\int_{\mathbb{R}^n} \left\langle L_{\mu_{[t]_N}^x}^x \varphi, (\lambda_{\mu_{[t]_N}^x}^x)_{]t[_N} \right\rangle \mu_{[t]_N}^N (dx) \right) dt.$$

Our first application of $(*)$ will be to prove that, for any $T > 0$, the family $\{\mu_t^N : t \in [0, T] \ \& \ N \geq 1\}$ is *tight* in the sense that

$$(**) \qquad \lim_{R \to \infty} \sup_{\substack{n \geq 0 \\ t \in [0, T]}} \mu_t^N \big(B_{\mathbb{R}^n}(0, R)\complement\big) = 0.$$

For this purpose, set $R_0 = 1$ and, for each $m \geq 1$, use (c) above to choose $R_m \geq 4 R_{m-1}$ so that $\mu_0\big(B_{\mathbb{R}^n}(0, R_m)\complement\big) \leq 2^{-m}$ and $M\big((x, \mu); B_{\mathbb{R}^n}(0, R_m)\complement\big) \leq 2^{-m}$ for all (x, μ). Next, choose $\psi \in C^\infty\big(\mathbb{R}^n; [0, 1]\big)$ so that $\psi \equiv 0$ on $B_{\mathbb{R}^n}(0, 1)$ and $\psi \equiv 1$ off $B_{\mathbb{R}^n}(0, 2)$, and define η_m so that $\eta_m(y) = \psi\big(R_m^{-1} y\big)$. Notice that if $m \geq 1$ and $|z| < \frac{1}{2} R_m$, then

$$L_\mu^x \eta_m(z) = \int_{\mathbb{R}^n} \eta_m(z+y) \, M\big((x, \mu), dy\big) \leq M\big((x, \mu), B_{\mathbb{R}^n}(0, R_{m-1})\complement\big) \leq 2^{1-m}.$$

At the same time, if $|z| \geq \frac{1}{2} R_m$, then $\eta_{m-1}(z) = 1$ and so

$$L_\mu^x \eta_m(z) \leq B \eta_{m-1}(z) \quad \text{where } B \equiv \sup_{(x, \mu)} \big\| L_\mu^x \eta_m \big\|_u < \infty.$$

Thus, $L_\mu^z \eta_m \leq 2^{1-m} + B \eta_{m-1}$ for each $m \geq 1$. Now set $f_M = \sum_{m=0}^M \eta_m$ for $M \in \mathbb{Z}^+$. Clearly $f_M \in C_b^2\big(\mathbb{R}^n; [0, \infty)\big)$. In addition, from the preceding, it is easy to check that $\langle f_M, \mu_0 \rangle \leq 2$ and $L_\mu^x f_M \leq A + B f_M$, where $A \equiv 2 + \sup_{(x, \mu)} \| L_\mu^x \eta_0 \|_u$. After plugging this estimate into $(*)$, we conclude that

$$\langle f_M, \mu_t^N \rangle \leq 2 + At + B \int_0^t \langle f_M, \mu_\tau^N \rangle \, d\tau,$$

which, by Gronwall's inequality (cf. Exercise 2.2.11 below), means that

$$\langle f_M, \mu_t^N \rangle \le K(t) \equiv 2e^{Bt} + A\frac{e^{Bt} - 1}{B} \quad \text{for all } M \ge 1.$$

Finally, note that $f_M \nearrow f$, where f is a smooth, strictly positive function which tends to ∞ as $|x| \to \infty$. Hence,

$$\sup_{t \in [0,T]} \sup_{N \ge 1} \mu_t^N\big(B_{\mathbb{R}^n}(0, R)\complement\big) \le \frac{K(T)}{\inf_{|x| \ge R} f(x)} \longrightarrow 0 \quad \text{as } R \to \infty,$$

which is the statement which we wanted.

We now use (**) to show that $\{\mu_\cdot^N : N \ge 1\}$ is a relatively compact as a sequence in $C\big([0,\infty); \mathbf{M}_1(\mathbb{R}^n)\big)$; and, by Lévy's continuity theorem, this is tantamount to showing that $\{\widehat{\mu_\cdot^N} : N \ge 1\}$ is relatively compact in $C\big([0,\infty) \times \mathbb{R}^n; \mathbb{C}\big)$. But, from (*) with $\varphi = \mathfrak{e}_\xi$, we know that

$$\sup_N \big|\widehat{\mu_{t_2}^N}(\xi) - \widehat{\mu_{t_1}^N}(\xi)\big| \le C(1 + |\xi|^2)|t_2 - t_1|, \quad \xi \in \mathbb{R}^n.$$

At the same time, since, for each $R > 0$,

$$\big|\widehat{\mu_t^N}(\xi') - \widehat{\mu_t^N}(\xi)\big| \le R|\xi' - \xi| + 2\mu_t^N\big(B_{\mathbb{R}^n}(0, R)\complement\big),$$

the estimate in (**) shows that, for each $T > 0$,

$$\lim_{\delta \searrow 0} \sup_{N \ge 1} \sup_{t \in [0,T]} \sup_{|\xi' - \xi| \le \delta} \big|\widehat{\mu_t^N}(\xi') - \widehat{\mu_t^N}(\xi)\big| = 0.$$

Finally, knowing that $\{\mu_\cdot^N : N \ge 1\}$ is relatively compact sequence in $C\big([0,\infty); \mathbf{M}_1(\mathbb{R}^n)\big)$, choose a convergent subsequence $\{\mu_\cdot^{N'} : N' \ge 1\}$, and let $t \rightsquigarrow \mu_t$ denote its limit. Because of (*), it is an easy application of standard facts about weak convergence (cf. §3.1, especially Lemma 3.1.10, in [36]) to check that $t \rightsquigarrow \mu_t$ must be an integral curve of Λ which starts from μ_0. $\square$

§2.2.2. Uniqueness for Affine Vector Fields: By essentially the same argument as we gave in the proof of Theorem 1.3.7, we have the following uniqueness criterion for integral curves of affine vector fields on $\mathbf{M}_1(\mathbb{R}^n)$.

THEOREM 2.2.6. *Let Λ be an affine vector field, and determine the operator L on $C_b^2(\mathbb{R}^n; \mathbb{R})$ accordingly, as in (2.2.2). If for each $T > 0$ and $\varphi \in C_c^\infty(\mathbb{R}^n; \mathbb{R})$ there exists a $u \in C_b^{1,2}\big([0,T] \times \mathbb{R}^n; \mathbb{R}\big)$ satisfying $u(0, \cdot) = \varphi$ and $\partial_t u = Lu$ on $(0,T) \times \mathbb{R}^n$, then for each $\mu \in \mathbf{M}_1(\mathbb{R}^n)$ there is at most one integral curve $t \rightsquigarrow \mu_t$ of Λ starting from μ. In fact, $\langle \varphi, \mu_T \rangle = \langle u(T, \cdot), \mu \rangle$ for each $T > 0$ and $\varphi \in C_c^\infty(\mathbb{R}^n; \mathbb{R})$. Thus, if in addition, the hypotheses of Theorem 2.2.5 hold (when $L_\mu = L$ for all μ), then for each μ there is precisely one integral curve of Λ starting from μ.*

PROOF: Clearly it suffices to check that $\langle u(0,\,\cdot\,),\mu_T\rangle = \langle u(T,\,\cdot\,),\mu\rangle$ when $u \in C_b^{1,2}\big([0,T] \times \mathbb{R}^n;\mathbb{R}\big)$ satisfies $\partial_t u = Lu$ on $(0,T) \times \mathbb{R}^n$ and $t \rightsquigarrow \mu_t$ is an integral curve of Λ. But, just as in the proof of Theorem 1.3.7, one sees that $\frac{d}{dt}\langle u(T-t,\,\cdot\,),\mu_t\rangle = 0$ for $t \in (0,T)$. $\quad\square$

§2.2.3. **The Markov Property and Kolmogorov's Equations:** Just as in §1.3.4, we can use the preceding to recover Kolmogorov's description of Markov flows. That is, suppose that Λ is an affine vector field to which Theorems 2.2.5 and 2.2.6 apply. For each $x \in \mathbb{R}^n$, use $t \rightsquigarrow P_t(x,\,\cdot\,)$ to denote the integral curve of Λ starting at x. Then (cf. Exercise 2.2.14 below) the map $(t,x) \rightsquigarrow P_t(x,\,\cdot\,)$ is continuous. Furthermore, by uniqueness, for any integral curve $t \rightsquigarrow \mu_t$ of Λ

$$(2.2.7) \qquad \mu_{s+t} = \int_{\mathbb{R}^n} P_t(x,\,\cdot\,)\,\mu_s(dx) \quad \text{for all } s,t \in [0,\infty).$$

In particular, this yields the *Chapman–Kolmogorov* equation

$$(2.2.8) \qquad P_{s+t}(x,\,\cdot\,) = \int_{\mathbb{R}^n} P_t(y,\,\cdot\,)\,P_s(x,dy), \quad (s,t,x) \in [0,\infty)^2 \times \mathbb{R}^n,$$

which expresses the *Markov property* in terms of the *transition probability function* $(t,x) \rightsquigarrow P_t(x,\,\cdot\,)$ determined by Λ.

Next let L be the operator on $C_b^2(\mathbb{R}^n;\mathbb{R})$ for which (2.2.2) holds. Then, because $\dot P_t(x,\,\cdot\,) = \Lambda_{P_t(x,\,\cdot\,)}$ is equivalent to

$$\partial_t \int_{\mathbb{R}^n} \varphi(y)\,P_t(x,dy) = \int_{\mathbb{R}^n} L\varphi(y)\,P_t(x,dy) \quad \text{for all } \varphi \in C_b^2(\mathbb{R}^n;\mathbb{R}),$$

we see that $(t,x) \rightsquigarrow P_t(x,\,\cdot\,)$ satisfies *Kolmogorov's forward equation*

$$(2.2.9) \qquad \partial_t P_t(x,\,\cdot\,) = L^* P_t(x,\,\cdot\,),$$

where L^* is the formal adjoint of L and its action is on the forward variable. To make all of this a little more concrete, assume for the moment that there is no Lévy measure term in L: that is,

$$L = \frac{1}{2}\sum_{i,j=1}^n a_{ij}(x)\partial_{x_i}\partial_{x_j} + \sum_{i=1}^n b_i(x)\partial_{x_i}.$$

In addition, assume that the coefficients in L are smooth and that $P_t(x,dy) = p_t(x,y)\,dy$ where $(t,x,y) \in (0,\infty) \times \mathbb{R}^n \times \mathbb{R}^n \longmapsto p_t(x,y) \in [0,\infty)$ is smooth. Then (2.2.9) becomes

$$\partial_t p_t(x,\,\cdot\,) = \frac{1}{2}\sum_{i,j=1}^n \partial_{x_i}\partial_{x_j}\big(a_{ij}p_t(\,\cdot\,,y)\big)(x) - \sum_{i=1}^n \partial_{x_i}\big(b_i p_t(\,\cdot\,,y)\big)(x).$$

In the physics and engineering communities, an equation of this form is called a *Fokker–Planck* equation.

Finally, the passage from Kolmogorov's forward equation to his backward equation is the same as it was in §1.3.4, only now we have to worry about differentiability. That is, let $\varphi \in C_{\mathrm{b}}(\mathbb{R}^n; \mathbb{R})$ be given, set $u(t, x) = \int \varphi(y) P_t(x, dy)$, and assume that $t \in (0, \infty) \longmapsto u(t, \cdot) \in C_{\mathrm{b}}^2(\mathbb{R}^n; \mathbb{R})$ is continuous. Then, because

$$u(t + h, x) = \int_{\mathbb{R}^n} u(t, y) \, P_h(x, dy) = u(t, x) + hLu(t, x) + o(h),$$

we find that u satisfies

$$\partial_t u(t, x) = Lu(t, x) \quad \text{on } (0, \infty) \times \mathbb{R}^n \quad \text{with } \lim_{t \searrow 0} u(t, x) = \varphi(x).$$

In particular, if $P_t(x, dy) = p_t(x, y) \, dy$ where $(t, x, y) \rightsquigarrow p_t(x, y)$ is smooth, then we arrive at *Kolmogorov's backward equation*

$$(2.2.10) \qquad\qquad \partial_t p_t(\cdot, y) = Lp_t(\cdot, y),$$

in which it is important to understand that L is acting on $p_t(x, y)$ as a function of the backward variable x.

§2.2.4. Exercises

EXERCISE 2.2.11. There is a whole family of inequalities, all of which are commonly attributed to Gronwall. In this exercise, we will investigate one of these. Namely, suppose that α, β, and u are bounded, almost everywhere continuous, $\mathbb{R}$-valued functions on $[0, T]$, and assume that β is non-negative. Under the condition that

$$u(t) \leq \alpha(t) + \int_0^t \beta(\tau)u(\tau) \, d\tau \quad \text{for } t \geq 0,$$

Gronwall's inequality is the statement that

$$(2.2.12) \qquad u(T) \leq \alpha(0)e^{B(T)} + \int_0^T e^{B(T) - B(t)} \, d\alpha(t), \quad t \geq 0,$$

where $B(t) \equiv \int_0^t \beta(\tau) \, d\tau$ and all the integrals are taken in the sense of Riemann.[8]

Hint: Set $U(t) = \int_0^t \beta(\tau)u(\tau) \, d\tau$, and observe that $\dot{U}(t) \leq \alpha(t)\beta(t) + \beta(t)U(t)$. Now use $e^{-B(t)}$ as an integrating factor to obtain $e^{-B(t)}U(t) \leq \int_0^t \alpha(\tau)\beta(\tau)e^{-B(\tau)} \, d\tau$, and plug this estimate back into the original inequality. Finally, integrate by parts.

---

[8] For those worried about the fact that α may not be of bounded variation, notice that B is absolutely continuous, and see Theorem 1.2.7 in [34].

EXERCISE 2.2.13. Let Λ be the translation invariant vector field determined by the Lévy system (a, b, M), and let f be given by (2.1.10). That is, for each $\mu \in \mathbf{M}_1(\mathbb{R}^n)$, $\Lambda_\mu = \int (\tau_x)_* \Lambda_{\delta_0}^{(a,b,M)} \mu(dx)$ and $f(\xi) = e^{\Lambda_{\delta_0} \varepsilon_\xi}$. Show that, for each $\mu \in \mathbf{M}_1(\mathbb{R}^n)$, the one and only integral curve of Λ starting from μ is the "straight line" $t \rightsquigarrow \lambda_t * \mu$, where $t \rightsquigarrow \lambda_t$ is the infinitely divisible flow such that $\hat{\lambda}_t(\xi) = e^{tf(\xi)}$.

EXERCISE 2.2.14. Suppose that Λ is a vector field which satisfies the hypotheses in Theorem 2.2.5, and let $\mathcal{I}(\Lambda)$ be the set of $t \rightsquigarrow \mu_t$ which are integral curves of Λ.

(i) Show that if K is a compact subset of $\mathbf{M}_1(\mathbb{R}^n)$, then, for each $T \in [0, \infty)$,

$$\{\mu_t : \mu_\cdot \in \mathcal{I}(\Lambda), \ t \in [0, T], \text{ and } \mu_0 \in K\} \text{ is a compact subset of } \mathbf{M}_1(\mathbb{R}^n).$$

(ii) Now assume that for each $\nu \in \mathbf{M}_1(\mathbb{R}^n)$ there is precisely one integral curve $t \rightsquigarrow \mu_t^\nu$ of Λ starting from ν, and show that $(t, \nu) \rightsquigarrow \mu_t^\nu$ is continuous. In particular, when Λ is affine and satisfies the conditions of Theorem 2.2.6, then the transition probability function $(t, x) \rightsquigarrow P_t(x, \cdot)$ is continuous.

EXERCISE 2.2.15. For the same reason as we needed (cf. §1.4.1) the analogous fact about $\mathbf{M}_1(\mathbb{Z}_n)$, it is important to realize that the introduction of time dependence requires very little change in our theory of affine vector fields on $\mathbf{M}_1(\mathbb{R}^n)$. In fact, there is a trick which enables one to reduce the time dependent case on $\mathbf{M}_1(\mathbb{R}^n)$ to the time independent case on $\mathbf{M}_1(\mathbb{R}^{n+1})$. To be precise, given a map taking each $t \in [0, \infty)$ into an affine vector $\Lambda(t)$ on $\mathbf{M}_1(\mathbb{R}^n)$, determine

$$t \rightsquigarrow L(t)\varphi(x) = \frac{1}{2} \sum_{i,j=1}^n a_{ij}(t, x) \partial_{x_i} \partial_{x_j} \varphi + \sum_{i=1}^n b_i(t, x) \partial_{x_i} \varphi$$

$$+ \int_{\mathbb{R}^n} \left(\varphi(x + y) - \varphi(x) - \frac{(y, \text{grad}\varphi(x))}{1 + |y|^2} \right) M\big((t, x), dy\big)$$

so that (cf. (2.2.2)) $\Lambda(t)_\mu \varphi = \langle L(t)\varphi, \mu \rangle$. Finally, define $\tilde{L}$ on $C_b^2(\mathbb{R} \times \mathbb{R}^n; \mathbb{C})$ by

$$\tilde{L}\tilde{\varphi}(t, x) = (\partial_t + L(t \vee 0))\tilde{\varphi}.$$

(i) Check that the equation $\tilde{\Lambda}_{\tilde{\mu}} \tilde{\varphi} = \langle \tilde{L}\tilde{\varphi}, \tilde{\mu} \rangle$ determines $\tilde{\Lambda}$ as an affine vector field on $\mathbf{M}_1(\mathbb{R}^{n+1})$. Further, given $(s, \mu) \in [0, \infty) \times \mathbf{M}_1(\mathbb{R}^n)$, show that $t \rightsquigarrow \tilde{\mu}_t$ is an integral curve of $\tilde{\Lambda}$ starting from $\delta_s \times \mu$ if and only if $\tilde{\mu}_t = \delta_{s+t} \times \mu_t^s$, where $t \rightsquigarrow \mu_t^s$ is an integral curve of $t \rightsquigarrow \Lambda(t)$ starting from μ at time s in the sense that $\partial_t \mu_t^s = \Lambda(s + t)_{\mu_t^s}$ with $\mu_0^s = \mu$.

(ii) Assume that the hypotheses of Theorem 2.2.5 hold when $\mathbb{R}^n$ and L_μ there are replaced by $\mathbb{R}^{n+1}$ and $\tilde{L}$, respectively. Show that for each $(s, \mu) \in [0, \infty) \times \mathbf{M}_1(\mathbb{R}^n)$ there is an integral curve of $t \rightsquigarrow \Lambda(t)$ starting from μ at time s.

(iii) Continue with the hypotheses in (ii). If $u \in C_b^{1,2}([s, s+T] \times \mathbb{R}^n; \mathbb{C})$ satisfies $\partial_t u + L(t)u = 0$, show that $\langle u(s, \cdot), \mu_T^s \rangle = \langle u(s+T, \cdot), \mu_0^s \rangle$ for any integral curve of $t \rightsquigarrow \Lambda(t)$ starting at time s. In particular, if for each $\varphi \in C_c^\infty(\mathbb{R}^n; \mathbb{R})$ and $(s, T) \in [0, \infty) \times (0, \infty)$ there exists a $u \in C_b^{1,2}([s, s+T]; \mathbb{C})$ satisfying $\partial_t u + L(t)u = 0$ and $u(s, \cdot) = \varphi$, conclude that, for each $(s, \mu) \in [0, \infty) \times \mathbf{M}_1(\mathbb{R}^n)$ there is at exactly one integral curve of $t \rightsquigarrow \Lambda(t)$ starting from μ at time s.

(iv) Continue in the setting at the end of (iii), and, for each $(s, x) \in [0, \infty) \times E$, let $t \rightsquigarrow P_t((s, x), \cdot)$ denote the integral curve of $t \rightsquigarrow \Lambda(t)$ starting from δ_x at time s. Show that $(s, x) \rightsquigarrow P_t((s, x), \cdot)$ is continuous. In addition, given any integral curve $t \rightsquigarrow \mu_t^s$ starting at time s, show that $\mu_t^s = \int P_t((s, x), \cdot) \mu_0^s(dx)$. In particular, arrive at the *time inhomogeneous Chapman–Kolmogorov equation*

$$P_{t_1+t_2}((s, x), \cdot) = \int_{\mathbb{R}^n} P_{t_2}((s+t_1, y), \cdot) P_{t_1}((s, x), dy).$$

2.3 Pathspace Realization, Preliminary Version

We now want to develop the analog in the present setting of the considerations in §1.4, and again we will begin with the approach, basically due to Kolmogorov, which is very general but fails to shed much light.

§2.3.1. **Kolmogorov's Construction:** The analog of Theorem 1.4.3 presents no difficulty. Namely, let $t \rightsquigarrow \mu_t$ be an integral curve of the vector field Λ on $\mathbf{M}_1(\mathbb{R}^n)$ given in terms of $\mu \rightsquigarrow L_\mu$ by (2.2.1), where L_μ is the operator corresponding to the Lévy system $x \rightsquigarrow (a((x, \mu)), b(x, \mu), M((x, \mu), \cdot))$ as in (2.1.13). Next, set $L(t) = L_{\mu_t}$ and determine the time dependent vector field $t \rightsquigarrow \Lambda(t)$ by $\Lambda(t)_\mu \varphi = \langle L(t)\varphi, \mu \rangle$.

THEOREM 2.3.1. *Referring to the preceding, assume that for each $(s, \mu) \in [0, \infty) \times \mathbf{M}_1(E)$ there is a unique integral curve of $t \rightsquigarrow \Lambda(t)$ (cf. Exercise 2.2.15) starting from μ at time s, and use $P_t((s, x), \cdot)$ to denote the integral curve starting from δ_x at time s. Then, for all $s \in [0, \infty)$ and $t > 0$,*

$$\mu_{s+t} = \int_{\mathbb{R}^n} P_t((s, x), \cdot) \mu_s(dx).$$

Furthermore, there exists a unique probability measure $\mathbb{P}$ on $(\mathbb{R}^n)^{[0,\infty)}$ with the property that (1.4.4) holds for all $m \geq 1$, $0 = t_0 < \cdots < t_m$, and

$\Gamma_1, \ldots, \Gamma_m \in \mathcal{B}_{\mathbb{R}^n}$. *Equivalently, the* $\mathbb{P}$-*distribution of* $p \rightsquigarrow p(0)$ *is* μ_0 *and, for each* $(s, t) \in [0, \infty)^2$, *the* $\mathbb{P}$-*conditional distribution of* $p \rightsquigarrow p(s + t)$ *given* $\sigma(\{p(\tau) : \tau \in [0, s]\})$ *is given by* $p \rightsquigarrow P_t((s, p(s)), \cdot)$. *In particular, for each* t, *the* $\mathbb{P}$-*distribution of* $p(t)$ *is* μ_t.

§2.3.2. Path Regularity: Just as in §1.4, where we started by putting our probability measures on $(\mathbb{Z}_n)^{[0, \infty)}$ and eventually (cf. Theorem 1.5.5 and Exercise 1.4.14) proved that they live on $D([0, \infty); \mathbb{Z}_n)$, so here we will want to know that the measure $\mathbb{P}$ discussed in Theorem 2.3.1 lives[9] on the space $D([0, \infty); \mathbb{R}^n)$ of right continuous paths $p : [0, \infty) \longrightarrow \mathbb{R}^n$ which possess a left limit at each $t \in (0, \infty)$.

In order to have something to which we will be able to compare Itô's approach, we will give in this section a brief review of an abstract approach to path regularity questions. In Chapters 3 and 4, we will develop the approach of It/, which is analogous to the one in §1.4.[10]

Because we are dealing here with Markov processes, whenever possible, it is desirable to state criteria for various properties in terms of the transition probability function; and, in the case of properties about path regularity, J.R. Kinney (cf. [18] and [8]) formulated one of the more useful criteria.

Kinney's criterion states that a probability measure $\mathbb{P}$ on $(\mathbb{R}^n)^{[0, \infty)}$ satisfying (1.4.4) lives on $D([0, \infty); E)$ (in the sense described in the preceding footnote) if, as $t \searrow 0$, $P_t((s, x), \cdot)$ assigns each ball $B_{\mathbb{R}^n}(x, r)$ measure locally uniformly close to 1. To be precise, his criterion is that, for each $R \in [0, \infty)$ and $r > 0$,

$$(2.3.2) \qquad \lim_{t \searrow 0} \sup_{(s, x) \in [0, R] \times \overline{B_{\mathbb{R}^n}(0, R)}} P_t((s, x), B_{\mathbb{R}^n}(x, r)\complement) = 0.$$

To check (2.3.2) for the measure $\mathbb{P}$ coming from Theorem 2.3.1, first choose an $\eta \in C^\infty(\mathbb{R}; [0, 1])$ so that $\eta \equiv 0$ on $(-\infty, \frac{1}{4}]$ and $\eta \equiv 1$ on $[1, \infty)$. Next, given $x \in \mathbb{R}^n$ and $r > 0$, we set $\psi_{x,r}(y) = \eta(r^{-2}|y - x|^2)$ and observe that, for each $R \in (0, \infty)$, $L(t)\psi_{x,r}(y)$ is bounded above, uniformly in $(t, x, y) \in [0, R] \times \overline{B_{\mathbb{R}^n}(0, R)} \times \mathbb{R}^n$, by some constant $K_r(R) < \infty$. Hence, for $(s, x) \in [0, R] \times \overline{B_{\mathbb{R}^n}(0, R)}$,

$$P_t((s, x), B_{\mathbb{R}^n}(x, r)\complement) \leq \langle \psi_{x,r}, P_t((s, x), \cdot) \rangle \rangle$$

$$= \int_0^t \langle L(s + \tau)\psi_{x,r}, P_\tau((s, x), \cdot) \rangle \rangle \, d\tau \leq K_r(R)t.$$

---

[9] Just as in §1.4, the technical meaning here of *lives* on $D([0, \infty); \mathbb{R}^n)$ is that it assigns $D([0, \infty); \mathbb{R}^n)$ outer measure 1 and therefore admits a unique restriction as a probability measure on that space.

[10] It must be admitted that Itô's theory is slightly deficient here. Namely, as we will see (cf.§3.2), Itô's theory does not cover the most general case.

Before recording our conclusions in a theorem, it is important to know that when, for each $t \geq 0$, the operator $L(t)$ is *local*, in the sense that $L(t)\psi(x) = 0$ for all $t \geq 0$ whenever ψ vanishes in a neighborhood of x, $D([0,\infty); \mathbb{R}^n)$ can be replaced (cf. Exercise 2.3.5 below) by $C([0,\infty); \mathbb{R}^n)$. In other words, when the associated Lévy measures $M((t,x), \cdot)$ are 0, $\mathbb{P}$ lives on $C([0,\infty); \mathbb{R}^n)$. To understand the relevance of locality, note that it implies that $L(s+\tau)\psi_{x,r} = 0$ on $B(x, \frac{r}{2})$ and therefore that, for $(s,x) \in [0,R] \times \overline{B_{\mathbb{R}^n}(0,R)}$,

$$P_t\big((s,x), B_{\mathbb{R}^n}(x,r)\complement\big)$$
$$\leq K_r(R) \int_0^t P_\tau\big((s,x), B_{\mathbb{R}^n}(x, \tfrac{r}{2})\complement\big)\, d\tau \leq \frac{K_r(R) K_{r/2}(R) t^2}{2}.$$

In particular, this means that

$$\lim_{t \searrow 0} \sup_{(s,x) \in [0,R] \times \overline{B_{\mathbb{R}^n}(0,R)}} t^{-1} P_t\big((s,x), B_{\mathbb{R}^n}(x,r)\big) = 0,$$

which, by another theorem of Kinney's (cf. *op. cit.*), is more than enough to guarantee that $\mathbb{P}$ gives $C([0,\infty); \mathbb{R}^n)$ outer measure 1.

We summarize our progress in the following statement.

THEOREM 2.3.3. *For each $t \geq 0$ set $\mathcal{B}_t = \sigma(\{p(\tau) : t \in [0,t]\})$, and take $\mathcal{B} = \bigvee_{t \geq 0} \mathcal{B}_t$. Given the integral curve $t \rightsquigarrow \mu_t$ of a vector field Λ on $\mathbf{M}_1(\mathbb{R}^m)$ for which the conditions in Theorem 2.3.1 hold, there is a unique probability measure[11] $\mathbb{P}$ on $D([0,\infty); \mathbb{R}^n)$ for which (1.4.4) holds. In fact, $\mathbb{P}$ is uniquely characterized by the properties that the $\mathbb{P}$-distribution of $p \rightsquigarrow p(0)$ is μ_0 and, for each $\varphi \in C_b(\mathbb{R}^n; \mathbb{R})$, (cf. Theorem 2.3.1),*

$$\left(\varphi\big(p(t)\big) - \int_0^t L(\tau)\varphi\big(p(\tau)\big)\, d\tau, \mathcal{B}_t, \mathbb{P} \right)$$

is a martingale. Finally, when, in addition, $L(t)$ is local for each $t \geq 0$, then $D([0,\infty); \mathbb{R}^n)$ can be replaced by $C([0,\infty); \mathbb{R}^n)$.

PROOF: The only assertion requiring comment is the characterization in terms of the asserted martingale property. However, given the last part of Theorem 2.3.1, its proof follows exactly the same line of reasoning as the proof that (1.4.5) implies the final assertion in Corollary 1.5.9. □

Recall the discussion following Corollary 1.5.9 in which we introduced the idea of a *martingale problem* for an operator L, and observe that the last part of Theorem 2.3.3 has the following corollary.

---

[11] From now on, we will be implicitly assuming that a probability measure on $D([0,\infty); \mathbb{R}^n)$ is defined on the σ-algebra $\mathcal{B}$. See Exercise 2.3.5 for further information about this choice of σ-algebra.

COROLLARY 2.3.4. *Assume that the Λ in Theorem 2.3.3 is affine, and let L be the associated operator on $C_b^2(\mathbb{R}^n; \mathbb{R})$. Then, for each $x \in \mathbb{R}^n$ there is precisely one solution $\mathbb{P}_x^L$ on $D([0, \infty); \mathbb{R}^n)$ to the martingale problem for L starting from x. In fact, if $(t, x) \rightsquigarrow P_t(x, \cdot)$ is the transition probability function determined by Λ, then $\mathbb{P}_x^L$ is the one and only $\mathbb{P}$ on $D([0, \infty); \mathbb{R}^n)$ which satisfies (1.4.1) with $\mu = \delta_x$.*

§2.3.3. Exercises

EXERCISE 2.3.5. The purpose of this exercise is to discuss some elementary properties of the spaces $C([0, \infty); \mathbb{R}^n)$ and $D([0, \infty); \mathbb{R}^n)$. Throughout, we give both of these spaces the topology of uniform convergence on compact subsets. See Remark 3.1.23 for further discussion of these matters.

(i) Show that[12]

$$\rho(p, q) \equiv \sum_{m=1}^{\infty} \frac{\|p - q\|_{[0,m]}}{2^m (1 + \|p - q\|_{[0,m]})}$$

is a complete metric for $D([0, \infty); \mathbb{R}^n)$, and check that $C([0, \infty); \mathbb{R}^n)$ is a separable, closed subset of $D([0, \infty); \mathbb{R}^n)$.

(ii) Let $\mathcal{B}$ be the σ-algebra over $D([0, \infty); \mathbb{R}^n)$ or $C([0, \infty); \mathbb{R}^n)$ generated by the maps $p \rightsquigarrow p(t)$ for $t \in [0, \infty)$. Given a separable, closed subset F of $D([0, \infty); \mathbb{R}^n)$, show that

$$p \in D([0, \infty); \mathbb{R}^n) \longmapsto \rho(p, F) \equiv \inf\{\rho(p, q) : q \in F\}$$

is a continuous, $\mathcal{B}$-measurable map, and conclude that every separable, closed subset of $D([0, \infty); \mathbb{R}^n)$ is $\mathcal{B}$-measurable. In particular, show that $C([0, \infty); \mathbb{R}^n)$ is a $\mathcal{B}$-measurable subset of $D([0, \infty); \mathbb{R}^n)$ and that the restriction of $\mathcal{B}$ to $C([0, \infty); \mathbb{R}^n)$ coincides with the Borel field there.

(iii) Show that every $\mathcal{B}$-measurable, $\mathbb{R}$-valued function on $D([0, \infty); \mathbb{R}^n)$ is the pointwise sequential limit of continuous, $\mathcal{B}$-measurable, $\mathbb{R}$-valued functions.

2.4 The Structure of Lévy Processes on $\mathbb{R}^n$

Let (a, b, M) be a Lévy system. As a special case of Theorem 2.3.3 applied to the translation invariant vector field determined by (a, b, M), we know that there exists a unique probability measure $\mathbb{P}^{(a,b,M)}$, the Lévy process determined by the Lévy system (a, b, M), on $D([0, \infty); \mathbb{R}^n)$ with the properties that $\mathbb{P}^{(a,b,M)}(p(0) = 0) = 1$ and, for all $s, t \in [0, \infty)$,

$$\mathbb{P}^{(a,b,M)}(p(s + t) - p(s) \in \Gamma \mid \mathcal{B}_s) = \lambda_t^{(a,b,M)}(\Gamma) \quad \mathbb{P}^{(a,b,M)}\text{-almost surely}$$

---

[12] Given $q : [0, \infty) \longrightarrow \mathbb{R}^n$, we will use $\|q\|_{[0,T]}$ to denote $\sup_{t \in [0,T]} |q(t)|$.

for each $\Gamma \in \mathcal{B}_{\mathbb{R}^n}$. The goal of this section is to give another construction of $\mathbb{P}^{(a,b,M)}$, one that completely avoids Kinney's theorem and, at the same time, reveals the structure of $\mathbb{P}^{(a,b,M)}$.

§**2.4.1. Construction:** In view of §1.4.2, one should suspect that $\mathbb{P}^{(0,0,M)}$ is basically a Poisson process. To confirm this suspicion, recall that a *compound Poisson process with jump distribution* $\mu \in \mathbf{M}_1(\mathbb{R}^n)$ *and intensity* $\lambda > 0$ on a probability space $(\Omega, \mathcal{F}, \mathbb{Q})$ is a family $\{X(t) : t \geq 0\}$ of $\mathbb{R}^n$-valued random variables which admit the representation

$$(2.4.1) \qquad X(t,\omega) = \sum_{\ell=1}^{N(t,\omega)} X_\ell(\omega) \, (\equiv 0 \text{ if } \mathbf{N}(t,\omega) = 0), \quad t \in [0,\infty),$$

where $\{\mathbf{N}(t) : t \geq 0\}$ is a (cf. §1.4.2) simple Poisson process with intensity λ (i.e., $\mathbb{E}[\mathbf{N}(t)] = t\lambda$) and the X_ℓ's are mutually independent, $\mathbb{R}^n \setminus \{0\}$-valued random variables which are independent of $\sigma(\{\mathbf{N}(t) : t \geq 0\})$ and have distribution μ. Equivalently, $X(0,\omega) = 0$, $t \rightsquigarrow X(t,\omega)$ is right continuous and piecewise constant, the times between the jumps of $X(\,\cdot\,,\omega)$ are mutually $\mathbb{Q}$-independent exponential random variables with mean λ^{-1}, and the jumps have distribution μ and are independent of both the jump times and each other. Observe that when the X_ℓ's are integrable,

$$\mathbb{E}^{\mathbb{Q}}[X(t)] = \mathbb{E}^{\mathbb{Q}}[\mathbf{N}(t)] \mathbb{E}^{\mathbb{Q}}[X_1] = \lambda t \int_{\mathbb{R}^n} y \, \mu(dy),$$

and

$$\mathbb{E}^{\mathbb{Q}}[|X(t)|^2] = \mathbb{E}^{\mathbb{Q}}\left[\left| \sum_1^{\mathbf{N}(t)} (X_\ell - \mathbb{E}^{\mathbb{Q}}[X_1]) \right|^2 \right]$$

$$+ 2\mathbb{E}^{\mathbb{Q}}\left[\mathbf{N}(t)\left(X(t), \mathbb{E}^{\mathbb{Q}}[X_1]\right)_{\mathbb{R}^n}\right] - \mathbb{E}^{\mathbb{Q}}\left[\mathbf{N}(t)^2\right] \left|\mathbb{E}[X_1]\right|^2$$

$$= \mathbb{E}^{\mathbb{Q}}[\mathbf{N}(t)] \mathbb{E}^{\mathbb{Q}}\left[\left|X_1 - \mathbb{E}^{\mathbb{Q}}[X_1]\right|^2\right] + \mathbb{E}^{\mathbb{Q}}\left[\mathbf{N}(t)^2\right]\left|\mathbb{E}^{\mathbb{Q}}[X_1]\right|^2,$$

which, because $\mathbb{E}^{\mathbb{Q}}[\mathbf{N}(t)] = t\lambda = \mathrm{var}\big(N(t)\big)$, leads to

$$\mathbb{E}^{\mathbb{Q}}\left[\left|X(t) - \mathbb{E}[X(t)]\right|^2\right] = \mathbb{E}^{\mathbb{Q}}\left[|X(t)|^2\right] - \left|\mathbb{E}^{\mathbb{Q}}[X(t)]\right|^2$$

$$= \mathbb{E}^{\mathbb{Q}}[\mathbf{N}(t)]\mathbb{E}^{\mathbb{Q}}\left[\left|X_1 - \mathbb{E}^{\mathbb{Q}}[X_1]\right|^2\right] + \mathrm{var}\big(\mathbf{N}(t)\big)\left|\mathbb{E}^{\mathbb{Q}}[X_1]\right|^2 = \lambda t \int |y|^2 \, \mu(dy).$$

That is, for an $\mathbb{R}^n$-valued compound Poisson process $\{X(t) : t \geq 0\}$ with intensity $\lambda > 0$ and jump distribution μ,

$$(2.4.2) \qquad
\begin{aligned}
\mathbb{E}^{\mathbb{Q}}[X(t)] &= t\lambda \int y \, \mu(dy) \\
\mathbb{E}^{\mathbb{Q}}\left[|X(t) - \mathbb{E}[X(t)]|^2\right] &= t\lambda \int |y|^2 \, \mu(dy)
\end{aligned}
\qquad \text{if } \int |y| \, \mu(dy) < \infty.$$

LEMMA 2.4.3. *Let M be a Lévy measure. If $M(\mathbb{R}^n) = 0$, then*

$$\mathbb{P}^{(0,0,M)}\big(p(t) = 0 \text{ for all } t \geq 0\big) = 1.$$

On the other hand, if $M(\mathbb{R}^n) \in (0,\infty)$ and $p \in D\big([0,\infty);\mathbb{R}^n\big) \longmapsto \tilde{p} \in D\big([0,\infty);\mathbb{R}^n\big)$ is given by

$$\tilde{p}(t) = p(t) + t \int \frac{1}{1+|y|^2} y\, M(dy),$$

then the $\mathbb{P}^{(0,0,M)}$-distribution of $p \rightsquigarrow \tilde{p}$ is that of a compound Poisson process with intensity $M(\mathbb{R}^n)$ and jump distribution $\frac{M}{M(\mathbb{R}^n)}$. In particular, when $\int(1+|y|)\, M(dy) < \infty$ and

$$\bar{p}(t) = p(t) - t \int \frac{|y|^2}{1+|y|^2} y\, M(dy) = \tilde{p}(t) - t \int y\, M(dy),$$

then

(2.4.4) $$\mathbb{P}^{(0,0,M)}\big(\|\bar{p}\|_{[0,T]} \geq R\big) \leq \frac{T}{R^2} \int |y|^2\, M(dy).$$

PROOF: To prove the first statement, let $\{X(t) : t \geq 0\}$ be given by (2.4.1) when $\lambda = M(\mathbb{R}^n)$ and $\mu = \frac{M}{M(\mathbb{R}^n)}$, and set

$$Y(t,\omega) = X(t,\omega) - t \int \frac{1}{1+|y|^2} y\, M(dy).$$

Because $\{\mathbf{N}(t) : t \geq 0\}$ has $\mathbb{Q}$-independent increments and is $\mathbb{Q}$-independent of the X_ℓ's, it is clear that both $\{X(t) : t \geq 0\}$ and $\{Y(t) : t \geq 0\}$ also have $\mathbb{Q}$-independent increments. In addition, because $\mathbf{N}(s+t) - \mathbf{N}(s)$ is a Poisson random variable with mean $tM(\mathbb{R}^n)$ and the X_ℓ's are identically distributed and are independent of both each other and of $\{\mathbf{N}(t) : t \geq 0\}$,

$$\mathbb{E}^{\mathbb{Q}}\Big[e^{\sqrt{-1}\big(\xi, X(s+t)-X(s)\big)_{\mathbb{R}^n}}\Big] = e^{-tM(\mathbb{R}^n)} \sum_{m=0}^{\infty} \frac{\big(tM(\mathbb{R}^n)\big)^m}{m!} \mathbb{E}^{\mathbb{Q}}\Big[e^{\sqrt{-1}(\xi, X_1)_{\mathbb{R}^n}}\Big]^m$$

$$= \exp\left(t \int_{\mathbb{R}^n} \Big(e^{\sqrt{-1}\,(\xi,y)_{\mathbb{R}^n}} - 1\Big) M(dy)\right),$$

and so

$$\mathbb{E}^{\mathbb{Q}}\Big[e^{\sqrt{-1}\big(\xi, Y(s+t)-Y(s)\big)_{\mathbb{R}^n}}\Big]$$

$$= \exp\left(t \int \Big(e^{\sqrt{-1}(\xi,y)_{\mathbb{R}^n}} - 1 - \frac{(\xi,y)_{\mathbb{R}^n}}{1+|y|^2}\Big) M(dy)\right).$$

But this means that the $\mathbb{Q}$-distribution of $\{Y(t) : t \geq 0\}$ is that of the Lévy process corresponding to $(0, 0, M)$, which is equivalent to the statement that the $\mathbb{P}^{(0,0,M)}$-distribution of $p \rightsquigarrow \tilde{p}$ is that of a compound Poisson process with jump measure $\frac{M}{M(\mathbb{R}^n)}$ and intensity $M(\mathbb{R}^n)$.

Given the first part, the rest is easy. In fact, all that one has to do is realize (cf. (2.4.2)) that the $\mathbb{P}^{(0,0,M)}$-distribution of $p \rightsquigarrow \bar{p}$ is the same as the $\mathbb{Q}$-distribution of $\omega \rightsquigarrow \bar{X}(\,\cdot\,, \omega)$, where $\bar{X}(t, \omega) \equiv X(t) - \mathbb{E}^{\mathbb{Q}}[X(t)]$. In particular $\{\bar{p}(t) : t \geq 0\}$ under $\mathbb{P}^{(0,0,M)}$ has independent, centered increments satisfying

$$\mathbb{E}^{\mathbb{P}^{(0,0,M)}}\left[|\bar{p}(s+t) - \bar{p}(s)|^2\right] = t \int |y|^2\, M(dy).$$

Hence by (cf. Theorem 1.4.5 in [36]) Kolmogorov's inequality, we now know that, for any $\ell \in \mathbb{Z}^+$,

$$\mathbb{P}^{(0,0,M)}\left(\max_{1 \leq k \leq \ell} |\bar{p}(\tfrac{kT}{\ell})| \geq R\right) \leq \frac{T}{R^2} \int |y|^2\, M(dy),$$

from which (2.4.4) follows immediately after one lets $\ell \to \infty$. (See (3.3.8) in [36] for more details about this sort of argument.) $\square$

LEMMA 2.4.5. *Given a Lévy measure M, choose $\{r_\ell\}_0^\infty \subseteq (0, 1]$ so that $r_\ell \searrow 0$ and*

$$\int_{B_{\mathbb{R}^n}(0, r_\ell)} |y|^2\, M(dy) \leq 2^{-\ell - 1},$$

and determine the Borel measures $\{M^\ell : \ell \in \mathbb{N}\}$ so that

$$M^\ell(dy) = \begin{cases} \mathbf{1}_{(r_0, \infty)}(|y|)\, M(dy) & \text{if } \ell = 0 \\ \mathbf{1}_{(r_\ell, r_{\ell-1}]}(|y|)\, M(dy) & \text{if } \ell \geq 1. \end{cases}$$

Next, set $\mathbb{Q} = \prod_{\ell \in \mathbb{N}} \mathbb{P}^{(0,0,M^\ell)}$ on $\Omega \equiv D([0, \infty); \mathbb{R}^n)^{\mathbb{N}}$ with the product σ-algebra $\mathcal{F} \equiv \mathcal{B}^{\mathbb{N}}$, and, for each $m \geq 0$, define

$$\omega = (p^0, \ldots, p^m, \ldots) \in \Omega \longmapsto q^{(m)}(\,\cdot\,, \omega) = \sum_{\ell=0}^{m} p^\ell \in D([0, \infty); \mathbb{R}^n).$$

Then there exists an $\mathcal{F}$-measurable map $\omega \in \Omega \longmapsto q(\,\cdot\,, \omega) \in D([0, \infty); \mathbb{R}^n)$ such that, for each $T \in [0, \infty)$, $\|q^{(m)}(\,\cdot\,, \omega) - q(\,\cdot\,, \omega)\|_{[0,T]} \longrightarrow 0$ for $\mathbb{Q}$-almost every ω. Moreover, $\mathbb{P}^{(0,0,M)}$ is the $\mathbb{Q}$-distribution of $\omega \rightsquigarrow q(\,\cdot\,, \omega)$.

PROOF: We begin by showing that if F is a finite subset of $\mathbb{N}$, $M^{(F)} \equiv \sum_{\ell \in F} M^\ell$ and $q^{(F)}(\,\cdot\,,\omega) \equiv \sum_{\ell \in F} p^\ell$, then $\mathbb{P}^{(0,0,M^{(F)})}$ is the $\mathbb{Q}$-distribution of $\omega \rightsquigarrow q^{(F)}(\,\cdot\,,\omega)$. Indeed, it is clear that $\mathbb{Q}(q^{(F)}(0) = 0) = 1$ and that $\{q^{(F)}(t) : t \geq 0\}$ has $\mathbb{Q}$-independent increments. Finally,

$$
\mathbb{E}^{\mathbb{Q}}\left[\exp\left(\sqrt{-1}(\xi, q^{(F)}(s+t) - q^{(F)}(s))_{\mathbb{R}^n}\right)\right]
$$
$$
= \prod_{\ell \in F} \mathbb{E}^{\mathbb{P}^{(0,0,M^\ell)}}\left[\exp\left(\sqrt{-1}(\xi, p(s+t) - p(s))_{\mathbb{R}^n}\right)\right]
$$
$$
= \prod_{\ell \in F} \exp\left(\int \left(e^{\sqrt{-1}(\xi,y)_{\mathbb{R}^n}} - 1 - \frac{\sqrt{-1}(\xi,y)_{\mathbb{R}^n}}{1 + |y|^2}\right) M^\ell(dy)\right)
$$
$$
= \exp\left(\int \left(e^{\sqrt{-1}(\xi,y)_{\mathbb{R}^n}} - 1 - \frac{\sqrt{-1}(\xi,y)_{\mathbb{R}^n}}{1 + |y|^2}\right) M^{(F)}(dy)\right).
$$

We next want to show that, for each $T > 0$,

(*) $$\lim_{m \to \infty} \mathbb{Q}\left(\exists N > m \; \|q^{(N)} - q^{(m)}\|_{[0,T]} \geq \tfrac{2T}{m}\right) = 0.$$

But, by the preceding paragraph and (2.4.4),

$$
\mathbb{Q}\left(\|q^{(N)} - q^{(m)}\|_{0,T]} \geq \tfrac{2T}{m}\right) \leq \mathbb{Q}\left(\|\bar{q}^{(N)} - \bar{q}^{(m)}\|_{0,T]} \geq \tfrac{T}{m}\right) \leq m^2 2^{-m},
$$

from which (*) is trivial.

Knowing (*), the existence of $\omega \rightsquigarrow q(\,\cdot\,,\omega)$ is easy. Namely, (cf. part **(i)** of Exercise 2.3.5) (*) means that there exists a $\mathbb{Q}$-null set $B \in \mathcal{B}$ such that $\{q^{(m)}(\,\cdot\,,\omega)\}_0^\infty$ converges uniformly on compact subsets for each $\omega \notin B$. Hence, we can take $q(\,\cdot\,,\omega) = \lim_{m \to \infty} q^{(m)}(\,\cdot\,,\omega)$ when $\omega \notin B$ and $q(\,\cdot\,,\omega) \equiv 0$ when $\omega \in B$.

Finally the identification of the $\mathbb{Q}$-distribution of $\omega \rightsquigarrow q(\,\cdot\,,\omega)$ as $\mathbb{P}^{(0,0,M)}$ follows easily from the above convergence plus the corresponding identification of the $\mathbb{Q}$-distribution of each $\omega \rightsquigarrow q^{(m)}(\,\cdot\,,\omega)$. $\square$

In connection with the claim made in the introduction to this section that we would give a construction of $\mathbb{P}^{(a,b,M)}$ on $D([0,\infty);\mathbb{R}^n)$ which avoids Kinney's theorem, it should be noted that so far we have given a construction of $\mathbb{P}^{(0,0,M)}$ on $D([0,\infty);\mathbb{R}^n)$ without reference to any of the material in §2.3. In order to complete the program, it suffices to know that $\mathbb{P}^{(0,0,M}$ exists on $D([0,\infty);\mathbb{R}^n)$ and $\mathbb{P}^{(a,b,0)}$ exists on $C([0,\infty);\mathbb{R}^n)$, since one can then produce $\mathbb{P}^{(a,b,M)}$ on $D([0,\infty);\mathbb{R}^n)$ as the $\mathbb{P}^{(a,b,0)} \times \mathbb{P}^{(0,0,M)}$-distribution of $(p^1, p^2) \in C([0,\infty);\mathbb{R}^n) \times D([0,\infty);\mathbb{R}^n) \longmapsto p^1 + p^2 \in D([0,\infty);\mathbb{R}^n)$.

Because N. Wiener is recognized as the first to have constructed such a measure, the measure $\mathbb{P}^{(a,b,0)}$ on $C([0,\infty);\mathbb{R}^n)$ is often called the *Wiener measure with diffusion coefficient a and drift b.* In order to produce $\mathbb{P}^{(a,b,0)}$, first observe that it suffices to construct $\mathbb{P}^{(1,0,0)}$ on $C([0,\infty);\mathbb{R})$. Indeed, to construct $\mathbb{P}^{(a,b,0)}$ on $C([0,\infty);\mathbb{R}^n)$ knowing the existence of $\mathbb{P}^{(1,0,0)}$ on $C([0,\infty);\mathbb{R})$, let σ denote the non-negative definite, symmetric square root of a and observe that when

$$X(t,p) = \sigma \begin{pmatrix} p_1(t) \\ \vdots \\ p_n(t) \end{pmatrix} + tb \quad \text{for } p = \begin{pmatrix} p_1(t) \\ \vdots \\ p_n(t) \end{pmatrix} \in C([0,\infty);\mathbb{R}^n)^n,$$

then $\mathbb{P}^{(a,b,0)}$ is the $(\mathbb{P}^{(1,0,0)})^n$-distribution of $p \in C([0,\infty);\mathbb{R}^n)^n \longmapsto X(\,\cdot\,,p) \in C([0,\infty);\mathbb{R}^n)$.

Thus, all that remains is to prove that $\mathbb{P}^{(1,0,0)}$ on $C([0,\infty);\mathbb{R})$ exists, which is what Wiener was the first to do. However, shortly after learning about Wiener's work, P. Lévy found a far more elementary argument, and it is basically Lévy's argument which we now present.

LEMMA 2.4.6. *Let $(\Omega, \mathcal{F}, \mathbb{Q})$ be a probability space on which there exist a doubly indexed sequence $\{Y_{m,N} : (m,N) \in \mathbb{N}^2\}$ of mutually independent, standard normal (i.e., centered Gaussian with variance 1) random variables. (For instance, one can take (cf. Theorem 1.1.9 in [36]) the unit interval with Lebesgue measure.) Set $B^0(0) = 0$, and, for each $m \in \mathbb{N}$, let $B^0(m+1) - B^0(m) = Y_{m,0}$ and make $t \in [m, m+1] \longmapsto B^0(t,\omega) \in \mathbb{R}$ linear. Next, for $N \geq 1$, determine $B^N(\,\cdot\,,\omega)$ so that*

$$B^N(t,\omega) = B^{N-1}(t,\omega) + 2^{-(N+1)/2}\left(1 - 2^N\left|t - \tfrac{2m+1}{2^N}\right|\right) Y_{m,N}$$

for $t \in [m2^{-N+1}, (m+1)2^{-N+1}]$. Then there exists an $\mathcal{F}$-measurable map $\omega \in \Omega \longmapsto B(\,\cdot\,,\omega) \in C([0,\infty);\mathbb{R})$ with the property that $B^N(\,\cdot\,,\omega) \longrightarrow B(\,\cdot\,,\omega)$ uniformly on compact subsets for $\mathbb{Q}$-almost every ω. Moreover, $\mathbb{P}^{(1,0,0)}$ is the $\mathbb{Q}$-distribution of $\omega \in \Omega \rightsquigarrow B(\,\cdot\,,\omega) \in C([0,\infty);\mathbb{R})$.

PROOF: Everything turns on the fact that if $\mathfrak{G}$ is the smallest closed linear subspace of $L^2(\mathbb{Q})$ containing $\{X_{m,N} : (m,N) \in \mathbb{N}^2\}$, then every element of $\mathfrak{G}$ is a centered Gaussian random variable; and therefore (cf. Exercise 4.2.39 in [36]), for any $S \subseteq \mathfrak{G}$, $\sigma(S)$ is $\mathbb{Q}$-independent of $\sigma(S^\perp \cap \mathfrak{G})$.

The first step is to show that, for each $N \in \mathbb{N}$, $\{B^N((m+1)2^{-N}) - B^N(m2^{-N}) : m \in \mathbb{N}\}$ is a family of mutually independent centered Gaussian random variables with variance 2^{-N}, and, in view of the preceding remark, this comes down to checking that

$$\mathbb{E}^{\mathbb{Q}}\left[\left(B^N((\ell+1)2^{-N}) - B^N(\ell2^{-N})\right)\left(B^N((m+1)2^{-N}) - B^N(m2^{-N})\right)\right]$$

equals 0 if $k \neq \ell$ and 2^{-N} if $k = \ell$.

The second step is to observe that, for each $N \geq 1$ and $m \in \mathbb{N}$,

$$\sup_{t \in [m2^{-N+1}, (m+1)2^{-N+1}]} |B^N(t) - B^{N-1}(t)| = 2^{-\frac{N+1}{2}} |Y_{m,N}|,$$

and conclude that, for any $M \geq 1$ and $q \in [1, \infty)$,

$$\|B^N - B^{N-1}\|_{[0,M]} = 2^{-\frac{N+1}{2}} \max_{0 \leq m < M2^{N-1}} |Y_{m,N}|$$

$$\leq 2^{-\frac{N+1}{2}} \left(\sum_{m=0}^{M2^{N-1}-1} |Y_{m,N}|^{2q} \right)^{\frac{1}{2q}}.$$

In particular, this means that there is a $K(q) < \infty$ for which

$$\mathbb{E}^{\mathbb{P}} \left[\|B^N - B^{N-1}\|_{[0,M]}^{2q} \right]^{\frac{1}{2q}} \leq K(q) M^{\frac{1}{2q}} 2^{-\frac{N}{2q'}},$$

where $q' = \frac{q}{q-1}$.

Starting from the estimate just derived and proceeding as in the last part of the proof of Lemma 2.4.5, one can easily show that there exists a measurable map $\omega \in \Omega \longmapsto B(\,\cdot\,, \omega) \in C([0, \infty); \mathbb{R})$ such that, for any $q \in [1, \infty)$,

$$(2.4.7) \qquad \mathbb{E}^{\mathbb{P}} \left[\|B - B^N\|_{[0,M]}^{2q} \right]^{\frac{1}{2q}} \leq K(q) M^{\frac{1}{2q}} 2^{-\frac{N}{2q'}}$$

for some (slightly different) constant $K(q) < \infty$. Further, for each $N \in \mathbb{N}$, $\{B((m+1)2^{-N}) - B(m2^{-N}) : m \in \mathbb{N}\}$ is a family of $\mathbb{Q}$-mutually independent centered Gaussian random variables with variance 2^{-N}, which, because we know that $B(\,\cdot\,, \omega) \in C([0, \infty); \mathbb{R})$ for all ω, makes it easy to check that $\mathbb{P}^{(1,0,0)}$ is the $\mathbb{Q}$-distribution of $\omega \rightsquigarrow B(\,\cdot\,, \omega)$. $\square$

§2.4.2. **Structure:** In the preceding section we saw that the measure $\mathbb{P}^{(a,b,M)}$ on $D([0, \infty); \mathbb{R}^n)$ can be built up by a superimposition procedure. Namely, $\mathbb{P}^{(a,b,M)}$ results from superimposing the Wiener measure $\mathbb{P}^{(a,b,0)}$ on $\mathbb{P}^{(0,0,M)}$, and $\mathbb{P}^{(0,0,M)}$ can be obtained by carefully superimposing compound Poisson processes on top of one another. In this subsection, we will go in the opposite direction. That is, we will show how to decompose the paths of a Lévy process into its constituent parts. The idea is simple. First, we will start with a path $p \in D([0, \infty); \mathbb{R}^n)$ and will remove its jumps in such a way that the remaining path $w(\,\cdot\,, p)$ is continuous. Second, we will show that the $\mathbb{P}^{(a,b,M)}$-distribution of $p \rightsquigarrow w(\,\cdot\,, p)$ is $\mathbb{P}^{(a,b,0)}$, that the $\mathbb{P}^{(a,b,M)}$-distribution of $p \rightsquigarrow p - w(\,\cdot\,, p)$ is $\mathbb{P}^{(0,0,M)}$, and that $\sigma(\{w(t,p) : t \geq 0\})$

is $\mathbb{P}^{(a,b,M)}$-independent of $\sigma(\{p(t) - w(t,p) : t \geq 0\})$. Although the idea is simple, its implementation is complicated by the fact that, in general, a path $p \in D([0,\infty); \mathbb{R}^n)$ does not admit a clean decomposition into a continuous part and purely jump part. The point is that p can have infinitely many very small jumps which, in composite, cannot be canonically distinguished from a continuous path.[13]

In the next lemma, and elsewhere, $p(t-) \equiv \lim_{s \nearrow t} p(s)$ will be used to denote the left limit at $t \in (0,\infty)$ of a path $p \in D([0,\infty); \mathbb{R}^n)$.

LEMMA 2.4.8. *For any Lévy system* (a,b,M),

$$\mathbb{P}^{(a,b,M)}(p(t) = p(t-)) = 1 \quad \text{for all } t \in (0,\infty).$$

In addition, if Δ *is a Borel subset of* $\mathbb{R}^n \setminus \{0\}$ *and* $M(\Delta) = 0$, *then*

$$\mathbb{P}^{(a,b,M)}(\exists t \in (0,\infty) \, p(t) - p(t-) \in \Delta) = 0.$$

PROOF: To prove the first part, note that, by Fatou's Lemma, for any $\epsilon > 0$,

$$\mathbb{P}^{(a,b,M)}(|p(t) - p(t-)| > \epsilon) \leq \lim_{s \nearrow t} \mathbb{P}^{(a,b,M)}(|p(t) - p(s)| > \epsilon)$$

$$= \lim_{s \nearrow t} \lambda_{t-s}^{(a,b,M)}(B_{\mathbb{R}^n}(0,\epsilon)\complement) = 0.$$

To prove the second assertion, remember that $\mathbb{P}^{(a,b,M)}$ is the $\mathbb{P}^{(a,b,0)} \times \mathbb{P}^{(0,0,M)}$-distribution of $(p^1, p^2) \rightsquigarrow p^1 + p^2$. Thus, since $\mathbb{P}^{(a,b,0)}$ lives on $C([0,\infty); \mathbb{R}^n)$, we may and will restrict our attention to $\mathbb{P}^{(0,0,M)}$. Second, observe that when $M(\mathbb{R}^n) < \infty$, the result follows immediately from the representation of $\mathbb{P}^{(0,0,M)}$ in Lemma 2.4.3. Finally, to handle general Lévy measures M, we use the representation of $\mathbb{P}^{(0,0,M)}$ given in Lemma 2.4.5. Namely, that lemma says that $\mathbb{P}^{(0,0,M)}$ is the $\mathbb{Q}$-distribution of $\omega \rightsquigarrow q(\cdot, \omega)$, where, for $\mathbb{Q}$-almost every $\omega = (p^1, \ldots, p^m, \ldots)$, $q(\cdot, \omega) = \sum_{\ell=0}^{\infty} p^\ell$ and the sum converges uniformly on compact subsets. Further, since $M^\ell(\Delta) \leq M(\Delta) = 0$ and $M^\ell(\mathbb{R}^n) < \infty$, we know that

$$\mathbb{Q}(\exists t \in (0,\infty) \, p^\ell(t) - p^\ell(t-) \in \Delta)$$

$$= \mathbb{P}^{(0,0,M^\ell)}(\exists t \in (0,\infty) \, p(t) - p(t-) \in \Delta) = 0.$$

Thus, we will be done once we show that, $\mathbb{Q}$-almost surely, $J(q(\cdot, \omega)) \subseteq \bigcup_{\ell=0}^{\infty} J(p^\ell)$, where, for $p \in D([0,\infty); \mathbb{R}^n)$, $J(p) \equiv \{p(t) - p(t-) : t \in (0,\infty)\}$, and, in turn, this will follow if we show that, $\mathbb{Q}$-almost surely, no two p^ℓ's

---

[13] An expanded account of this point is given in §3.2 of [36].

jump at the same time. But, because $\mathbb{P}^{M^\ell}(p(t) \neq p(t-)) = 0$ for all $t \in (0, \infty)$ and $\ell \in \mathbb{N}$, it is easy (cf. Exercise 2.4.14 below) to see that, $\mathbb{Q}$-almost surely, no two p^ℓ's share a jump time. $\square$

To state the next result, we need to introduce the map taking $(t, p) \in [0, \infty) \times D([0, \infty); \mathbb{R}^n)$ into the Borel measure $\eta(t, \cdot, p)$ on $\mathbb{R}^n \setminus \{0\}$ given by

$$(2.4.9) \qquad \eta(t, \Gamma, p) = \sum_{\tau \leq t} 1_\Gamma \big(p(\tau) - p(\tau-)\big).$$

Observe that, because $p \in D([0, \infty); \mathbb{R}^n)$, the sum on the right hand side of (2.4.9) makes sense, since, for each $\epsilon > 0$, $p \restriction [0, t]$ can have at most finitely many jumps of size at least ϵ. That is, $\eta(t, \cdot, p) \restriction \mathbb{R}^n \setminus B_{\mathbb{R}^n}(0, \epsilon)$ is a finite, $\mathbb{N}$-valued measure for every $\epsilon > 0$.

LEMMA 2.4.10. *For any Lévy system (a, b, M) and a Borel set $\Delta \subseteq \mathbb{R}^n \setminus \{0\}$ with $M(\Delta) < \infty$, the $\mathbb{P}^{(a,b,M)}$-distribution of $p \rightsquigarrow \eta(\cdot, \Delta, p)$ is that of a simple Poisson process with intensity $M(\Delta)$. Moreover, if $M^\Delta(dy) = 1_\Delta(y) M(dy)$ and*

$$p^\Delta(t) \equiv \int_\Delta y \eta(\cdot, dy, p) - t \int_\Delta \frac{1}{1 + |y|^2} y\, M(dy),$$

then the $\mathbb{P}^{(a,b,M)}$-distribution of $p \in D([0, \infty); \mathbb{R}^n) \longmapsto (p^\Delta, p - p^\Delta) \in D([0, \infty); \mathbb{R}^n)^2$ is $\mathbb{P}^{(0,0,M^\Delta)} \times \mathbb{P}^{(a,b,M-M^\Delta)}$.

PROOF: Like the analogous results in §1.4.2, these results are most easily proved by making a construction and using uniqueness. Here is an outline. A more detailed account of a similar argument can be found in the proof of Corollary 3.3.17 in [36].

Note that $\mathbb{P}^{(a,b,M)}$ is the distribution of $(p^1, p^2) \in D([0, \infty); \mathbb{R}^n)^2 \longmapsto p^1 + p^2 \in D([0, \infty); \mathbb{R}^n)$ under $\mathbb{Q} \equiv \mathbb{P}^{(0,0,M^\Delta)} \times \mathbb{P}^{(a,b,M-M^\Delta)}$. In particular, we know (cf. Exercise 2.4.14) that p^1 and p^2 never jump at the same time. Hence, by the second part of Lemma 2.4.8,

$$\eta(\cdot, \Delta, p^1 + p^2) = \eta(t, \mathbb{R}^n, p^1) \quad \mathbb{Q}\text{-almost surely},$$

which, by the first part of Lemma 2.4.3, means that the $\mathbb{P}^{(a,b,M)}$-distribution of $p \rightsquigarrow \eta(\cdot, \Delta, p)$ is that of a simple Poisson process with intensity $M(\Delta)$. At the same time, by that same lemma,

$$p^1(t) + t \int_\Delta \frac{1}{1 + |y|^2} y\, M(dy) = \int_\Delta y \eta(t, dy, p^1 + p^2) \quad \mathbb{Q}\text{-almost surely},$$

and so $p \rightsquigarrow (p^\Delta, p - p^\Delta)$ has the same distribution under $\mathbb{P}^{(a,b,M)}$ as (p^1, p^2) have under $\mathbb{Q}$, which is the assertion that we wanted to prove. $\square$

We can now prove that the paths of an $\mathbb{R}^n$-valued independent process admit a decomposition which mimics the components making up the associated Lévy system.

THEOREM 2.4.11. *Let* (a, b, M) *be a Lévy system, and set*

$$\tilde{\eta}(t, dy, p) = \eta(t, dy, p) - \frac{t}{1 + |y|^2} M(dy).$$

Next, choose $1 \geq r_m \searrow 0$ *so that* $\int_{B_{\mathbb{R}^n}(0, r_m)} |y|^2 M(dy) \leq 2^{-m}$. *Then, for* $\mathbb{P}^{(a,b,M)}$-*almost every* $p \in D([0, \infty); \mathbb{R}^n)$, *the limit*

$$(2.4.12) \qquad\qquad \tilde{p}(t) \equiv \lim_{m \to \infty} \int_{|y| \geq r_m} y \tilde{\eta}^M(t, dy, p)$$

exists, uniformly with respect to t *in finite intervals; and therefore* $t \rightsquigarrow \tilde{p}(t)$ *is* $\mathbb{P}^{(a,b,M)}$-*almost surely an element of* $D([0, \infty); \mathbb{R}^n)$. *In fact, the* $\mathbb{P}^{(a,b,M)}$-*distribution of* $\tilde{p}$ *is* $\mathbb{P}^{(0,0,M)}$. *Furthermore, if* $w(\cdot, p) \equiv p - \tilde{p}$, *then,* $w(\cdot, p) \in C([0, \infty); \mathbb{R}^n)$ $\mathbb{P}^{(a,b,M)}$-*almost surely,* $\sigma(\{w(t, p) : t \geq 0\})$ *is* $\mathbb{P}^{(a,b,M)}$-*independent of* $\sigma(\{\tilde{p}(t) : t \geq 0\})$, *and the* $\mathbb{P}^{(a,b,M)}$-*distribution of* $p \rightsquigarrow w(\cdot, p)$ *is* $\mathbb{P}_0^{(a,b,0)}$. *In particular, this means that* $\sigma(\{w(\cdot, p) : t \geq 0\})$ *is* $\mathbb{P}^{(a,b,M)}$-*independent of* $\sigma(\{\eta(t, \Gamma, p) : t \geq 0 \ \& \ \Gamma \in \mathcal{B}_{\mathbb{R}^n \setminus \{0\}}\})$.

PROOF: Set $M^m(dy) = \mathbf{1}_{[r_m, \infty)}(|y|) M(dy)$, and

$$\tilde{p}^m(t) = \int_{|y| \geq r_m} y \, \tilde{\eta}(t, dy, p).$$

    In order to prove that the asserted limit exists in the required sense, set $\Delta^{(m,m')} = \{y \in \mathbb{R}^n : r_{m'} < |y| \leq r_m\}$, observe that, for $0 \leq m < m'$, (cf. the notation in Lemma 2.4.10)

$$\tilde{p}^m(t) - \tilde{p}^{m'}(t) = \bar{p}^{\Delta^{m,m'}}(t) - t\rho^{m,m'}, \text{ where}$$

$$\bar{p}^{\Delta^{(m,m')}}(t) = p^{\Delta^{(m,m')}}(t) - t \int_{\Delta^{(m,m')}} y \, M(dy) \text{ and}$$

$$\rho^{m,m'} \equiv \int_{\Delta^{(m,m')}} \frac{|y|^2}{1 + |y|^2} y \, M(dy).$$

Thus, by Lemma 2.4.10 and (2.4.4),

$$\mathbb{P}^{(a,b,M)}\left(\|\tilde{p}^{m'} - \tilde{p}^m\|_{[0,T]} \geq R + T2^{-m}\right) \leq \frac{T}{R^2} \int_{\Delta^{m,m'}} |y|^2 M(dy).$$

from which the required convergence result is immediate. Furthermore, because, by Lemma 2.4.10, the $\mathbb{P}^{(a,b,M)}$-distribution of $\tilde{p}^m$ is $\mathbb{P}^{(0,0,M^m)}$, it is an easy matter to check that $\mathbb{P}^{(0,0,M)}$ is the $\mathbb{P}^{(a,b,M)}$-distribution of $\tilde{p}$.

Next, set $\tilde{q}^m = p - \tilde{p}^m$. By Lemma 2.4.10, we know that, under $\mathbb{P}^{(a,b,M)}$, $\sigma(\{\tilde{q}^m(t) : t \geq 0\})$ is independent of $\sigma(\{\tilde{p}^m : t \geq 0\})$ and that the distribution of $\tilde{q}^m$ is $\mathbb{P}_0^{(a,b,M-M^m)}$. In particular, by Lemmas 2.4.8,

$$\mathbb{P}^{(a,b,M)}\big(\exists t \in [0,\infty) \,\big|q^m(t) - q^m(t-)\big| > r_m\big) = 0.$$

Hence, $w(\,\cdot\,,p) \in C\big([0,\infty);\mathbb{R}^n\big)$ $\mathbb{P}^{(a,b,M)}$-almost surely, and the required independence result as well as the identification of the $\mathbb{P}^{(a,b,M)}$-distribution of $w(\,\cdot\,,p)$ follow from the same sort of elementary limit argument which we just applied to $\{\tilde{p}^m\}_1^\infty$. $\square$

REMARK 2.4.13. It is important to recognize that the preceding contains an example of *stochastic integration*. Namely, one might be inclined to write

$$\tilde{p}(t) = \int_{\mathbb{R}^n \setminus \{0\}} y \, \tilde{\eta}(t, dy, p).$$

However, the integral on the right is *not* well defined in any completely classical sense. Indeed, the whole reason why we went to the trouble of replacing $\eta(t, dy, p)$ by $\tilde{\eta}(t, dy, p)$ was to produce the cancellation properties (resulting from centering and independence) which would allow us to show that, for almost every p, the right hand side exists as the limit of the classically defined integrals $\int_{|y| \geq r_m} y \, \tilde{\eta}(t, dy, p)$.

§2.4.3. Exercises

EXERCISE 2.4.14. Begin by showing that any $p \in D\big([0,\infty);\mathbb{R}^n\big)$ has at most countably many jumps. Next, suppose that $\mathbb{P}^1$ and $\mathbb{P}^2$ are a pair of probability measures on $D\big([0,\infty);\mathbb{R}^n\big)$ with the property that $\mathbb{P}^i\big(p(t) \neq p(t-)\big) = 0$ for $t \in (0,\infty)$ and $i \in \{1,2\}$. Show that $\mathbb{P}^1 \times \mathbb{P}^2$-almost surely, p^1 and p^2 never jump at the same time.
Hint: Let $\mathcal{T}(p) = \{t \in (0,\infty) : p(t) \neq p(t-)\}$, and note that

$$\mathbb{P}^1 \times \mathbb{P}^2\big(\{(p^1,p^2) : \mathcal{T}(p^1) \cap \mathcal{T}(p^2) \neq \emptyset\}\big)$$

$$\leq \int \left(\sum_{t \in \mathcal{T}(p^1)} \mathbb{P}^2\big(p^2(t) \neq p^2(t-)\big) \right) \mathbb{P}^1(dp^1).$$

EXERCISE 2.4.15. Let (a,b,M) be a Lévy system, and define the operator $L^{(a,b,M)}$ accordingly, as in (2.1.3). We know that

$$\left(\varphi(p(t)) - \int_0^t L^{(a,b,M)} \varphi(p(\tau)) \, d\tau, \mathcal{B}_t, \mathbb{P}^{(a,b,M)} \right)$$

is a martingale for every $\varphi \in C_b^2(\mathbb{R}^n; \mathbb{R})$.

Next, suppose that that $\int_{\mathbb{R}^n} |y|^2 \, M(dy) < \infty$. Show that

$$\mathbb{E}^{\mathbb{P}^{(a,b,M)}}[p(t)] = t\bar{b}, \quad \text{where } \bar{b} \equiv b + \int_{\mathbb{R}^n} \frac{|y|^2}{1+|y|^2} \, y M(dy)$$

and

$$\mathbb{E}^{\mathbb{P}^{(a,b,M)}}[\bar{p}(t) \otimes \bar{p}(t)] = t\bar{a} \quad \text{where}$$

$$\bar{p}(t) \equiv p(t) - t\bar{b} \text{ and } \bar{a} \equiv a + \int_{\mathbb{R}^n} y \otimes y \, M(dy).$$

Conclude that both $\big(\bar{p}(t), \mathcal{B}_t, \mathbb{P}^{(a,b,M)}\big)$ and $\big(\bar{p}(t) \otimes \bar{p}(t) - t\bar{a}, \mathcal{B}_t, \mathbb{P}^{(a,b,M)}\big)$ are martingales.

EXERCISE 2.4.16. Let everything be as in Lemma 2.4.6. Without much effort, one can improve on the conclusion made there and get an estimate on the Hölder modulus of continuity of Brownian paths.

(i) Note that, $\mathbb{P}$-almost surely, B^N is the polygonal path which one obtains by replacing the limit path B by its own linearization on each of the intervals $[m2^{-N}, (m+1)2^{-N}]$, and conclude that, $\mathbb{P}$-almost surely,

$$|B(t) - B(s)| \leq 2^N (t-s) \max_{0 \leq m < 2^N} |B((m+1)2^{-N}) - B(m2^{-N})| + 2\|B - B^N\|_{[0,1]}$$

for all $0 \leq s < t \leq 1$. Now, given $2^{-N-1} < t - s \leq 2^{-N}$, show first that

$$\frac{|B(t) - B(s)|}{(t-s)^\alpha} \leq 2^\alpha 2^{N\alpha} \left(\sum_{m=0}^{2^N-1} |B((m+1)2^{-N}) - B(m2^{-N})|^{2q} \right)^{\frac{1}{2q}}$$

$$+ 2^{1+\alpha} 2^{N\alpha} \|B - B^N\|_{[0,1]},$$

and then that

$$\mathbb{E}^{\mathbb{P}} \left[\sup_{\substack{0 \leq s < t \leq 1 \\ \frac{1}{2} \leq 2^N(t-s) < 1}} \left| \frac{|B(t) - B(s)|}{(t-s)^\alpha} \right|^{2q} \right]^{\frac{1}{2q}} \leq K(q) 2^{N(\alpha - \frac{1}{2q'})}$$

for an appropriate $K(q) < \infty$. In particular, if $\alpha \in (0, \frac{1}{2})$, use this to arrive at

$$\mathbb{E}^{\mathbb{P}} \left[\sup_{0 \leq s < t \leq 1} \left| \frac{|B(t) - B(s)|}{(t-s)^\alpha} \right|^{2q} \right]^{\frac{1}{2q}} < \infty$$

for all $q \in [1, \infty)$ with $\alpha < \frac{1}{2q'}$.

(ii) Given $T > 0$, set $B_T(t) = T^{-\frac{1}{2}} B(tT)$, and show that the $\mathbb{P}$-distribution $\omega \rightsquigarrow B_T(\cdot, \omega)$ is the same for all T. This is sometimes called the *Brownian scaling invariance* property. Finally, use this together with the result in (i) to show that, for each $\alpha \in (0, \frac{1}{2})$ and $q \in [1, \infty)$ there is a $K(\alpha, q) < \infty$ such that

$$\mathbb{E}^{\mathbb{P}^{(I,0,0)}} \left[\sup_{0 \le s < t \le T} \left| \frac{|p(t) - p(s)|}{(t-s)^\alpha} \right|^q \right]^{\frac{1}{q}} \le \sqrt{n} K(\alpha, q) T^{\frac{1}{2} - \alpha}$$

for all $T > 0$.

EXERCISE 2.4.17. The line of reasoning in the preceding really has very little to do with Brownian motion and can be easily generalized to give a proof of *Kolmogorov's continuity criterion*. (See Theorem 2.1.6 and Exercise 2.4.1 in [36] for an approach which is both more elegant and more general.) Namely, suppose that $\{X(t) : t \in [0, T]\}$ is a family of random variables on $(\Omega, \mathcal{F}, \mathbb{P})$ which take their values in some Banach space $\mathbf{X}$. Further, assume that, for some $q \in [1, \infty)$ and $C_q < \infty$,

$$\mathbb{E}^{\mathbb{P}} \left[\|X(t) - X(s)\|_{\mathbf{X}}^{2q} \right]^{\frac{1}{2q}} \le C_q (t-s)^{\frac{1}{2}} \quad \text{for all } 0 \le s < t \le T,$$

where $\| \cdot \|_{\mathbf{X}}$ is used to denote the Banach norm on $\mathbf{X}$. Prove Kolmogorov's result which says there is a family $\{\tilde{X}(t) : t \in [0, T]\}$ of $\mathbf{X}$-valued random variables with the properties that $t \in [0, T] \longmapsto \tilde{X}(t, \omega) \in \mathbf{X}$ is continuous for each $\omega \in \Omega$ and $\tilde{X}(t) = X(t)$ $\mathbb{P}$-almost surely for each $t \in [0, T]$. Furthermore, if $\alpha \in \left(0, \frac{q-1}{2q}\right)$, show there is a $K(\alpha, q) < \infty$, such that

$$\mathbb{E}^{\mathbb{P}} \left[\sup_{0 \le s < t \le T} \left(\frac{\|\tilde{X}(t) - \tilde{X}(s)\|_{\mathbf{X}}}{(t-s)^\alpha} \right)^{2q} \right]^{\frac{1}{2q}} \le K(\alpha, q) T^{\frac{1}{2} - \alpha}.$$

In fact, for each $\alpha \in (0, \frac{1}{2})$, the constant $K(\alpha, q)$ can be chosen to depend only on α, q, and the constant C_q.

Hint: Begin by rescaling to the case when $T = 1$, and, for $Y : [0, 1] \longrightarrow \mathbf{X}$, use $\|Y\|_{[0,1],\mathbf{X}}$ to denote $\sup_{t \in [0,1]} \|Y(t)\|_{\mathbf{X}}$. Next, for each $N \in \mathbb{N}$, let $X^N(\cdot, \omega)$ be the polygonal path obtained from $X(\cdot, \omega)$ by linearization between the dyadic points $\{m2^{-N} : 0 \le m \le 2^N\}$, check that

$$\|X^N(\cdot) - X^{N-1}(\cdot)\|_{[0,1],\mathbf{X}} \le \left(\sum_{m=1}^{2^N} \|X(m2^{-N}) - X((m-1)2^{-N})\|_{\mathbf{X}}^{2q} \right)^{\frac{1}{2q}},$$

conclude that,

$$\mathbb{E}^{\mathbb{P}}\left[\left\|X^N(\,\cdot\,) - X^{N-1}(\,\cdot\,)\right\|_{[0,1],\mathbf{X}}^{2q}\right]^{\frac{1}{2q}} \le C_q 2^{-\frac{N}{2q'}},$$

and use this to produce **X**-valued random variables $\{\tilde{X}(t) : t \in [0,1]\}$ such that $t \rightsquigarrow \tilde{X}(t,\omega)$ is continuous for each ω and

$$\mathbb{E}^{\mathbb{P}}\left[\left\|\tilde{X}(\,\cdot\,) - X^N(\,\cdot\,)\right\|_{[0,1],\mathbf{X}}^{2q}\right]^{\frac{1}{2q}} \le \left(1 - 2^{-\frac{1}{2q'}}\right)^{-1} C_q 2^{-\frac{N}{2q'}}.$$

In particular, this is more than enough to see that $\tilde{X}(t) = X(t)$ $\mathbb{P}$-almost surely first for each dyadic $t \in [0,1]$ and then for all $t \in [0,1]$. In addition, just as in (ii) of the preceding, it is a simple matter to pass from here to the asserted Hölder continuity estimate.

CHAPTER 3

Itô's Approach in the Euclidean Setting

In §2.3 we used Kolmogorov's approach to lift the integral curve of a vector field on $\mathbf{M}_1(\mathbb{R}^n)$ to a measure on the pathspace $D([0, \infty); \mathbb{R}^n)$ (cf. Theorems 2.3.1 and 2.3.3), and we then showed in §2.4 that the measure associated with a Lévy process admits a much more transparent construction. Our goal in this chapter is, following Itô, to show how to build the paths corresponding to more general affine vector fields from Lévy processes. In other words, we want to implement here the analog for $\mathbb{R}^n$ of what we did for $\mathbb{Z}_n$ in §§1.5 and 1.6.

There are two major challenges which make Itô's theory in $\mathbb{R}^n$ more complicated than it is (cf. §1.5) in $\mathbb{Z}_n$. The first of these is the fact that, as we saw in §2.4, there is a much richer variety of infinitely divisible laws in $\mathbb{R}^n$ than there was in $\mathbb{Z}_n$. The second is that the most interesting of these laws correspond to Lévy processes for which the paths are very far from piecewise constant. Indeed, their paths need not even have locally bounded variation, which means (cf. Remark 2.4.13) that the analog of integrals like the one on the right hand side of (1.6.1) cannot be interpreted à la either Lebesgue or Riemann.

3.1 Itô's Basic Construction

Given a map taking $x \in \mathbb{R}^n$ into a Lévy system $\big(a(x), b(x), M(x, \cdot)\big)$, our goal is to develop a method for transforming an appropriate Lévy process into a process which solves the martingale problem for the operator L given on $C_b^2(\mathbb{R}^n; \mathbb{R})$ by

$$L\varphi(x) \equiv \frac{1}{2} \sum_{i,j=1}^{n} a_{ij}(x) \partial_i \partial_j \varphi(x) + \sum_{i=1}^{n} b_i(x) \partial_i \varphi(x)$$

(3.1.1)
$$+ \int_{\mathbb{R}^{n'}} \left(\varphi(x+y) - \varphi(x) - \frac{(y, \operatorname{grad}_x \varphi)_{\mathbb{R}^n}}{1 + |y|^2} \right) M(x, dy).$$

The procedure which we will use is basically Itô's adaptation of Euler's approximation scheme, this time carried out path by path.

§3.1.1. Transforming Lévy Processes: In order to get going, it is important to see how to proceed in the trivial case when the Lévy system does not depend on x. That is, we want to understand how one might go about transforming one Lévy process into another.

The Lévy processes out of which we want to build other processes will, in general, take values in $\mathbb{R}^{n'}$, where n' need not equal n, and will be associated with a Lévy system of the form $(I, 0, M')$ for some $M' \in \mathfrak{M}(\mathbb{R}^{n'})$; and for this reason we try to make the notation less cumbersome by taking $\mathbb{P}^{M'} \equiv \mathbb{P}^{(I,0,M')}$, In particular (cf. the notation in Theorem 2.4.11), the $\mathbb{P}^{M'}$-distribution of $p' \rightsquigarrow w(\,\cdot\,, p')$ will be that of a standard, $\mathbb{R}^{n'}$-valued Wiener process (a.k.a. an $\mathbb{R}^{n'}$-valued Brownian motion). Now suppose that we are given a Lévy system (a, b, M) for an infinitely divisible law on $\mathbb{R}^n$. How can we go about transforming a path $p' \in D([0,\infty); \mathbb{R}^{n'})$ into a path $X(\,\cdot\,, p') \in D([0,\infty); \mathbb{R}^n)$ in such a way that the $\mathbb{P}^{M'}$-distribution of $p' \rightsquigarrow X(\,\cdot\,, p')$ will be $\mathbb{P}^{(a,b,M)}$?

To answer this question, we treat one component of (a, b, M) at a time. The diffusion coefficient a and drift b are easy. Indeed, as long as $n' \geq n$, we can find a $\sigma \in \mathrm{Hom}(\mathbb{R}^{n'}; \mathbb{R}^n)$ such that $a = \sigma\sigma^\top$, in which case the $\mathbb{P}^{M'}$-distribution of $p' \rightsquigarrow X(\,\cdot\,, p')$, where $X(t, p') \equiv \sigma w(t, p') + tb$, will be $\mathbb{P}^{(a,b,0)}$. To handle the Lévy measure, we will need the following analog of the considerations in §1.5.1.

LEMMA 3.1.2. *Given a Borel measurable function $F : \mathbb{R}^{n'} \longrightarrow \mathbb{R}^n$ which vanishes in a neighborhood of the origin, define (cf. (2.4.9))*

$$p' \in D([0,\infty); \mathbb{R}^{n'}) \longmapsto X_F(\,\cdot\,, p') \equiv \int F(y')\, \eta(\,\cdot\,, dy', p') \in D([0,\infty); \mathbb{R}^n).$$

Then, for each $t \in [0,\infty)$, $p' \rightsquigarrow X_F(t, p')$ is $\mathcal{B}'_t \equiv \sigma(\{p'(\tau) : \tau \in [0,t]\})$-measurable. Moreover, if $F_* M'$ is the Lévy measure on $\mathbb{R}^n$ given by

$$(3.1.3) \qquad F_* M'(\Gamma) = M'(\{y' \in \mathbb{R}^{n'} : F(y') \in \Gamma \setminus \{0\}\}), \quad \Gamma \in \mathcal{B}_{\mathbb{R}^n},$$

then either $F_* M' = 0$, and $X_F(\,\cdot\,, p) \equiv 0$ $\mathbb{P}^{M'}$-almost surely, or $F_* M'(\mathbb{R}^n) \in (0,\infty)$ and $\{X_F(t) : t \in [0,\infty)\}$ under $\mathbb{P}^{M'}$ is a compound Poisson process with intensity $F_* M'(\mathbb{R}^n)$ and jump measure $\frac{F_* M'}{F_* M'(\mathbb{R}^n)}$. Equivalently, $\mathbb{P}^{(0,0,F_* M')}$ is the $\mathbb{P}^{M'}$-distribution of $p' \rightsquigarrow \tilde{X}_F(\,\cdot\,, p')$ when

$$\tilde{X}_F(t, p) \equiv X_F(t, p) - t \int \frac{1}{1 + |F(y')|^2} F(y')\, M'(dy').$$

Finally, if $\int_{\mathbb{R}^{n'}} |F(y')|\, M'(dy') < \infty$ and

$$\bar{X}_F(t, p') \equiv X_F(t, p') - t \int_{\mathbb{R}^{n'}} F(y')\, M'(dy')$$

$$= \tilde{X}_F(t, p) - t \int_{\mathbb{R}^{n'}} \frac{|F(y')|^2}{1 + |F(y')|^2} F(y')\, M'(dy'),$$

then

$$(3.1.4) \quad R^2 \mathbb{P}^{M'} \left(\|\bar{X}_F\|_{[0,T]} \geq R \right) \leq \mathbb{E}^{\mathbb{P}^{M'}} \left[|\bar{X}_F(T)|^2 \right] = T \int_{\mathbb{R}^{n'}} |F(y')|^2\, M'(dy').$$

PROOF: Because

$$X_F(s, p') = \sum_{\sigma \leq s} F\big(p'(\sigma) - p'(\sigma-)\big),$$

it is clear that $X_F(\cdot, p') \in D\big([0, \infty); \mathbb{R}^n\big)$ and that $p' \rightsquigarrow X_F(t, p')$ is $\mathcal{B}'_t$-measurable. In addition, one sees that $X_{F_m}(s, p') \longrightarrow X_F(s, p')$ if $\{F_m\}_1^\infty$ is a sequence of functions, all of which vanish is a fixed neighborhood of the origin, which tends pointwise to F as $m \to \infty$. Thus, throughout, we may and will assume that F is bounded.

Now suppose that F is simple in the sense that $F(y') = \sum_1^m z_\ell \mathbf{1}_{\Delta_\ell}(y')$ for some choice of $\{z_\ell\}_1^m \subseteq \mathbb{R}^n$ and mutually disjoint $\{\Delta_\ell\}_1^m \subseteq \mathcal{B}_{\mathbb{R}^{n'}}$ with $0 \notin \overline{\Delta}_\ell$. Then

$$X_F(t, p') \equiv \int F(y')\, \eta(t, dy', p') = \sum_{\ell=1}^m z_\ell \eta(t, \Delta_\ell, p'),$$

from which it is obvious (cf. the first part of Lemma 2.4.10) that, under $\mathbb{P}_0^{(a', b', M')}$, $X_F(s + t, p') - X_F(s, p')$ is independent of $\mathcal{B}'_s$ and that its distribution is the same as that of $X_F(t, p')$. That is,

$$\mathbb{E}^{\mathbb{P}^{M'}} \left[e^{\sqrt{-1}\, \big(\xi, X_F(s+t) - X_F(s)\big)_{\mathbb{R}^n}} \,\Big|\, \mathcal{B}'_s \right] = \mathbb{E}^{\mathbb{P}^{M'}} \left[e^{\sqrt{-1}\, \big(\xi, X_F(t)\big)_{\mathbb{R}^n}} \right].$$

At the same time, by Lemma 2.4.10,

$$\mathbb{E}^{\mathbb{P}^{M'}} \left[e^{\sqrt{-1}\, \big(\xi, X_F(t)\big)_{\mathbb{R}^n}} \right] = \prod_1^m \exp\left(t M'(\Delta_\ell) \left(e^{\sqrt{-1}\, (\xi, z_\ell)_{\mathbb{R}^n}} - 1 \right) \right)$$

$$= \exp\left(t \int \left(e^{\sqrt{-1}\, \big(\xi, F(y)\big)_{\mathbb{R}^n}} - 1 \right) M'(dy) \right).$$

Hence, we have now completed the case when F is simple.

To handle general F's which are bounded, choose a uniformly bounded sequence $\{F_m\}_1^\infty$ of simple functions, all of which vanish on some fixed neighborhood of the origin, so that $F_m \longrightarrow F$ in $L^1(M'; \mathbb{R}^{n'})$. Because $X_{F_{m'}} - X_{F_m} = X_{F_{m'} - F_m}$, either $(F_{m'})_* M' = (F_m)_* M'$, and therefore $X_{F_{m'}}(\,\cdot\,, p') = X_{F_m}(\,\cdot\,, p')$ $\mathbb{P}^{M'}$-almost surely, or $(F_{m'})_* M' \neq (F_m)_* M'$ and the $\mathbb{P}^{M'}$-distribution of $X_{F_{m'}} - X_{F_m}$ is that of a compound Poisson process with intensity $(F_{m'} - F_m)_* M'(\mathbb{R}^{n'})$ and jump distribution $\frac{(F_{m'} - F_m)_* M'}{(F_{m'} - F_m)_* M'(\mathbb{R}^{n'})}$. In either case,

$$\|X_{F_{m'}}(t) - X_{F_m}(t)\|_{L^1(\mathbb{P}^{M'}; \mathbb{R}^n)} = t\|F_{m'} - F_m\|_{L^1(M'; \mathbb{R}^{n'})}.$$

Hence, for each $t \geq 0$, $X_{F_m}(t) \longrightarrow X_F(t)$ in $\mathbb{P}^{M'}$-measure; and so, for any $A' \in \mathcal{B}'_s \equiv \sigma(\{p'(\sigma) : \sigma \in [0, s]\})$ with $\mathbb{P}^{M'}(A') > 0$,

$$\mathbb{E}^{\mathbb{P}^{M'}}\left[e^{\sqrt{-1}\left(\xi, X_F(s+t) - X_F(s)\right)_{\mathbb{R}^n}} \,\Big|\, A'\right]$$

$$= \lim_{m\to\infty} \mathbb{E}^{\mathbb{P}^{M'}}\left[e^{\sqrt{-1}\left(\xi, X_{F_m}(s+t) - X_{F_m}(s)\right)_{\mathbb{R}^n}} \,\Big|\, A'\right]$$

$$= \exp\left(t \int \left(e^{\sqrt{-1}\left(\xi, F(y)\right)_{\mathbb{R}^n}} - 1\right) M'(dy)\right).$$

Knowing that the $\mathbb{P}^{M'}$-distribution of $p' \rightsquigarrow X_F(\,\cdot\,, p')$ is that of a compound Poisson process, we get (3.1.4) as a particular case of (2.4.4). □

Lemma 3.1.2, together with the remark preceding it, already tells us how to transform paths of the Lévy process with distribution $\mathbb{P}^{M'}$ into the Lévy process with distribution $\mathbb{P}^{(a,b,M)}$ under the condition that (cf. (3.1.3)) $M = F_* M'$ for some F which vanishes near the origin. Furthermore, by mimicking the line of reasoning which we used in the proof of Lemma 2.4.5, we could replace this vanishing condition by the condition that there exists an $r > 0$ for which $\int_{B_{\mathbb{R}^{n'}}(0,r)} |F(y')|^2\, dy' < \infty$. However, because such extensions will be incorporated into the program which follows, we will not bother with them here.

§3.1.2. **Hypotheses and Goals:** Throughout the rest of this section, we will be dealing with the situation described in hypotheses **(H1)**–**(H4)** below.

(H1) M' will be a Lévy measure on $\mathbb{R}^{n'}$ and $\mathbb{P}^{M'}$ will be the probability measure $\mathbb{P}^{(I,0,M')}$ on $D([0, \infty); \mathbb{R}^{n'})$.

(H2) $F : \mathbb{R}^n \times \mathbb{R}^{n'} \longrightarrow \mathbb{R}^n$ will be a Borel measurable map which takes $\mathbb{R}^n \times \{0\}$ to 0 and satisfies the estimates

(a) $$\lim_{r \searrow 0} \sup_{x \in \mathbb{R}^n} \frac{1}{1 + |x|^2} \int_{B_{\mathbb{R}^{n'}}(0,r)} |F(x, y')|^2\, M'(dy') = 0,$$

and, for each $R \in (0, \infty)$:

(b)
$$\sup_{x \in \mathbb{R}^n} \frac{1}{1 + |x|^2} \int_{B_{\mathbb{R}^{n'}}(0,R)} |F(x, y')|^2 \, M'(dy') < \infty,$$

and

(c)
$$\sup_{x_1 \neq x_2} \frac{1}{|x_2 - x_1|^2} \int_{B_{\mathbb{R}^{n'}}(0,R)} |F(x_2, y') - F(x_1, y')|^2 \, M'(dy') < \infty.$$

(H3) $\sigma : \mathbb{R}^n \longrightarrow \mathrm{Hom}(\mathbb{R}^{n'}, \mathbb{R}^n)$ and $b : \mathbb{R}^n \longrightarrow \mathbb{R}^n$ will be continuous maps which satisfy the condition[1]

$$\sup_{x_2 \neq x_1} \frac{\|\sigma(x_2) - \sigma(x_1)\|_{\mathrm{H.S.}} \vee |b(x_2) - b(x_1)|}{|x_2 - x_1|} < \infty,$$

and $a : \mathbb{R}^n \longrightarrow \mathrm{Hom}(\mathbb{R}^n, \mathbb{R}^n)$ will be used to denote the symmetric, non-negative definite matrix-valued function $\sigma \sigma^\top$.

(H4) Using (a) of **(H2)**, choose $1 \geq r_N \searrow 0$ so that

$$\sup_{x \in \mathbb{R}^n} \frac{1}{1 + |x|^2} \int_{B_{\mathbb{R}^{n'}}(0,r_N)} |F(x, y')|^2 \, M'(dy') \leq 2^{-N}.$$

Then, for $N \in \mathbb{N}$,

$$(x, p') \in \mathbb{R}^n \times D\big([0, \infty); \mathbb{R}^{n'}\big) \longmapsto X^N(\,\cdot\,, x, p') \in D\big([0, \infty); \mathbb{R}^n\big)$$

is defined inductively so that $X^N(0, x, p') = x$ and (cf. the notation in Theorem 2.4.11)

$$X^N(t, x, p') - X^N\big([t]_N, x, p'\big)$$
$$= \sigma\big(X^N([t]_N, x, p')\big)\big(w(t, p') - w([t]_N, p')\big) + b\big(X^N([t]_N, x, p')\big)(t - [t]_N)$$
$$+ \int_{|y'| \geq r_N} F\big(X^N([t]_N, x, p'), y'\big)\big(\bar{\eta}(t, dy', p') - \bar{\eta}([t]_N, dy', p')\big),$$

where $[t]_N \equiv 2^{-N}[2^N t]$ ([t] being the integer part of t) and (cf. (2.4.9))
$$\bar{\eta}(t, dy, p') \equiv \eta(t, dy, p') - t\mathbf{1}_{[0,1)}(|y'|)\, M(dy').$$

---

[1] We use $\|\cdot\|_{\mathrm{H.S.}}$ to denote the Hilbert–Schmidt norm.

Our main goals in the rest of this chapter will be to prove the following:

(G1) There exists a measurable map $p' \in D\big([0,\infty); \mathbb{R}^{n'}\big) \longmapsto X(\,\cdot\,, x, p') \in D\big([0,\infty); \mathbb{R}^n\big)$ to which $\{X^N(\,\cdot\,, x, p') : N \geq 0\}$ converges uniformly on finite intervals for $\mathbb{P}^{M'}$-almost every p'. In particular, for each $s \geq 0$, $p' \rightsquigarrow X(s, x, p')$ is measurable with respect to the $\mathbb{P}^{M'}$-completion $\overline{\mathcal{B}'_s}$ (cf. the last part of Remark 3.1.23 below) of $\mathcal{B}'_s \equiv \sigma\big(\{p'(\sigma) : \sigma \in [0, s]\}\big)$.

(G2) For every $\varphi \in C^2_{\mathrm{b}}(\mathbb{R}^n; \mathbb{R})$ and $x \in \mathbb{R}^n$,

$$\int_{\mathbb{R}^{n'}} \bigg| \varphi\big(x + F(x, y')\big) - \varphi(x)$$

$$- \mathbf{1}_{[0,1)}(|y'|)\big(F(x, y'), \mathrm{grad}_x \varphi\big)_{\mathbb{R}^{n'}} \bigg| \, M'(dy') < \infty,$$

and

(3.1.5) $$L\varphi \equiv \frac{1}{2} \sum_{i,j=1}^n a_{ij} \partial_i \partial_j \varphi + \sum_{i=1}^n b_i \partial_i \varphi + K\varphi$$

is a continuous where $K\varphi(x)$ is given by

$$\int_{\mathbb{R}^{n'}} \Big(\varphi\big(x + F(\,\cdot\,, y')\big) - \varphi(x) - \mathbf{1}_{[0,1)}(|y'|)\big(F(x, y'), \mathrm{grad}_x \varphi\big)_{\mathbb{R}^n} \Big) \, M'(dy').$$

Moreover, if $\varphi \in C^2_{\mathrm{c}}(\mathbb{R}^n; \mathbb{R})$, then $L\varphi$ is bounded and

$$\bigg(\varphi\big(X(t, x, p')\big) - \int_0^t L\varphi\big(X(\tau, x, p')\big) \, d\tau, \overline{\mathcal{B}'_t}, \mathbb{P}^{M'} \bigg)$$

is a martingale. That is, the $\mathbb{P}^{M'}$-distribution of $p' \rightsquigarrow X(\,\cdot\,, x, p')$ solves the martingale problem starting at x for the operator L on $C^2_{\mathrm{c}}(\mathbb{R}^n; \mathbb{R})$.

REMARK 3.1.6. The reader may well be asking why the L in (3.1.5) does not look more like the L in (3.1.1). The answer is that it has the form in (3.1.1), but in disguise. Namely, define $\hat{b} : \mathbb{R}^n \longrightarrow \mathbb{R}^n$ so that

(3.1.7)
$$\hat{b}(x) = b(x) - \int_{\mathbb{R}^{n'}} \bigg(\frac{\mathbf{1}_{[0,1)}(|y'|)|F(x, y')|^2}{1 + |F(x, y')|^2}$$

$$- \mathbf{1}_{[1,\infty)}(|y'|) \bigg) F(x, y') \, M'(dy')$$

and $x \rightsquigarrow M(x, \cdot)$ is a Lévy measure valued map so that

(3.1.8) $\quad M(x, \Gamma) = F(x, \cdot)_* M'(\Gamma) \equiv M'(\{y' \in \mathbb{R}^{n'} : F(x, y') \in \Gamma \setminus \{0\}\}).$

It is then an easy matter to check that the action of the L in (3.1.5) is given by

(3.1.9)
$$L\varphi \equiv \frac{1}{2} \sum_{i,j=1}^{n} a_{ij} \partial_i \partial_j \varphi + \sum_{i=1}^{n} \hat{b}_i \partial_i \varphi$$
$$+ \int_{\mathbb{R}^n} \left(\varphi(\cdot + y) - \varphi - \frac{(y, \operatorname{grad}\varphi)_{\mathbb{R}^n}}{1 + |y|^2} \right) M(x, dy).$$

REMARK 3.1.10. The hypotheses **(H1)–(H3)** do not guarantee that $L\varphi$ is bounded for every $\varphi \in C_b^2(\mathbb{R}^n; \mathbb{R})$. Thus, if we insist on a vector field possessing the boundedness properties which we have imposed heretofore, we will not in general be able to determine a vector field Λ on $\mathbf{M}_1(\mathbb{R}^n)$ via the formula $\Lambda_\mu \varphi = \langle L\varphi, \mu \rangle$.

Be that as it may, suppose for the moment that we add to **(H1)–(H3)** the assumption that σ, b, and $x \rightsquigarrow \int_{\mathbb{R}^{n'}} \frac{|F(x,y')|^2}{1+|F(x,y')|^2} M'(dy')$ are all uniformly bounded in $x \in \mathbb{R}^n$. Then it is easy to check that $\|L\varphi\|_u \leq C\|\varphi\|_{C_b^2(\mathbb{R}^n;\mathbb{R})}$ for some $C < \infty$, and so we can determine the vector affine field Λ on $\mathbf{M}_1(\mathbb{R}^n)$ by $\langle \varphi, \Lambda\mu \rangle = \langle L\varphi, \mu \rangle$. In addition, by an easy cutoff argument, one sees that the martingale property in **(G2)** extends to all $\varphi \in C_b^2(\mathbb{R}^n; \mathbb{R})$. Thus, the $\mathbb{P}^{M'}$-distribution of $p' \rightsquigarrow X(\cdot, x, p')$ solves the martingale problem for L starting from x. In particular, if we assume that there is only one integral curve of Λ starting from each $\mu \in \mathbf{M}_1(\mathbb{R}^n)$, then the last part of Theorem 2.3.3 tells us that the $\mathbb{P}^{M'}$-distribution of $p' \rightsquigarrow X(\cdot, x, p')$ is the measure $\mathbb{P}$ described in that theorem. Hence, if $P_t(x, \cdot)$ denotes the $\mathbb{P}^{M'}$-distribution of $p' \rightsquigarrow X(t, x, p')$, then

$$\mathbb{P}^{M'}\big(X(t_0, x) \in \Gamma_0, \ldots, X(t_m, x) \in \Gamma_m\big)$$

is given by the right hand side of (1.4.1) with $\mu_0 = \delta_x$.

§**3.1.3. Important Preliminary Observations:** Perhaps the most crucial observation, the one on which everything else turns, is that we already understand $p' \rightsquigarrow X^N(\cdot, x, p')$ locally. To wit, define $\hat{b}^N : \mathbb{R}^n \longrightarrow \mathbb{R}^n$ by

$$\hat{b}^N(x) = b(x) - \int_{\mathbb{R}^{n'}} \left(\frac{\mathbf{1}_{[r_N, 1)}(|y'|)|F(x, y')|^2}{1 + |F(x, y')|^2} - \mathbf{1}_{[1,\infty)}(|y'|) \right) F(x, y') \, M'(dy')$$

and a Lévy measure valued map $x \rightsquigarrow M^N(x, \cdot)$ so that

$$M^N(x, \Gamma) = M'(\{|y'| \geq r_N : F(x, y') \in \Gamma \setminus \{0\}\}).$$

Then, by the considerations in §3.1.1, for each $m \in \mathbb{N}$, the conditional $\mathbb{P}^{M'}$-distribution of

$$p' \rightsquigarrow \left(X^N(\,\cdot\, + m2^{-N}, x, p') - X^N(m2^{-N}, x, p')\right) \restriction [0, 2^{-N}] \text{ given } \mathcal{B}'_{m2^{-N}}$$

is the same as that of the Lévy process corresponding to the Lévy system

$$\left(a\big(X^N(m2^{-N}, x, p')\big), \hat{b}^N\big(X^N(m2^{-N}, x, p')\big), M^N\big(X^N(m2^{-N}, x, p'), \,\cdot\,\big)\right)$$

restricted to the same time interval.

LEMMA 3.1.11. *Assume that*

$$\int_{\mathbb{R}^{n'}} |F(x, y')|^2\, M'(dy') + \|\sigma(x)\|^2_{\text{H.S.}} + |b(x)|^2 \le C\big(1 + |x|^2\big)$$

for some $C < \infty$ *and all* $x \in \mathbb{R}^n$. *Then there is a* $K < \infty$, *which depends only on* C, $M'\big(\mathbb{R}^{n'} \setminus B_{\mathbb{R}^{n'}}(0, 1)\big)$, *and* n, *such that*

$$\mathbb{E}^{\mathbb{P}^{M'}}\big[\|X^N(\,\cdot\,, x)\|^2_{[0,T]}\big] \le \big(1 + |x|^2\big)e^{K(T+1)} \text{ for } (T, x) \in [0, \infty) \times \mathbb{R}^n,$$

(3.1.12) $\mathbb{E}^{\mathbb{P}^{M'}}\big[|X^N(m2^{-N} + t, x) - X^N(m2^{-N}, x)|^2\,\big|\,\mathcal{B}'_{m2^{-N}}\big]$

$$\le K(1 + |x|^2)te^{Km2^{-N}} \text{ for } m \in \mathbb{N} \ \& \ (t, x) \in [0, 2^{-N}] \times \mathbb{R}^n.$$

Moreover, this K *can be chosen so that for any* $\varphi \in C^2(\mathbb{R}^n; \mathbb{R})$ *with*

$$\|\varphi\|^{(2)}_u \equiv \sup\{\|\partial^2_\xi \varphi\|_u : \xi \in \mathbb{S}^{n-1}\} < \infty,$$

(3.1.13)
$$\int_{\mathbb{R}^n} \Big|\varphi\big(x_1 + F(x_2, y')\big) - \varphi(x_1)$$

$$- \mathbf{1}_{[0,1)}(|y'|)\big(F(x_2, y'), \text{grad}_{x_1}\varphi\big)_{\mathbb{R}^n}\Big|\, M'(dy')$$

$$\le K\big(|\text{grad}_0\varphi| + \|\varphi\|^{(2)}_u\big)\big(1 + |x_1|^2 + |x_2|^2\big)$$

for all $(x_1, x_2) \in (\mathbb{R}^n)^2$. *Thus, if* $L^N\varphi(t, x, p')$ *is given by*

$$\frac{1}{2}\sum_{i,j=1}^n a_{ij}\big(X^N([t]_N, x, p')\big)\partial_i\partial_j\varphi\big(X^N(t, x, p')\big)$$

$$+ \sum_{i=1}^n b_i\big(X^N([t]_N, x, p')\big)\partial_i\varphi\big(X^N(t, x, p')\big)$$

$$+ \int_{|y'| \ge r_N} \Big(\varphi\big(X^N(t, x, p') + F(X^N([t]_N, x, p'), y')\big)\Big) - \varphi\big(X^N(t, x, p')\big)$$

$$- \mathbf{1}_{[0,1)}(|y'|)\big(F(X^N([t]_N, x, p')), \text{grad}_{X^N(t,x,p')}\varphi\big)_{\mathbb{R}^n}\Big)\, M'(dy'),$$

then, for a $K < \infty$ depending only on C and n,

$$(3.1.14) \quad \begin{aligned} |L^N \varphi(t, x, p')| &\leq K\big(|\text{grad}_0 \varphi| + \|\varphi\|_u^{(2)}\big) \\ &\quad \times \big(1 + |X^N([t]_N, x, p')|^2 + |X^N(t, x, p')|^2\big). \end{aligned}$$

Finally, for every $\varphi \in C^2(\mathbb{R}^n; \mathbb{R})$ with $\|\varphi\|_u^{(2)} < \infty$,

$$\left(\varphi(X^N(t, x, p')) - \int_0^t L^N \varphi(\tau, x, p') \, d\tau, \mathcal{B}_t', \mathbb{P}^{M'} \right)$$

is a martingale. In particular, if

$$\bar{B}^N(t, x, p') \equiv \int_0^t \bar{b}(X^N([\tau]_N, x, p')) \, d\tau$$

$$\text{where } \bar{b}(x) \equiv b(x) - \int_{|y'| \geq 1} F(x, y') \, M'(dy')$$

and $\bar{X}^N(t, x, p') \equiv X^N(t, x, p') - \bar{B}^N(t, x, p')$, then $\big(\bar{X}^N(t, x, p'), \mathcal{B}_t', \mathbb{P}^{M'}\big)$ is a square integrable, $\mathbb{R}^n$-valued martingale.

PROOF: First observe that we already know $t \rightsquigarrow \|X^N(t, x)\|_{L^2(\mathbb{P}^{M'}; \mathbb{R}^n)}$ is locally bounded. Indeed, from, the preceding description of $p' \rightsquigarrow X^N(\cdot, x, p')$ locally in terms of Lévy processes, (3.1.4), and the boundedness hypotheses under which we are presently working, we know that

$$(*) \quad \begin{aligned} \mathbb{E}^{\mathbb{P}^{M'}} &\left[|X^N(m2^{-N} + t, x) - X^N(m2^{-N}, x)|^2 \,\middle|\, \mathcal{B}_{m2^{-N}}' \right](p') \\ &\leq At\big(1 + |X^N(m2^{-N}, x, p')|^2\big), \quad t \in [0, 2^{-N}], \end{aligned}$$

for some $A < \infty$ depending only on C and n. Hence, boundedness of $t \rightsquigarrow \|X^N(t, x)\|_{L^2(\mathbb{P}^{M'}; \mathbb{R}^n)}$ on $[0, m2^{-N}]$ follows by induction on $m \in \mathbb{N}$.

The next step is to check the estimate in (3.1.13). But, by Taylor's theorem,

$$\int_{\mathbb{R}^n} \left| \varphi(x_1 + F(x_2, y')) - \varphi(x_1) - \mathbf{1}_{[0,1)}(|y'|)(F(x_2, y'), \text{grad}_{x_1} \varphi)_{\mathbb{R}^n} \right| M'(dy')$$

$$\leq \int_{\mathbb{R}^n} \left| \varphi(x_1 + F(x_2, y')) - \varphi(x_1) - (F(x_2, y'), \text{grad}_{x_1} \varphi)_{\mathbb{R}^n} \right| M'(dy')$$

$$+ \int_{|y'| \geq 1} \left| (F(x_2, y'), \text{grad}_{x_1} \varphi)_{\mathbb{R}^n} \right| M'(dy')$$

$$\leq \frac{\|\varphi\|_u^{(2)}}{2} \int_{\mathbb{R}^n} |F(x_2, y')|^2 \, M'(dy')$$

$$+ |\text{grad}_{x_1} \varphi| M'(\mathbb{R}^{n'} \setminus B_{\mathbb{R}^{n'}}(0, 1))^{\frac{1}{2}} \left(\int_{\mathbb{R}^n} |F(x_2, y')|^2 \, M'(dy') \right)^{\frac{1}{2}}.$$

Hence, since $|\text{grad}_{x_1}\varphi| \leq |\text{grad}_0\varphi| + \|\varphi\|_u^{(2)}|x_1|$, (3.1.13) is now proved.

Given (3.1.13), (3.1.14) is easy. Moreover, knowing (3.1.14), one can prove the asserted martingale property as follows. First suppose that $\varphi \in C_b^2(\mathbb{R}^n; \mathbb{R})$. By the preceding description of the $\mathbb{P}^{M'}$-distribution of $p' \rightsquigarrow X^N(\cdot, x, p')$ in terms of Lévy processes combined with the characterization, in Corollary 2.3.4, of a Lévy process in terms of a martingale problem, we know that

$$\mathbb{E}^{\mathbb{P}^{M'}}\left[\varphi\big(X^N(t_2, x)\big) - \varphi\big(X^N(t_1, x)\big) \,\Big|\, \mathcal{B}'_{t_1}\right] = \mathbb{E}^{\mathbb{P}^{M'}}\left[\int_{t_1}^{t_2} L^N\varphi(\tau, x, p')\, d\tau \,\Big|\, \mathcal{B}'_{t_1}\right]$$

for $m2^{-N} \leq t_1 < t_2 \leq (m+1)2^{-N}$ and each $m \in \mathbb{N}$. Thus, because of (3.1.14) and our opening comment about the local boundedness of $t \rightsquigarrow \|X^N(t, x)\|_{L^2(\mathbb{P}^{M'};\mathbb{R}^n)}$, it is easy to justify

$$\mathbb{E}^{\mathbb{P}^{M'}}\left[\varphi\big(X^N(t_2, x)\big) - \varphi\big(X^N(t_1, x)\big) \,\Big|\, \mathcal{B}'_{t_1}\right]$$

$$= \sum_{m=0}^{\infty} \mathbb{E}^{\mathbb{P}^{M'}}\left[\varphi\big(X^N(t_2 \wedge (m+1)2^{-N}, x)\big) - \varphi\big(X^N(t_1 \wedge m2^{-N}, x)\big) \,\Big|\, \mathcal{B}'_{t_1}\right]$$

$$= \sum_{m=0}^{\infty} \mathbb{E}^{\mathbb{P}^{M'}}\left[\int_{t_1 \wedge m2^{-N}}^{t_2 \wedge (m+1)2^{-N}} L^N\varphi(\tau, x, p')\, d\tau \,\Big|\, \mathcal{B}'_{t_1}\right]$$

$$= \mathbb{E}^{\mathbb{P}^{M'}}\left[\int_{t_1}^{t_2} L^N\varphi(\tau, x, p')\, d\tau \,\Big|\, \mathcal{B}'_{t_1}\right].$$

To handle general $\varphi \in C^2(\mathbb{R}^n; \mathbb{R})$ with bounded second order derivatives, choose a $\psi \in C^\infty(\mathbb{R}; [0, 1])$ which is identically 1 on $(-\infty, 1]$ and vanishes off $[2, \infty)$, and set $\varphi_R(y) = \psi(R^{-2}|y|^2)\varphi(y)$ for $R \geq 1$. Then, because the second derivatives of φ_R can be uniformly bounded with a bound that does not depend on $R \geq 1$, it is an easy matter to apply the preceding to φ_R and then use (3.1.14) and Lebesgue's dominated convergence Theorem to show that the equality survives the passage to limit as $R \to \infty$.

Turning to (3.1.12), apply the preceding to $\varphi(y) = |y|^2$ and, using (3.1.14), conclude that, for a $K < \infty$ with the required dependence,

$$u(t) \leq 1 + |x|^2 + K \int_0^t u(\tau)\, d\tau \quad \text{where } u(t) \equiv 1 + \sup_{\tau \in [0,t]} \mathbb{E}^{\mathbb{P}^{M'}}\left[|X^N(\tau, x)|^2\right].$$

Now apply Gronwall's inequality to arrive at

$$(**)\qquad\qquad \mathbb{E}^{\mathbb{P}^{M'}}\left[|X^N(t, x)|^2\right] \leq (1 + |x|^2)e^{Kt},$$

and then use (*) and (**) to get the second line of (3.1.12). Finally, to pass from (**) to the first line of (3.1.12), we apply Doob's inequality (cf. Theorem 7.1.8 in [36]) to get

$$\mathbb{E}^{\mathbb{P}^{M'}}\left[\|\bar{X}^N(\,\cdot\,,x)\|_{[0,T]}^2\right] \le 4\mathbb{E}^{\mathbb{P}^{M'}}\left[|\bar{X}^N(T,x)|^2\right]$$

$$\le 8\mathbb{E}^{\mathbb{P}^{M'}}\left[|X^N(T,x)|^2\right] + 8\mathbb{E}^{\mathbb{P}^{M'}}\left[\|\bar{B}^N(\,\cdot\,,x)\|_{[0,T]}^2\right].$$

Thus,

$$\mathbb{E}^{\mathbb{P}^{M'}}\left[\|X^N(\,\cdot\,,x)\|_{[0,T]}^2\right] \le 2\mathbb{E}^{\mathbb{P}^{M'}}\left[\|\bar{X}^N(\,\cdot\,,x)\|_{[0,T]}^2\right] + 2\mathbb{E}^{\mathbb{P}^{M'}}\left[\|\bar{B}^N(\,\cdot\,,x)\|_{[0,T]}^2\right]$$

$$\le 16\mathbb{E}^{\mathbb{P}^{M'}}\left[|X^N(T,x)|^2\right] + 18\mathbb{E}^{\mathbb{P}^{M'}}\left[\|\bar{B}^N(\,\cdot\,,x)\|_{[0,T]}^2\right].$$

But

$$\|\bar{B}^N(t,x,p')\|_{[0,T]}^2 \le \left(\int_0^T \left|\bar{b}(X^N([t]_N,x,p'))\right|dt\right)^2$$

$$\le T\int_0^T \left|\bar{b}(X^N([t]_N,x,p'))\right|^2 dt,$$

and

$$|\bar{b}(x)|^2 \le 2|b(x)|^2 + 2M'\big(B_{\mathbb{R}^{n'}}(0,1)\complement\big)\int_{\mathbb{R}^{n'}} |F(x,y')|^2 M'(dy') \le K(1+|x|^2)$$

for a $K < \infty$ with the required dependence. Hence, after combining these together and applying (**), the first line of (3.1.12) follows. $\square$

A second observation is one which will allow us to do most of our calculations in the case dealt with in Lemma 3.1.11.

LEMMA 3.1.15. *Given $R \ge 1$, set*

$$p'_R(t) = p(t) - \int_{|y'|\ge R} y'\eta(t,dy') \quad \text{and} \quad X^N_R(\,\cdot\,,x,p') = X^N(\,\cdot\,,x,p'_R).$$

Then $p' \rightsquigarrow X^N_R(\,\cdot\,,x,p')$ is the map described in (H4) when $M'(dy')$ in (H1) is replaced by $\mathbf{1}_{[0,R)}(|y'|)M'(dy')$. Furthermore, if

$$A(T,R) = \{p' : \eta(T, B_{\mathbb{R}^{n'}}(0,R)\complement, p') = 0\},$$

then

$$\mathbb{P}^{M'}\big(A(T,R)\big) = e^{-TM'(B_{\mathbb{R}^{n'}}(0,R)\complement)}$$

and the $\mathbb{P}^{M'}$-conditional distribution of $\{X^N(t,x) : (t,x) \in [0,T] \times \mathbb{R}^n\} \cup \{p'(t) : t \in [0,T]\}$ given $A(T,R)$ is the same as the (unconditional) $\mathbb{P}^{M'}$-distribution of $\{X^N_R(t,x) : (t,x) \in [0,T] \times \mathbb{R}^n\} \cup \{p'_R(t) : t \in [0,T]\}$.

PROOF: The first assertion and the computation of $\mathbb{P}^{M'}(A(T,R))$ are covered by Lemma 2.4.10. As for the second part, first observe that $X^N(t,x,p')$ $= X^N(t,x,p'_R)$ if $(t,p') \in [0,T] \times A(T,R)$. Second, apply Lemma 2.4.10 to see that $\sigma(\{p'_R(t) : t \geq 0\})$ is $\mathbb{P}^{M'}$-independent of $A(T,R)$, and notice that $p' \rightsquigarrow X_R^N(\,\cdot\,,x,p') \upharpoonright [0,T]$ is $\sigma(\{p'_R(t) : t \in [0,T]\})$-measurable for all $x \in \mathbb{R}^n$. Hence, if $0 \leq t_1 < \cdots < t_\ell$, $\{x_k\}_1^\ell \subseteq \mathbb{R}^n$, and $\Phi : (\mathbb{R}^n \times \mathbb{R}^{n'})^\ell \longrightarrow \mathbb{R}$ is bounded and measurable, then

$$\mathbb{E}^{\mathbb{P}^{M'}}\left[\Phi\Big(\big(X^N(t_1,x_1,p'),p'(t_1)\big),\dots\big(X^N(t_\ell,x_\ell,p'),p'(t_\ell)\big)\Big)\,\Big|\,A(T,R)\right]$$

$$= \mathbb{E}^{\mathbb{P}^{M'}}\left[\Phi\Big(\big(X_R^N(t_1,x_1,p'),p'_R(t_1)\big),\dots\big(X_R^N(t_\ell,x_\ell,p'),p'_R(t_\ell)\big)\Big)\,\Big|\,A(T,R)\right]$$

$$= \mathbb{E}^{\mathbb{P}^{M'}}\left[\Phi\Big(\big(X_R^N(t_1,x_1,p'),p'_R(t_1)\big),\dots\big(X_R^N(t_\ell,x_\ell,p'),p'_R(t_\ell)\big)\Big)\right]. \quad \square$$

§3.1.4. The Proof of Convergence: We prove here the convergence result asserted in **(G1)**. Although our proof is a little tedious, it is much less so than Itô's original proof, which was carried out without the benefit of martingale technology. Specifically, for Itô the difficult part of the proof came when he wanted to show that the convergence is uniform on finite time intervals. Itô handled this by making repeated applications of Kolmogorov's continuity criterion (cf. Exercise 2.4.17). Our lives are made easier because we have (cf. the last part of the proof of Lemma 3.1.11) Doob's inequality available to us.

LEMMA 3.1.16. *In addition to the assumptions made in Lemma 3.1.11, replace* (c) *in* **(H2)** *by the assumption that*

$$\int_{\mathbb{R}^{n'}} \big|F(x_2,y') - F(x_1,y')\big|^2\, M'(dy') \leq C^2 |x_2 - x_1|^2.$$

Then there exists a nondecreasing map $T \in [0,\infty) \longmapsto K(T) \in [0,\infty)$, depending only on the C and n, such that

$$(3.1.17)\quad \mathbb{E}^{\mathbb{P}^{M'}}\left[\big\|X^{N+1}(\,\cdot\,,x) - X^N(\,\cdot\,,x)\big\|_{[0,T]}^2\right] \leq \frac{K(T)^2}{2^N}\big(1 + |x|^2\big), \quad N \in \mathbb{N}.$$

PROOF: Using the notation introduced in the last part of the statement of Lemma 3.1.11, note that $\big(\bar{X}^{N+1}(t,x,p') - \bar{X}^N(t,x,p'),\mathcal{B}'_t,\mathbb{P}^{M'}\big)$ is a square-integrable martingale and that

$$\mathbb{E}^{\mathbb{P}^{M'}}\left[\big\|X^{N+1}(\,\cdot\,,x) - X^N(\,\cdot\,,x)\big\|_{[0,T]}^2\right]$$

$$\leq 2\mathbb{E}^{\mathbb{P}^{M'}}\left[\big\|\bar{X}^{N+1}(\,\cdot\,,x) - \bar{X}^N(\,\cdot\,,x)\big\|_{[0,T]}^2\right]$$

$$\quad + 2\mathbb{E}^{\mathbb{P}^{M'}}\left[\big\|\bar{B}^{N+1}(\,\cdot\,,x) - \bar{B}^N(\,\cdot\,,x)\big\|_{[0,T]}^2\right].$$

Hence, by Doob's inequality,

(*)
$$\begin{aligned}
&\mathbb{E}^{\mathbb{P}^{M'}}\left[\|X^{N+1}(\,\cdot\,,x) - X^N(\,\cdot\,,x)\|_{[0,T]}^2\right]\\
&\leq 8\mathbb{E}^{\mathbb{P}^{M'}}\left[|\bar{X}^{N+1}(T,x) - \bar{X}^N(T,x)|^2\right]\\
&\quad + 2\mathbb{E}^{\mathbb{P}^{M'}}\left[\|\bar{B}^{N+1}(\,\cdot\,,x) - \bar{B}^N(\,\cdot\,,x)\|_{[0,T]}^2\right].
\end{aligned}$$

Next, another application of the martingale property yields

$$\mathbb{E}^{\mathbb{P}^{M'}}\left[|\bar{X}^{N+1}(2^k,x) - \bar{X}^N(2^k,x)|^2\right]$$

$$= \sum_{m=0}^{2^{N+1+k}-1} \mathbb{E}^{\mathbb{P}^{M'}}\left[|(\bar{X}^{N+1}((m+1)2^{-N-1},x) - \bar{X}^{N+1}(m2^{-N-1},x))\right.$$

$$\left. - (\bar{X}^N((m+1)2^{-N-1},x) - \bar{X}^N(m2^{-N-1},x))|^2\right]$$

$$\leq \frac{3}{2^{N+1}} \sum_{m=0}^{2^{N+1+k}-1} \mathbb{E}^{\mathbb{P}^{M'}}\left[\|\sigma(X^{N+1}(m2^{-N-1},x))\right.$$

$$\left. - \sigma(X^N([m2^{-N-1}]_N,x))\|_{\text{H.S.}}^2\right]$$

$$+ \frac{3}{2^{N+1}} \sum_{m=0}^{2^{N+1+k}-1} \mathbb{E}^{\mathbb{P}^{M'}}\left[\iint_{|y'|\geq r_N} |F(X^{N+1}(m2^{-N-1},x),y')\right.$$

$$\left. - F(X^N([m2^{-N-1}]_N,x),y')|^2\, M'(dy')\right]$$

$$+ \frac{3}{2^{N+1}} \sum_{m=0}^{2^{N+1+k}-1} \mathbb{E}^{\mathbb{P}^{M'}}\left[\int_{r_{N+1}\leq|y'|<r_N} |F(X^{N+1}(m2^{-N-1},x),y')|^2\, M'(dy')\right],$$

where we have used the equality part of (3.1.4) as well as the fact that, because $p' \rightsquigarrow w(s+t,p') - w(s,p')$ under $\mathbb{P}^{M'}$ is a centered Gaussian which has covariance $tI_{\mathbb{R}^n}$ and is independent of $\mathcal{B}'_s$,

$$\mathbb{E}^{\mathbb{P}^{M'}}\left[|(\sigma(X^{N+1}([s]_{N+1},x,p')) - \sigma(X^N([s]_N,x,p')))(w(t,p') - w(s,p'))|^2\right]$$

$$= (t-s)\mathbb{E}^{\mathbb{P}^{M'}}\left[\|\sigma(X^{N+1}([s]_{N+1},x)) - \sigma(X^N([s]_N,x))\|_{\text{H.S.}}^2\right].$$

Now set

$$u^N(j) \equiv \mathbb{E}^{\mathbb{P}^{M'}}\left[\|X^{N+1}(t,x) - X^N(t,x)\|_{[0,j2^{-N}]}^2\right].$$

By plugging the preceding into (*), taking into account our Lipschitz continuity assumptions, and applying the second part of (3.1.12), we see that, for any $k \in \mathbb{Z}$ there is a $C(k) < \infty$ such that

$$u^N(j) \leq C(k)(1+|x|^2)2^{-N} \sum_{i=0}^{j-1} (u^N(i) + 2^{-N}), \quad 0 \leq j \leq 2^{N+1+k}.$$

Finally, this is just the sort of relation to which the discrete version of Gronwall's inequality (cf. Exercise 3.1.30 below) applies and gives (3.1.17). $\square$

LEMMA 3.1.18. *Under the conditions in Lemma 3.1.16, there exists for each $x \in \mathbb{R}^n$ a map $p' \rightsquigarrow X(\,\cdot\,, x, p')$ of the sort described in* (G1). *In fact, there exists an nondecreasing map $T \in [0, \infty) \longmapsto K(T) \in [0, \infty)$ such that*

$$(3.1.19) \qquad \mathbb{E}^{\mathbb{P}^{M'}} \left[\|X(\,\cdot\,, x) - X^N(\,\cdot\,, x)\|_{[0,T]}^2 \right] \leq K(T)2^{-N}(1+|x|^2).$$

PROOF: By (3.1.17) and Minkowski's inequality,

$$\mathbb{E}^{\mathbb{P}^{M'}} \left[\sup_{N' > N} \|X^{N'}(\,\cdot\,, x) - X^N(\,\cdot\,, x)\|_{[0,T]}^2 \right]$$

$$\leq K(T)^2 (1+|x|^2) \left(\sum_{N' \geq N} 2^{-\frac{N'}{2}} \right)^2 \leq \left(\frac{2^{\frac{1}{2}} K(T)}{2^{\frac{1}{2}} - 1} \right)^2 (1+|x|^2)2^{-N},$$

where $K(T)$ is the constant in (3.1.17). Hence (cf. Exercise 2.3.5) $p' \rightsquigarrow X(\,\cdot\,, x, p')$ exists and (3.1.19) follows by a simple application of Fatou's lemma plus the redefinition of $K(T)$ as $\left(\frac{2^{\frac{1}{2}} K(T)}{2^{\frac{1}{2}} - 1} \right)^2$. $\square$

THEOREM 3.1.20. *Under the conditions in* (H1)–(H4), *there exists, for each $x \in \mathbb{R}^n$, a map $p' \rightsquigarrow X(\,\cdot\,, x, p')$ of the sort described in* (G1). *In fact,*

$$\lim_{N \to \infty} \sup_{x \in \mathbb{R}^n} \mathbb{P}^{M'} \left(\sup_{N' \geq N} \|X(\,\cdot\,, x) - X^N(\,\cdot\,, x)\|_{[0,T]} \geq \sqrt{1+|x|^2}\,\epsilon \right) = 0$$

for all $\epsilon > 0$ and $T \in [0, \infty)$.

PROOF: Refer to the notation in Lemma 3.1.15, and use Lemma 3.1.18 to produce, for each $R \in [1, \infty)$, a $p' \rightsquigarrow X_R(\,\cdot\,, x, p')$ to which $\{X_R^N(\,\cdot\,, x, p')\}_{N=0}^{\infty}$ $\mathbb{P}^{M'}$-almost surely converges uniformly on compacts. Because $X^N(\,\cdot\,, x, p') \restriction [0, T] = X_R^N(\,\cdot\,, x, p') \restriction [0, T]$ for $p' \in A(T, R)$ and

$$\mathbb{P}^{M'}(A(T, R)) = e^{-TM'(B_{\mathbb{R}^{n'}}(0,R)^{\complement})} \longrightarrow 1 \quad \text{as } R \to \infty,$$

we can construct a measurable $p' \rightsquigarrow X(\,\cdot\,, x, p')$ so that, for each $T \in [0, \infty)$, $X(\,\cdot\,, x, p') \upharpoonright [0, T] = X_\ell(\,\cdot\,, x, p') \upharpoonright [0, T]$ for all $\ell \in \mathbb{Z}^+$ and $p' \in A(T, \ell)$. Furthermore, it is clear that, for, $\mathbb{P}^{M'}$-almost every p', $X^N(\,\cdot\,, x, p') \longrightarrow X(\,\cdot\,, x, p')$ uniformly on compacts. Finally, because

$$\mathbb{P}^{M'}\left(\sup_{N' > N} \|X(\,\cdot\,, x) - X^N(\,\cdot\,, x)\|_{[0,T]} \geq \epsilon\right)$$

$$\leq \mathbb{P}^{M'}\left(\sup_{N' > N} \|X_\ell(\,\cdot\,, x) - X_\ell^N(\,\cdot\,, x)\|_{[0,T]} \geq \epsilon\right) + \mathbb{P}^{M'}(A(T, \ell)),$$

the final assertion follows from (3.1.19) applied to $\{X_\ell^N - X_\ell\}_{N+0}^\infty$. □

REMARK 3.1.21. Working by analogy with (1.6.1), one is tempted to find an "integral equation" which describes the transformation $p' \rightsquigarrow X(\,\cdot\,, x, p')$ under consideration. More precisely, one might hope that the path $X(\,\cdot\,, x, p')$ can be thought of as the solution to the equation

(3.1.22)
$$X(t, x, p') = x + \int_0^t \tilde{F}(X(\tau-, x, p'), p'(\tau) - p'(\tau-)) \, dp'(\tau)$$
$$+ \int_0^t \tilde{b}(X(\tau, s, p')) \, d\tau,$$

where $\tilde{F} : \mathbb{R}^n \times \mathbb{R}^{n'} \longrightarrow \mathrm{Hom}(\mathbb{R}^{n'}; \mathbb{R}^n)$ is given by

$$\tilde{F}(X, y') \equiv \begin{cases} \sigma(X) & \text{if } y' = 0 \\ F(X, y') \otimes \frac{y'}{|y'|^2} & \text{if } y' \neq 0, \end{cases}$$

and

$$\tilde{b}(X) = b(X) - \int_{|y'| < 1} F(X, y') \, M'(dy').$$

Indeed, if one passes to the limit as $N \to \infty$ and is sufficiently credulous about what is happening, this is precisely the equation to which the prescription in **(H4)** appears to be leading. Of course, after a little reflection, it becomes apparent that something is quite wrong. For one thing, the integral in the second term of the expression for $\tilde{b}$ will not be finite in general. This divergence is, of course, inextricably connected with the fact (cf. Remark 2.4.13) that the integral

$$\int_0^t \mathbf{1}_{\mathbb{R}^{n'} \setminus \{0\}}(p(\tau) - p(\tau-)) \tilde{F}(X(\tau-, x, p'), p'(\tau) - p'(\tau-)) \, dp'(\tau)$$

$$= \iint_{[0,t] \times \mathbb{R}^{n'}} F(X(\tau-, x, p'), y') \, \eta(dt \times dy', p')$$

will usually not exist in any classical sense. Further complicating the interpretation of (3.1.22) is the fact that the meaning of

$$\int_0^t \mathbf{1}_{\{0\}}\big(p(\tau) - p(\tau-)\big)\tilde{F}\big(X(\tau-,x,p'),p'(\tau) - p'(\tau-)\big)\,dp'(\tau)$$

$$= \int_0^t \sigma\big(X(\tau-,x,p')\big)\,dw(\tau,p')$$

is up for grabs. Namely, it is well known (cf. Theorem 4.1.10 and Exercise 4.1.11 in [36]) that, for $\mathbb{P}^{M'}$-almost every p', $w(\,\cdot\,,p')$ has unbounded variation on every non-trivial interval. Hence, it is highly unlikely that the meaning of such integrals will be readily supplied by either of Messieurs Riemann or Lebesgue. Instead, to the extent to which they can be rationalized as integrals, it was Itô who provided the rationalization in what is now called Itô's theory of stochastic integration. See §3.3 below to find out why one might care and Chapters 5–8 for a partial explanation of what he did.

REMARK 3.1.23. A second, more technical and less interesting, remark about the preceding is the role of uniform convergence on compacts to produce a topology on $D\big([0,\infty);\mathbb{R}^n\big)$. As we saw in part (i) of Exercise 2.3.5, this topology admits a complete metric ρ. However, there is a very good reason why this notion of convergence is not the one probabilists usually use for $D\big([0,\infty);\mathbb{R}^n\big)$. For one thing, in many applications, uniform convergence is just too demanding. For example, there are many circumstances in which one would like to say the paths $p_k = \mathbf{1}_{[k^{-1},\infty)}$ are converging in $D\big([0,\infty);\mathbb{R}\big)$ to the path $\mathbf{1}_{[0,\infty)}$, but clearly $\rho(p_k,p_\ell) = \frac{1}{2}$ for $k \neq \ell$. Another manifestation of the same problem is that $D\big([0,\infty);\mathbb{R}^n\big)$ is hideously nonseparable in the topology induced by ρ. (Indeed, $\rho\big(\mathbf{1}_{[\alpha,\infty)},\mathbf{1}_{[\beta,\infty)}\big) = \frac{1}{2}$ for all $\alpha,\beta \in [0,1)$ with $\alpha \neq \beta$.) Thus, if uniform convergence on compacts is used to induce its topology, there are continuous functions which fail to be measurable with respect to the σ-algebra $\mathcal{B}$ introduced in part (ii) of Exercise 2.3.5. Similarly, the standard theory of weak convergence (cf. §3.1 in [36]) of measures on $D\big([0,\infty);\mathbb{R}^n\big)$ falls apart. To overcome these objections, A.V. Skorohod (cf. [27]) introduced a clever, but somewhat less than pleasant, notion of convergence in $D\big([0,\infty);\mathbb{R}^{n'}\big)$, and probabilists have usually adopted one of several variations on Skorohod's topology when they study refined convergence results on this space. The reason why we are able to avoid Skorohod's topology is that the convergence results which we are proving here involve sequences of paths in which the jumps are changing in a tame manner. To wit, for any given p', the jumps of all the $X^N(\,\cdot\,,x,p')$'s are tightly controlled by those of p': as N increases, the only new jumps which appear are either very small, very unlikely, or very nearly equal to jumps which occur at an earlier stage.

In this connection, it may be helpful to make a comment about the proof of Lemma 3.1.18 and its implications for the statement in **(G1)**. In that proof we were dealing with the sequence $\{X^N(\,\cdot\,,x)\}_1^\infty$ of $D\big([0,\infty);\mathbb{R}^n\big)$-valued random variables, and the heart of our argument was our proof that, for p' outside of a $\mathbb{P}^{M'}$-null set $\mathcal{N}$, $\{X^N(\,\cdot\,,x,p')\}_1^\infty$ satisfies Cauchy's convergence criterion (relative to uniform convergence on compact intervals). We then applied Cauchy's criterion to assert the existence of a $D\big([0,\infty);\mathbb{R}^n\big)$-valued random variable $X(\,\cdot\,,x)$ to which $\{X^N(\,\cdot\,,x)\}_1^\infty$ converges (uniformly on compact intervals) $\mathbb{P}^{M'}$-almost surely. Of course, what was meant, but not said, was that $X(\,\cdot\,,x,p') = \lim_{N\to\infty} X^N(\,\cdot\,,x,p')$ for $p' \notin \mathcal{N}$ and $X(\,\cdot\,,x,p')$ is anything convenient (e.g., $X(\,\cdot\,,x,p') \equiv x$) for $p' \in \mathcal{N}$. Although this sort of procedure is completely standard in all measure theoretic reasoning, it leads to a technical problem here. Namely, even though $p' \rightsquigarrow X(\,\cdot\,,x,p')$ is $\mathcal{B}'$-measurable, it, as distinguished from the approximating $p' \rightsquigarrow X^N(\,\cdot\,,x,p')$, will not, in general, be *adapted* to the family $\{\mathcal{B}'_t : t \geq 0\}$. That is, just because, for each $N \geq 0$, $p' \rightsquigarrow X^N(t,x,p')$ is $\mathcal{B}'_t$-measurable for all $t \geq 0$, there is no way to guarantee that $p' \rightsquigarrow X(t,x,p')$ will be $\mathcal{B}_t$-measurable. The point is that the definition of $p' \rightsquigarrow X(t,x,p')$ depends on the set $\mathcal{N}$, and $\mathcal{N}$ is not $\mathcal{B}_t$-measurable. On the other hand, because it is a $\mathbb{P}^{M'}$-null set, $\mathcal{N}$ is in the $\mathbb{P}^{M'}$-completion $\overline{\mathcal{B}'_t}$ of $\mathcal{B}'_t$, and this is the reason for our having to introduce these completed σ-algebras. For a different approach to the same problem, see Exercise 4.6.8 in [41], where it was shown that one can avoid completions if one is willing to have $X(\,\cdot\,,x,p')$ right continuous everywhere but in $D\big([0,\infty);\mathbb{R}^n\big)$ only $\mathbb{P}^{M'}$-almost everywhere.

§3.1.5. Verifying the Martingale Property in (G2): It remains to prove that the $p' \rightsquigarrow X(\,\cdot\,,x,p')$ produced in Theorem 3.1.20 has the property described in **(G2)**. However, before doing so, we should check that $L\varphi \in C_b(\mathbb{R}^n;\mathbb{R})$ when $\varphi \in C_c(\mathbb{R}^n;\mathbb{R})$.

LEMMA 3.1.24. *There exists a $K < \infty$, depending only on*

$$\sup_{x\in\mathbb{R}^n} \frac{1}{1+|x|^2} \left(\|\sigma(x)\|_{\mathrm{H.S.}}^2 \vee |b(x)|^2 \vee \int_{B_{\mathbb{R}^{n'}}(0,1)} |F(x,y')|^2 \, M'(dy') \right)$$

and $M'\big(B_{\mathbb{R}^{n'}}(0,1)\complement\big)$, such that, for each $\varphi \in C_b^2(\mathbb{R}^n;\mathbb{R})$, $L\varphi \in C(\mathbb{R}^n;\mathbb{R})$ and

$$|L\varphi(x)| \leq K(1+|x|^2)\|\varphi\|_{C_b^2(\mathbb{R}^n;\mathbb{R})}.$$

Moreover, if $\varphi \in C_c^2(\mathbb{R}^n; \mathbb{R})$ and φ is supported in the ball $B_{\mathbb{R}^n}(0, R)$, then

$$\|L\varphi\|_u \leq K(3 + 8R^2)\|\varphi\|_{C_b^2(\mathbb{R}^n;\mathbb{R})} + K'\|\varphi\|_u \quad \text{where}$$

$$K' \equiv \sup_{x \in \mathbb{R}^n} M'(\{y' : |F(x, y')| \geq \tfrac{1}{2}\sqrt{1 + |x|^2}\}) < \infty.$$

PROOF: First observe that

$$M'(\{y' : |F(x, y')| \geq \tfrac{1}{2}\sqrt{1 + |x|^2}\})$$

$$\leq \frac{4}{1 + |x|^2} \int_{B_{\mathbb{R}^{n'}}(0,1)} |F(x, y')|^2 \, M'(dy') + M'(B_{\mathbb{R}^{n'}}(0, 1)\complement),$$

which, by **(b)** in **(H2)**, means that K' is indeed finite.

Assume that $\varphi \in C_b^2(\mathbb{R}^n; \mathbb{R})$, and observe that, by **(a)** of **(H2)**,

$$\int_{\mathbb{R}^{n'}} \Big(\varphi(x + F(x, y')) - \varphi(x) - \mathbf{1}_{[0,1)}(|y'|)(F(x, y'), \mathrm{grad}_x\varphi)_{\mathbb{R}^n}\Big) M'(dy')$$

$$= \lim_{r \searrow 0} \int_{|y'| \geq r} \Big(\varphi(x + F(x, y')) - \varphi(x)$$

$$- \mathbf{1}_{[0,1)}(|y'|)(F(x, y'), \mathrm{grad}_x\varphi)_{\mathbb{R}^n}\Big) M'(dy')$$

uniformly for x in compacts. At the same time, it is easy to draw from **(H3)** and **(c)** in **(H2)** the conclusion that

$$\int_{|y'| \geq r} \Big(\varphi(x + F(x, y')) - \varphi(x) - \mathbf{1}_{[0,1)}(|y'|)(F(x, y'), \mathrm{grad}_x\varphi)_{\mathbb{R}^n}\Big) M'(dy')$$

is a continuous function of x for each $r > 0$. Hence, the continuity of $L\varphi$ is now proved.

Turning to the asserted bounds, note that the first estimate is an easy consequence of (cf. the notation in (3.1.13))

$$\left| \int_{\mathbb{R}^{n'}} \Big(\varphi(x + F(x, y')) - \varphi(x) - \mathbf{1}_{[0,1)}(|y'|)(F(x, y'), \mathrm{grad}_x\varphi)_{\mathbb{R}^n}\Big) M'(dy') \right|$$

$$\leq \frac{1}{2} \left(\int_{B_{\mathbb{R}^n}(0,1)} |F(x, y')|^2 \, M'(dy') \right) \|\varphi\|_u^{(2)} + 2M'(B_{\mathbb{R}^{n'}}(0, 1)\complement)\|\varphi\|_u.$$

Next, assume that φ is supported in $B_{\mathbb{R}^n}(0, R)$. Then the preceding gives

$$|L\varphi(x)| \leq K(3 + 8R^2)\|\varphi\|_{C_b^2(\mathbb{R}^n;\mathbb{R})} \quad \text{for } x \in \overline{B_{\mathbb{R}^n}(0, 2R + 1)}.$$

On the other hand, if $|x| \geq 2R + 1$, then $x + F(x, y') \notin \mathrm{supp}(\varphi)$ unless $|F(x, y')| \geq \tfrac{1}{2}\sqrt{1 + |x|^2}$. Hence,

$$|L\varphi(x)| = \left| \int_{|F(x,y')| \geq \frac{1}{2}\sqrt{1+|x|^2}} \varphi(x + F(x, y')) \, M'(dy') \right|$$

$$\leq M'(\{y' : |F(x, y')| \geq \tfrac{1}{2}\sqrt{1 + |x|^2}\})\|\varphi\|_u$$

when $x \notin B_{\mathbb{R}^n}(0, 2R + 1)$. $\square$

LEMMA 3.1.25. *Under the conditions in Lemma 3.1.16,* $\|X(\cdot, x)\|_{[0,T]} \in$ $L^2(\mathbb{P}^{M'})$ *for all* $(T, x) \in [0, \infty) \times \mathbb{R}^n$. *Moreover, for each* $\varphi \in C^2(\mathbb{R}^n; \mathbb{R})$ *with bounded second order derivatives,* $\sup_{x \in \mathbb{R}^n} (1 + |x|^2)^{-1} |L\varphi(x)| < \infty$ *and*

$$\left(\varphi\big(X(t, x, p')\big) - \int_0^t L\varphi\big(X(\tau, x, p')\big) \, d\tau, \overline{\mathcal{B}'_t}, \mathbb{P}^{M'} \right)$$

is a martingale.

PROOF: The first assertion follows from the first line of (3.1.12) together with (3.1.19), and the second assertion is covered by (3.1.13). Hence, all that remains is to check the asserted martingale property. Thus let $\varphi \in C^2(\mathbb{R}^n; \mathbb{R})$ with bounded second order derivatives be given. By (3.1.19), it is clear that

$$\varphi\big(X^N(t, x)\big) \longrightarrow \varphi\big(X(t, x)\big) \quad \text{in } L^1(\mathbb{P}^{M'})$$

for each $(t, x) \in [0, \infty) \times \mathbb{R}^n$. In addition, by continuity, (3.1.19), and (3.1.12), we know that

$$\mathbb{P}^{M'} \left(\lim_{N \to \infty} \int_0^T \big| L\varphi\big(X(t, x)\big) - L\varphi\big(X^N(t, x)\big) \big| \, dt = 0 \right) = 1$$

for all $(T, x) \in [0, \infty) \times \mathbb{R}^n$. At the same time, because $\sup_{x \in \mathbb{R}^n} (1 + |x|^2)^{-1} |L\varphi(x)| < \infty$, (3.1.19) plus Lebesgue's dominated convergence Theorem[2] tells us that the preceding almost everywhere convergence implies convergence in $L^1(\mathbb{P}^{M'})$. That is,

$$\lim_{N \to \infty} \mathbb{E}^{\mathbb{P}^{M'}} \left[\int_0^T \big| L\varphi\big(X(t, x)\big) - L\varphi\big(X^N(t, x)\big) \big| \, dt \right] = 0.$$

Finally, by Lemma 3.1.11, we know that

$$\left(\varphi\big(X^N(t, x, p')\big) - \int_0^t L^N \varphi(\tau, x, p') \, d\tau, \mathcal{B}'_t, \mathbb{P}^{M'} \right)$$

is a martingale for each $N \geq 1$. Hence, we need only show that

$$\lim_{N \to \infty} \mathbb{E}^{\mathbb{P}^{M'}} \left[\int_0^T \big| L\varphi\big(X^N(t, x)\big) - L^N \varphi(t, x) \big| \, dt \right] = 0.$$

But, by the second line of (3.1.12), (c) in (H2), and (H3), we know that

$$\int_0^T \big| L\varphi\big(X^N(t, x)\big) - L^N \varphi(t, x) \big| \, dt \longrightarrow 0$$

in $\mathbb{P}^{M'}$-measure. Thus, by (3.1.14), our bound on $|L\varphi(x)|$, (3.1.19), and Lebesgue's dominated convergence Theorem, we have our result. $\square$

---

[2] The version we are using here is the one which says that if $f_n \to 0$ almost everywhere or in measure and if $|f_n| \leq g_n$ where $\{g_n\}_1^\infty$ is an L^1-convergent sequence, then $f_n \to 0$ in L^1.

THEOREM 3.1.26. *For each $\varphi \in C_c^2(\mathbb{R}^n; \mathbb{R})$, $L\varphi \in C_b(\mathbb{R}^n; \mathbb{R})$ and*

$$\left(\varphi\big(X(t,x,p')\big) - \int_0^t L\varphi\big(X(\tau,x,p')\big)\,d\tau, \overline{\mathcal{B}'_t}, \mathbb{P}^{M'}\right)$$

is a martingale. Hence, the $\mathbb{P}^{M'}$-distribution of $p' \rightsquigarrow X(\,\cdot\,, x, p')$ solves the martingale problem for L on $C_c^2(\mathbb{R}^n; \mathbb{R})$ starting from x.

PROOF: In view of Lemma 3.1.24, we need only prove the martingale property. To this end, let $R \geq 1$ be given and, referring to Lemma 3.1.15, apply Lemma 3.1.25 to see that

$$\left(\varphi\big(X_R(t,x,p')\big) - \int_0^t L_R\varphi\big(X_R(t,x,p')\big)\,d\tau, \overline{\mathcal{B}'_t}, \mathbb{P}^{M'}\right)$$

is a martingale, where L_R is the operator described in **(G2)** when $M'(dy')$ is replaced by $\mathbf{1}_{[0,R)}(|y'|)\, M'(dy')$. In particular, by Lemma 3.1.15, this means that, for any $0 \leq t_1 < t_2$ and $A \in \overline{\mathcal{B}'_{t_1}}$,

$$\mathbb{E}^{\mathbb{P}^{M'}}\left[\varphi\big(X(t_2,x)\big) - \varphi\big(X(t_1,x)\big),\, A \cap A(t_2, R)\right]$$
$$= \mathbb{E}^{\mathbb{P}^{M'}}\left[\int_{t_1}^{t_2} L_R\varphi\big(X(\tau,x)\big)\,d\tau,\, A \cap A(t_2, R)\right].$$

Hence, because $L_R\varphi \longrightarrow L\varphi$ pointwise and boundedly while $\mathbb{P}^{M'}\big(A(t_2, R)\big) \longrightarrow 1$ as $R \to \infty$, the desired conclusion follows. $\square$

We close this section by giving the analog for the present context of the uniqueness result in Theorem 2.2.6, and, as a consequence, find a condition which guarantees that the $\mathbb{P}^{M'}$-distribution of $p' \rightsquigarrow X(\,\cdot\,, x, p')$ is Markov.

COROLLARY 3.1.27. *Suppose that for each $\varphi \in C_c^\infty(\mathbb{R}^n; \mathbb{R})$ there is a bounded, $u_\varphi \in C^{1,2}\big([0, \infty) \times \mathbb{R}^n; \mathbb{R}\big)$ which solves*

$$(3.1.28) \qquad \partial_t u_\varphi(t,x) = L u_\varphi(t,x) \quad \text{with} \lim_{t \searrow 0} u_\varphi(t,x) = \varphi(x).$$

Then the $\mathbb{P}^{M'}$-distribution of $p' \rightsquigarrow X(\,\cdot\,, x, p')$ is the one and only solution $\mathbb{P}_x^L$ to the martingale problem for L on $C_c^2(\mathbb{R}^n; \mathbb{R})$ starting from x. In fact, if $P_t(x, \,\cdot\,)$ denotes the $\mathbb{P}^{M'}$-distribution of $p' \rightsquigarrow X(t, x, p')$, then $(t, x) \in [0, \infty) \times \mathbb{R}^n \longmapsto P_t(x, \,\cdot\,) \in \mathbf{M}_1(\mathbb{R}^n)$ is continuous and

$$\mathbb{P}^{M'}\big(X(s + t) \in \Gamma \,\big|\, \overline{\mathcal{B}'_s}\big) = P_t\big(X(s, x), \Gamma\big) \quad \mathbb{P}^{M'}\text{-almost surely.}$$

In particular, $(t, x) \rightsquigarrow P_t(x, \,\cdot\,)$ is a transition probability function, and $\mathbb{P}_x^L$ is the distribution of the Markov process determined (cf. (1.4.1)) by this transition probability function.

PROOF: We already know that the $\mathbb{P}^{M'}$-distribution $\mathbb{Q}$ of $p' \rightsquigarrow X(\,\cdot\,, x, p')$ solves the martingale problem for L on $C_c^2(\mathbb{R}^n; \mathbb{R})$ starting from x.

Next suppose that $\mathbb{P}$ is any solution to the martingale problem for L on $C_c^2(\mathbb{R}^n; \mathbb{R})$. We will now show that if $v \in C_b^{1,2}([0, T] \times \mathbb{R}^n; \mathbb{R})$ vanishes off of $[0, T] \times B_{\mathbb{R}^n}(0, R)$ for some $R > 0$, then

$$\left(v\big(t \wedge T, p(t \wedge T)\big) - \int_0^t (\partial_\tau + L) v\big(\tau, p(\tau)\big), \mathcal{B}_t, \mathbb{P} \right)$$

is a martingale. To this end, first notice that it suffices to handle smooth v and then observe that

$$\mathbb{E}^{\mathbb{P}}\left[v\big(t_2, p(t_2)\big) - v(t_1, p(t_1)), A \right]$$
$$= \mathbb{E}^{\mathbb{P}}\left[\int_{t_1}^{t_2} \Big(\partial_t v\big(t, p(t_2)\big) + Lv\big(t_1, p(t)\big) \Big) \, dt, A \right]$$

for $0 \le t_1 < t_2$ and $A \in \mathcal{B}_{t_1}$. At the same time,

$$\mathbb{E}^{\mathbb{P}}\left[\int_{t_1}^{t_2} \Big(\partial_t v\big(t, p(t_2)\big) + Lv\big(t_1, p(t)\big) \Big) \, dt, A \right]$$
$$= \mathbb{E}^{\mathbb{P}}\left[\int_{t_1}^{t_2} \Big(\partial_t v\big(t, p(t)\big) + Lv\big(t, p(t)\big) \Big) \, dt, A \right]$$
$$+ \mathbb{E}^{\mathbb{P}}\left[\iint_{t_1 \le \tau < t \le t_2} \Big(\partial_\tau Lv\big(\tau, p(t)\big) - L\partial_t v\big(\tau, p(t)\big) \Big) \, d\tau dt, A \right].$$

Hence, since $\partial_\tau Lv = L\partial_\tau v$, this leads to the desired conclusion.

Next suppose that u_φ is a solution of (3.1.28) which is bounded, and again let $\mathbb{P}$ be as in the preceding paragraph. Given $T > 0$, set $v(t, x) = u_\varphi(T - t, x)$ for $(t, x) \in [0, T] \times \mathbb{R}^n$. Then $v \in C_b^{1,2}([0, T] \times \mathbb{R}^n; \mathbb{R})$, $v(T, \,\cdot\,) = \varphi$, and $(\partial_t + L)v = 0$ on $(0, \infty) \times \mathbb{R}^n$. Now choose $\psi \in C_c^\infty(\mathbb{R}^n; [0, 1])$ so that $\psi \equiv 1$ on $B(0, 1)$, and set $v_R(t, x) = \psi(R^{-1}x)v(t, x)$. By the preceding, we know that

$$\left(v_R\big(t \wedge T, p(t \wedge T)\big) - \int_0^{t \wedge T} (\partial_\tau + L) v_R\big(\tau, p(\tau)\big) \, d\tau, \mathcal{B}_t, \mathbb{P} \right)$$

is a martingale. Thus, if

$$\zeta_R(p) \equiv \inf \left\{ t \ge 0 : \sup_{\tau \in [0, t]} |p(t)| \ge R \right\} \wedge T,$$

then $p \rightsquigarrow \zeta_R(p)$ is a bounded stopping time relative to $\{\mathcal{B}_t : t \geq 0\}$ (cf. §7.1 in [36]) and therefore, by Doob's stopping time theorem (cf. Corollary 7.1.15 in [36]),

$$\left(v\left(t \wedge \zeta_R, p(t \wedge \zeta_R)\right), \mathcal{B}_t, \mathbb{P}\right)$$

is also a martingale. In particular, this means that

$$\mathbb{E}^{\mathbb{P}}\left[u_\varphi\left(0, p(T \wedge \zeta_R)\right) \,\middle|\, \mathcal{B}_s\right] = u_\varphi\left((T - s) \wedge \zeta_R, p(s \wedge \zeta_R)\right)$$

$\mathbb{P}$-almost surely for all $s \in [0, T]$ and $R > 0$. Thus, since u_φ is bounded and continuous and $\zeta_R \longrightarrow \infty$ pointwise as $R \to \infty$, we conclude that

$$\mathbb{E}^{\mathbb{P}}\left[\varphi\left(p(T)\right) \,\middle|\, \mathcal{B}_s\right] = u_\varphi(T - s, p(s)) \quad \mathbb{P}\text{-almost surely.}$$

To complete the proof from here, let $P_t(x, \cdot)$ be the distribution of $\mathbb{P}^{M'}$-distribution of $p' \rightsquigarrow X(t, x, p')$. Then, by the preceding $\langle \varphi, P_t(x, \cdot) \rangle = u_\varphi(t, x)$ for $\varphi \in C_c^\infty(\mathbb{R}^n; \mathbb{R})$, and from this it follows that $(t, x) \rightsquigarrow P_t(x, \cdot)$ is continuous. In addition, we now know that, for any $\mathbb{P}$ which solves the martingale problem for L on $C_c^2(\mathbb{R}^n; \mathbb{R})$,

$$\langle \varphi, P_t(p(s), \cdot) \rangle = u_\varphi(t, x) = \mathbb{E}^{\mathbb{P}}\left[\varphi(p(s + t)) \,\middle|\, \mathcal{B}_s\right] \quad \text{for all } \varphi \in C_c^\infty(\mathbb{R}^n; \mathbb{R}).$$

Hence, $P_t(p(s), \cdot)$ is the $\mathbb{P}$-conditional distribution of $p \rightsquigarrow p(s + t)$ given $\mathcal{B}_s$. In particular, this, together with $\mathbb{P}(p(0) = x) = 1$, implies that $\mathbb{P}$ satisfies (1.4.1) with $\mu = \delta_x$, and so we have now proved that $\mathbb{Q}$ is indeed the one and only solution $\mathbb{P}_x^L$ to the martingale problem for L on $C_c^2(\mathbb{R}^n; \mathbb{R})$ starting from x.

Finally, to check that $P_t(x, \cdot)$ satisfies the Chapman–Kolmogorov equation, and is therefore a transition probability function, we use

$$\langle \varphi, P_{s+t}(x, \cdot) \rangle = \mathbb{E}^{\mathbb{P}_x^L}\left[\varphi(p(s + t))\right]$$
$$= \mathbb{E}^{\mathbb{P}_x^L}\left[\langle \varphi, P_t(p(s), \cdot) \rangle\right] = \int \langle \varphi, P_t(y, \cdot) \rangle \, P_s(x, dy). \quad \square$$

§3.1.6. Exercises

EXERCISE 3.1.29. Suppose that $x \rightsquigarrow \mathbb{P}_x$ is a map taking an $x \in \mathbb{R}^n$ to a probability measure $\mathbb{P}_x$ on $\left(D([0, \infty); \mathbb{R}^n), \mathcal{B}\right)$, and assume that $x \rightsquigarrow \mathbb{E}^{\mathbb{P}_x}[\Phi]$ is continuous for all $\mathcal{B}$-bounded continuous Φ's. Using part (iii) of Exercise 2.3.5, show that $x \rightsquigarrow \mathbb{E}^{\mathbb{P}_x}[\Phi]$ is measurable for each $\mathcal{B}$-measurable Φ which is either bounded or non-negative.

EXERCISE 3.1.30. Here is the discrete version of Gronwall's inequality which we used to end the proof of Lemma 3.1.16. Let $J \in \mathbb{Z}^+$, and suppose that $\{u_j\}_0^J \subseteq [0, \infty)$ satisfies the condition that $u_j \leq A \sum_{i=0}^{j-1} u_i + B$ for each $1 \leq j \leq J$. Show that $u_J \leq (Au_0 + B)(1 + A)^{J-1}$.

Hint: Set $U_j = \sum_{i=0}^{j-1} u_i$, and get $U_j \leq (1 + A)U_{j-1} + B$ for $1 \leq j \leq J$. Working by induction, conclude that

$$U_j \leq (1+A)^j u_0 + B \sum_{i=0}^{j-1}(1+A)^i = (1+A)^j u_0 + \frac{(1+A)^j - 1}{A} B \text{ for } 1 \leq j \leq J.$$

Finally, plug this into $u_J \leq AU_{J-1} + B$ to arrive at the asserted estimate.

EXERCISE 3.1.31. Let $p' \rightsquigarrow X(\cdot, x, p')$ be the map described in **(G1)**.

(i) Show that $p' \rightsquigarrow X(\cdot, x, p')$ has no fixed points of discontinuity in the sense that

$$\mathbb{P}^{M'}\big(X(t, x) \neq X(t-, x)\big) = 0 \quad \text{for all } (t, x) \in [0, \infty) \times \mathbb{R}^n.$$

(ii) Show that

$$\int_{\mathbb{R}^n} |F(x, y')|^2 M'(dy') \equiv 0$$
$$\implies \mathbb{P}^{M'}\big(\{p' : X(\cdot, x, p') \in C([0, \infty); \mathbb{R}^n)\}\big) = 1.$$

(iii) For each $T \in (0, \infty)$, show that

$$\lim_{R \to \infty} \sup_{x \in \mathbb{R}^n} \mathbb{P}^{M'}\big(\|X(\cdot, x)\|_{[0,T]} \geq R\sqrt{1 + |x|^2}\big) = 0.$$

EXERCISE 3.1.32. There is one case in which (3.1.22) presents no problems and does indeed determine $X(\cdot, x, p')$. Namely, consider the case in which $n' = n$ and $M' = 0$, and take $\sigma \equiv I$ and $F \equiv 0$. Note, for $\mathbb{P}^0$-almost every p', $p' = w(\cdot, p')$ and therefore p' is continuous. Thus, we may and will think of $\mathbb{P}^0$ as a probability measure on $C([0, \infty); \mathbb{R}^n)$, in which case $p' = w(\cdot, p')$ for every p'. Show that, for each $p' \in C([0, \infty); \mathbb{R}^n)$, $X^N(\cdot, x, p')$ converges uniformly on compacts to the one and only solution $X(\cdot, x, p')$ to the integral equation

$$X(t, x, p') = x + p'(t) + \int_0^t b\big(X(\tau, x, p')\big) d\tau.$$

Further, check that $p' \rightsquigarrow X(\cdot, x, p')$ is continuous as a map of $C([0, \infty); \mathbb{R}^n)$ to itself with the topology of uniform convergence on compacts.

3.2 When Does Itô's Theory Work?

In §3.1 we assumed that the quantities in **(H1)**–**(H3)** had been handed to us and proceeded to show how Itô's method led to a construction of a solution to the martingale problem for the operator L in (3.1.5). However, in practice, someone hands you the operator L, and your first problem is to construct functions σ, b, and F and find a Lévy measure M' so that they not only give rise to the operator L but also satisfy the conditions in **(H1)**–**(H3)**. That is, besides satisfying **(H1)**–**(H3)**, we need that $a(x) = \sigma(x)\sigma(x)^\top$ and (3.1.8) holds. The purpose of the section is to shed some light on this problem.

§3.2.1. The Diffusion Coefficients: By far the most satisfactory solution to the problem just raised applies to the diffusion coefficients. Namely, what we are seeking are conditions under which a symmetric, non-negative definite, operator-valued function $x \in \mathbb{R}^n \longmapsto a(x) \in \mathrm{Hom}(\mathbb{R}^n; \mathbb{R}^n)$ admits the representation $a(x) = \sigma(x)\sigma(x)^\top$ for some Lipschitz continuous function $\sigma : \mathbb{R}^n \longrightarrow \mathrm{Hom}(\mathbb{R}^{n'}; \mathbb{R}^n)$, and this is a question to which the answer is quite well understood.

In what follows, $a : \mathbb{R}^n \longrightarrow \mathrm{Hom}(\mathbb{R}^n; \mathbb{R}^n)$ will be a symmetric, non-negative definite operator-valued function, and we will use $a^{\frac{1}{2}}(x)$ to denote the symmetric, non-negative definite square of $a(x)$.

Lemma 3.2.1. *Assume that a is positive definite at x, and let $\lambda_{\min}(x)$ be the smallest eigenvalues of $a(x)$. Then $a^{\frac{1}{2}}$ is differentiable at x and*

$$\left\| (\partial_\xi)_x a^{\frac{1}{2}} \right\|_{\mathrm{op}} \leq \frac{\left\| (\partial_\xi)_x a \right\|_{\mathrm{op}}}{2\sqrt{\lambda_{\min}(x)}}$$

for all $\xi \in \mathbb{R}^n$

Proof: Use α and β to denote $\lambda_{\min}(x)$ and $\|a(x)\|_{\mathrm{op}}$, respectively, and set

$$m(y) = \frac{a(y)}{\beta} - I, \quad y \in \mathbb{R}^n.$$

Then, because $y \rightsquigarrow \lambda_{\max}(y)$ and $y \rightsquigarrow \lambda_{\min}(y)$ are continuous, we can find an $r > 0$ such that $\theta \equiv \sup\{\|m(y)\|_{\mathrm{op}} : |y - x| \leq r\} < 1$. Thus, for $y \in B_{\mathbb{R}^n}(x, r)$,

$$(*) \qquad\qquad a^{\frac{1}{2}}(y) = \beta^{\frac{1}{2}} \sum_{\ell=0}^{\infty} \binom{\frac{1}{2}}{\ell} m(y)^\ell,$$

where

$$\binom{\frac{1}{2}}{\ell} \equiv \begin{cases} 1 & \text{if } \ell = 0 \\ \frac{\frac{1}{2}(\frac{1}{2}-1)\cdots(\frac{1}{2}-\ell+1)}{\ell!} & \text{if } \ell \geq 1 \end{cases}$$

is the coefficient of η^ℓ in the Taylor's expansion of $(1 + \eta)^{\frac{1}{2}}$ for $|\eta| < 1$. To check the validity of (*), first observe that the right hand side is symmetric and then act the right hand side twice on any eigenvector of $a(y)$.

Knowing (*), one sees that, for any $\xi \in \mathbb{R}^n$, $(\partial_\xi)_x a^{\frac{1}{2}}$ exists and is given by

$$(\partial_\xi)_x a^{\frac{1}{2}} = \beta^{-\frac{1}{2}} \sum_{\ell=1}^{\infty} \ell \binom{\frac{1}{2}}{\ell} m(x)^{\ell-1} (\partial_\xi)_x a.$$

Hence, since $(-1)^{\ell-1} \binom{\frac{1}{2}}{\ell} \geq 0$ for $\ell \geq 1$ and $\sum_1^\infty \ell \binom{\frac{1}{2}}{\ell}(-x)^{\ell-1} = \frac{d}{dx}(1-x)^{\frac{1}{2}}$ for $|x| < 1$,

$$\left\| (\partial_\xi)_x a^{\frac{1}{2}} \right\|_{\mathrm{op}} \leq \beta^{-\frac{1}{2}} \|(\partial_\xi)_x a\|_{\mathrm{op}} \sum_{\ell=1}^{\infty} \ell \binom{\frac{1}{2}}{\ell} \left(-\|m(x)\| \right)^{\ell-1}$$

$$= \frac{\beta^{-\frac{1}{2}} \|(\partial_\xi)_x a\|_{\mathrm{op}}}{2(1 - \|m(x)\|_{\mathrm{op}})^{\frac{1}{2}}}.$$

Finally, because $0 \leq I - \frac{a(x)}{\beta} \leq \left(1 - \frac{\alpha}{\beta}\right)I$, $\|m(x)\|_{\mathrm{op}} \leq 1 - \frac{\alpha}{\beta}$, and the required estimate follows. $\square$

Obviously, Lemma 3.2.1 is very satisfactory as long as a is bounded below by a positive multiple of the identity. That is, in the terminology of partial differential equations, when $\sum_{i,j=1}^n a_{ij}(x)\partial_i\partial_j$ is a uniformly elliptic operator. However, one of the virtues of the probabilistic approach to partial differential equations is that it has the potential to handle elliptic operators which are degenerate. Thus, it is important to understand what can be said about the Lipschitz continuity of $a^{\frac{1}{2}}$ when a becomes degenerate.

The basic fact which allows us to deal with degenerate a's is the following elementary observation about non-negative functions. Namely, if $f \in C^2(\mathbb{R}; [0, \infty))$, then

(3.2.2) $$|f'(t)| \leq \sqrt{2\|f''\|_\mathrm{u} f(t)}, \quad t \in \mathbb{R}.$$

To see this, use Taylor's theorem to write $0 \leq f(t + h) \leq f(t) + hf'(t) + \frac{h^2}{2}\|f''\|_\mathrm{u}$ for all $h \in \mathbb{R}$. Hence $|f'(t)| \leq h^{-1}f(t) + \frac{h}{2}\|f''\|_\mathrm{u}$ for all $h > 0$, and so (3.2.2) results when one minimizes with respect to h.

LEMMA 3.2.3. Assume that a is twice differentiable and that

$$K = \sup\left\{ \left\|(\partial_\xi)_x^2 a\right\|_{\mathrm{op}} : x \in \mathbb{R}^n \text{ and } \xi \in \mathbb{S}^{n-1} \right\} < \infty.$$

Then

$$\left\| a^{\frac{1}{2}}(y) - a^{\frac{1}{2}}(x) \right\|_{\mathrm{H.S.}} \leq n\sqrt{2K} \, |y - x|.$$

PROOF: First observe that it suffices to handle a's which are uniformly positive definite. Indeed, given the result in that case, we can prove the result in general by replacing a with $a + \epsilon I$ and then letting $\epsilon \searrow 0$. Thus, we will, from now on, assume that $a(x) \geq \epsilon I$ for some $\epsilon > 0$ and all $x \in \mathbb{R}^n$. In particular, by Lemma 3.2.1, this means that $a^{\frac{1}{2}}$ is differentiable and that the required estimate will follow once we show that

$$(*) \qquad\qquad \left\| (\partial_\xi)_x a^{\frac{1}{2}} \right\|_{\text{H.S.}} \leq n\sqrt{2K}$$

for all $x \in \mathbb{R}^n$ and $\xi \in \mathbb{S}^{n-1}$.

To prove (*), let x be given, and work with an orthonormal coordinate system in which $a(x)$ is diagonal. Then, from $a = a^{\frac{1}{2}} a^{\frac{1}{2}}$ and Leibnitz's rule, one obtains

$$(\partial_\xi)_x a_{ij} = \left((\partial_\xi)_x a_{ij}^{\frac{1}{2}} \right) \left(\sqrt{a_{ii}(x)} + \sqrt{a_{jj}(x)} \right).$$

Hence, because $\sqrt{\alpha} + \sqrt{\beta} \geq \sqrt{\alpha + \beta}$ for all $\alpha, \beta \geq 0$,

$$\left| (\partial_\xi)_x a_{ij}^{\frac{1}{2}} \right| \leq \frac{|(\partial_\xi)_x a_{ij}|}{\sqrt{a_{ii}(x) + a_{jj}(x)}}.$$

To complete the proof of (*), set

$$f_\pm(t) = a_{ii}(x + t\xi) \pm 2a_{ij}(x + t\xi) + a_{jj}(x + t\xi),$$

and apply (3.2.2) to get

$$\left| (\partial_\xi)_x a_{ij} \right| \leq \frac{|f'_+(0)| + |f'_-(0)|}{4} \leq \sqrt{2K} \sqrt{a_{ii}(x) + a_{jj}(x)},$$

which, in conjunction with the preceding, leads first to

$$\left| (\partial_\xi)_x a_{ij}^{\frac{1}{2}} \right| \leq \sqrt{2K}$$

and then to (*). $\square$

REMARK 3.2.4. The preceding results might incline one to believe that there is little or no reason to take any square root other than the symmetric, non-negative definite one. Indeed, if either a is uniformly positive definite or all one cares about is Lipschitz continuity, then there really is no reason to consider other square roots. However, if one is dealing with degenerate a's and needs actual derivatives, then one can sometimes do better by looking at other square roots. For example, consider a bounded, smooth $a : \mathbb{R}^2 \longrightarrow$

$\mathrm{Hom}(\mathbb{R}^2; \mathbb{R}^2)$ such that $a(x) = (x_1^2 + x_2^2) I_{\mathbb{R}^2}$ for $x \in B_{\mathbb{R}^2}(0, 1)$. Clearly $a^{\frac{1}{2}}(x) = \sqrt{x_1^2 + x_2^2}\, I$ for $x \in B_{\mathbb{R}^2}(0, 1)$, and so, although $a^{\frac{1}{2}}$ is Lipschitz, it is not continuously differentiable at the origin. On the other hand, $a(x) = \sigma(x)\sigma(x)^\top$ on $B_{\mathbb{R}^2}(0, 1)$ if $\sigma(x) \equiv \begin{pmatrix} x_1 & x_2 \\ -x_2 & -x_1 \end{pmatrix}$ there, and σ is smooth on $B_{\mathbb{R}^2}(0, 1)$.

Finally, it should be recognized that, even if a is real analytic as a function of x, it will not in general be possible to find a smooth σ such that $a = \sigma\sigma^\top$. The reasons for this have their origins in classical algebraic geometry. Indeed, D. Hilbert showed that it is not possible to express every non-negative polynomial as a finite sum of squares of polynomials. After combining this fact with Taylor's theorem, one realizes that it rules out the existence of a smooth choice of σ. Of course, the problems arise only at the places where a degenerates: away from degeneracies, as the proof of Lemma 3.2.1 shows, the entries of $a^{\frac{1}{2}}$ are analytic functions of the entries of a.

§3.2.2. The Lévy Measure: Although, as we have just seen, Itô's theory handles the diffusion term in L with ease, his theory has a lot of problems with the Lévy measure term. Namely, although, as we are about to show, it is always possible to represent a measurable map taking $x \in \mathbb{R}^n$ into a Lévy measure $M(x, \cdot)$ as $M(x, \cdot) = F(x, \cdot)_* M'$ for appropriate choices of Lévy measure M' and measurable map F, I know of no general criterion in terms of the smoothness of $x \rightsquigarrow M(x, \cdot)$ which guarantees that this representation can be given by an F satisfying **(H2)**.

THEOREM 3.2.5. *Assume that $n' \geq 2$, and let M' be the Lévy measure on $\mathbb{R}^{n'}$ given by $M'(dy) = \mathbf{1}_{\mathbb{R}^{n'} \setminus \{0\}} |y|^{-n-1}\, dy$. Given a measurable map $x \rightsquigarrow M(x, \cdot)$ from $x \in \mathbb{R}^n$ to Lévy measures $M(x, \cdot)$ on $\mathbb{R}^n$, there exists a measurable map $F : \mathbb{R}^n \times \mathbb{R}^{n'} \longrightarrow \mathbb{R}^n$ with the property that $M(x, \cdot) = F(x, \cdot)_* M'$ for each $x \subset \mathbb{R}^n$.*

PROOF: Because this result will not be used in the sequel, only an outline of its proof will be given here.

The technical basis on which the proof rests is the following. Given a Polish space Ω (i.e., a complete separable metric space), think (cf. §3.1 in [36]) of $\mathbf{M}_1(\Omega)$ as a Polish space with the topology of weak convergence. Next, suppose that X and Y are a pair of Polish spaces, and set $\Omega = X \times Y$. Then there is a measurable map $(P, y) \in \mathbf{M}_1(\Omega) \times Y \longmapsto P_y \in \mathbf{M}_1(X)$ with the property that, for all $\mathcal{B}_\Omega$-measurable functions $\varphi : \Omega \longrightarrow [0, \infty)$,

$$\int_\Omega \varphi(\omega)\, P(d\omega) = \int_Y \left(\int_X \varphi(x, y)\, P_y(dx) \right) P_Y(dy),$$

where P_Y is the marginal distribution of P on Y (i.e., $P_Y(dy) = P(X \times dy)$). This result follows rather easily from the well known result which guarantees the existence of regular conditional probability distributions for probability measures on a Polish space. Namely, set $\Sigma = \{X \times \Gamma : \Gamma \in \mathcal{B}_Y\}$. Then Theorem 5.1.15 in [36] says that, for each $P \in \mathbf{M}_1(\Omega)$, there is a Σ-measurable map $\omega \in \Omega \longmapsto P_\omega \in \mathbf{M}_1(\Omega)$ with the property that, for each φ, $\omega \rightsquigarrow \mathbb{E}^{P_\omega}[\varphi]$ is a conditional expectation of φ given Σ. Furthermore, a careful look at the proof reveals that $(P, \omega) \in \mathbf{M}_1(\Omega) \times \Omega \longmapsto P_\omega \in \mathbf{M}_1(\Omega)$ is measurable. In addition, because Σ is countably generated, for each P, there is a P-null set $\mathcal{N} \in \Sigma$ such that $P_\omega(A) = \mathbf{1}_A(\omega)$ for all $A \in \Sigma$ and $\omega \notin \mathcal{N}$. Hence, all that we have to do is choose and fix an $x_0 \in X$ and take P_y equal to the marginal distribution $\left(P_{(x_0,y)}\right)_X$ of $P_{(x_0,y)}$ on X.

As our first application of the preceding, we show that, for each $n \geq 1$, there is a measurable map $(\mu, t) \in \mathbf{M}_1(\mathbb{R}^n) \times [0,1) \longmapsto f_n(\mu, t) \in \mathbb{R}^n$ such that $\mu = f_n(\mu, \cdot)_* \lambda_{[0,1)}$, where $\lambda_{[0,1)}$ is Lebesgue measure on $[0,1)$. When $n = 1$, we can take $f_1(\mu, t) = \sup\{x \in \mathbb{R} : \mu([x, \infty)) \geq t\}$. In order to handle $n \geq 2$, we need to know that there exists a pair of independent, uniformly distributed $[0,1)$-valued random variables U_1 and U_2 on $\left([0,1), \mathcal{B}_{[0,1)}, \lambda_{[0,1)}\right)$. For example, given $t \in [0,1)$, determine $\{\epsilon_m(t) : m \geq 1\} \subseteq \{0,1\}$ so that

$$t - \sum_{m=1}^n \epsilon_m(t)2^{-m} < 2^{-n} \quad \text{for all } n \geq 1.$$

It is easy to see (cf. Lemma 1.1.5 and Exercise 1.1.19 in [36]) that the ϵ_m's are mutually independent Bernoulli random variables and that $U_1 \equiv \sum_{m=0}^\infty \epsilon_{2m+1}2^{-m}$ and $U_2 \equiv \sum_{m=1}^\infty \epsilon_{2m}2^{-m}$ are random variables of the required sort. Now let $n \geq 2$ be given, assume the result for $n-1$, take $X = \mathbb{R}$ and $Y = \mathbb{R}^{n-1}$ in the preceding, and define $(\mu, y) \rightsquigarrow \mu_y$ accordingly. It is then an easy matter to check that f_n can be taken so that

$$f_n(\mu, t) = \left(f_1\big(\mu_t, U_2(t)\big), f_{n-1}\big(\mu_Y, U_1(t)\big) \right) \quad \text{where } \mu_t \equiv \mu_{f_{n-1}(\mu_Y, U_1(t))}.$$

In preparation for the next step, note that if μ is supported on $\mathbb{S}^{n-1}$, then we may and will take $t \in [0,1) \longmapsto f_n(\mu, t) \in \mathbb{S}^{n-1}$. Indeed, because, in any case, $f_n(\mu, t)$ will be in $\mathbb{S}^{n-1}$ for $\lambda_{[0,1)}$-almost every t, we can replace $f_n(\mu, t)$ by $\frac{f_n(\mu,t)}{|f_n(\mu,t)|}$ when $f_n(\mu, t) \neq 0$ and $(1, 0, \ldots, 0)$ when $f_n(\mu, t) = 0$.

Now suppose that $x \rightsquigarrow M(x, \cdot)$ is a measurable map of $x \in \mathbb{R}^n$ into Lévy measures on $\mathbb{R}^n$, and define $\bar{M}(x \cdot) \in \mathbf{M}_1(\mathbb{R}^n \setminus \{0\})$ so that

$$\bar{M}(x, \Gamma) = \int_\Gamma \frac{|y|^2}{1 + |y|^2} M(x, dy).$$

Next, set $m(x) = \bar{M}(x, \mathbb{R}^n \setminus \{0\})$, and define $x \in \mathbb{R}^n \longmapsto \mu(x, \cdot) \in \mathbf{M}_1\big((0, \infty)$ $\times \mathbb{S}^{n-1}\big)$ so that

$$\mu(x, \Gamma) = \frac{1}{m(x)} \int_{\mathbb{R}^n \setminus \{0\}} \mathbf{1}_\Gamma\big(|y|, \tfrac{y}{|y|}\big) \bar{M}(x, dy)$$

if $m(x) > 0$ and $\mu(x, \cdot) = \delta_{(1,0,\dots,0)}$ if $m(x) = 0$. By another application of the our first result, we can now find a measurable map $(x, \omega) \in \mathbb{R}^n \times \mathbb{S}^{n-1} \longrightarrow$ $\mu(x, \omega, \cdot) \in \mathbf{M}_1\big((0, \infty)\big)$ so that, for all measurable $\varphi : \mathbb{R}^n \setminus \{0\} \longrightarrow [0, \infty)$,

$$\int_{\mathbb{R}^n \setminus \{0\}} \varphi(y) \, M(x, dy)$$
$$= m(x) \int_{\mathbb{S}^{n-1}} \left(\int_{(0,\infty)} \varphi(r\omega) \frac{1 + r^2}{r^2} \mu(x, \omega, dr) \right) \mu_{\mathbb{S}^{n-1}}(x, d\omega),$$

where $\mu_{\mathbb{S}^{n-1}}(x, \cdot)$ denotes the marginal distribution of $\mu(x, \cdot)$ on $\mathbb{S}^{n-1}$. Thus, if we take (in the following, the supremum over the empty set is taken to be 0)

$$\rho(x, \omega, r) \equiv \sup \left\{ \rho \in [0, \infty) : m(x) \int_\rho^\infty \frac{1 + s^2}{s^2} \mu(x, \omega, ds) \geq \frac{1}{r} \right\}$$

and $G : \mathbb{R}^n \times (0, \infty) \times [0, 1) \longrightarrow \mathbb{R}^n$ is given by

$$G(x, r, t) = \rho\Big(x, f_n\big(\mu_{\mathbb{S}^{n-1}}(x, \cdot), t\big), r\Big) f_n\big(\mu_{\mathbb{S}^{n-1}}(x, \cdot), t\big),$$

then G is measurable and

$$\int_{\mathbb{R}^n \setminus \{0\}} \varphi(y) \, M_x(dy) = \int_{[0,1)} \left(\int_{(0,\infty)} \varphi(G(x, r, t)) \frac{dr}{r^2} \right) dt.$$

To complete the proof from here, let $n' \geq 2$, use $\lambda_{\mathbb{S}^{n'-1}}$ to denote surface measure on $\mathbb{S}^{n'-1}$, and define $\psi : [-1, 1) \longrightarrow [0, 1)$ by $\psi(s) = \frac{1}{w_{n'-1}} \lambda_{\mathbb{S}^{n'-1}}(\{\omega : \omega_1 \leq s\})$, where $w_{n'-1} \equiv \lambda_{\mathbb{S}^{n'-1}}(\mathbb{S}^{n'-1})$. If we take $F(x, r\omega) = G\big((x, r, \psi(\omega_1))\big)$ for $(r, \omega) \in (0, \infty) \times \mathbb{S}^{n'-1}$, then $F(x, \cdot)_* M_0 = w_{n'-1} M(x, \cdot)$. Thus, all that remains is to make the construction of G for $x \rightsquigarrow \frac{1}{w_{n'-1}} M(x, \cdot)$ instead of $x \rightsquigarrow M(x, \cdot)$. $\square$

Although the preceding shows that representations of $x \rightsquigarrow M(x, \cdot)$ in form $F(x, \cdot)_* M_0 = M(x, \cdot)$ are available, it leaves open the problem of finding one with an F which satisfies **(H2)**. On the other hand, it does indicate how one might proceed in good situations. To be more precise, assume that

$$(3.2.6) \quad \int_{\mathbb{R}^n \setminus \{0\}} \varphi(y) \, M(x, dy) = \int_{\mathbb{S}^{n-1}} \left(\int_{(0,\infty)} \varphi(r\omega) \beta(x, \omega, r) \, dr \right) \mu(d\omega)$$

for some $\mu \in \mathbf{M}_1(\mathbb{S}^{n-1})$ and measurable $\beta : \mathbb{R}^n \times \mathbb{S}^{n-1} \times (0, \infty) \longrightarrow (0, \infty)$ which satisfies

$$(3.2.7) \quad \begin{aligned} &\int_{(0,\infty)} \frac{r^2 \beta(x, \omega, r)}{1 + r^2} \, dr < \infty \quad \text{and} \\ &\int_{(0,\infty)} \beta(x, \omega, r) \, dr = \infty \quad \text{for all } (x, \omega) \in \mathbb{R}^n \times \mathbb{S}^{n-1}. \end{aligned}$$

Next, determine $F : \mathbb{R}^n \times \mathbb{R}^n \longrightarrow \mathbb{R}^n$ so that

$$(3.2.8) \quad \begin{aligned} &F(x, 0) = 0 \text{ and } F(x, r\omega) = \rho(x, \omega, r)\omega \quad \text{for } (r, \omega) \in (0, \infty) \times \mathbb{S}^{n-1} \\ &\text{where } \rho \text{ is determined by } \int_{\rho(x,\omega,r)}^{\infty} \beta(x, \omega, s) \, ds = \frac{1}{r}. \end{aligned}$$

It is then clear that F is measurable and that $M(x, \cdot) = F(x, \cdot)_* M_\mu$, where

$$(3.2.9) \quad M_\mu(\Gamma) = \int_{\mathbb{S}^{n-1}} \left(\int_{(0,\infty)} \mathbf{1}_\Gamma(r\omega) \frac{dr}{r^2} \right) \mu(d\omega).$$

THEOREM 3.2.10. *Let $\beta : \mathbb{R}^n \times \mathbb{S}^{n-1} \times (0, \infty) \longrightarrow (0, \infty)$ be a measurable map for which (3.2.7) holds, and assume that $\beta(\cdot, \omega, r) \in C^1(\mathbb{R}^n; \mathbb{R})$ for each $(\omega, r) \in \mathbb{S}^{n-1} \times (0, \infty)$. Further, assume that*

$$\int_{(0,\infty)} \frac{r^2 \beta(x, \omega, r)}{1 + r^2} \, dr$$

and, for each $R \in (0, \infty)$,

$$\int_{(0,R]} \frac{\left(\int_{(r,\infty)} |\mathrm{grad}_x \beta(\cdot, \omega, s)| \, ds \right)^2}{\beta(x, \omega, r)} \, dr$$

are bounded uniformly in $(x, \omega) \in \mathbb{R}^n \times \mathbb{S}^{n-1}$. Finally, assume that

$$\lim_{r \searrow 0} \int_{(0,r)} s^2 \beta(x, \omega, r) \, ds = 0 = \lim_{R \nearrow \infty} \int_{[R,\infty)} \beta(x, \omega, s) \, ds$$

*uniformly in $(x, \omega) \in \mathbb{R}^n \times \mathbb{S}^{n-1}$. If $F : \mathbb{R}^n \times \mathbb{R}^n \longrightarrow \mathbb{R}^n$ is defined as in (3.2.8), then for any $\mu \in \mathbf{M}_1(\mathbb{S}^{n-1})$, F satisfies **(H2)** with $n' = n$ and $M' = M_\mu$, the Lévy measure in (3.2.9).*

PROOF: First note that because

$$\lim_{R \nearrow \infty} \sup_{(x,\omega) \in \mathbb{R}^n \times \mathbb{S}^{n-1}} \int_{[R,\infty)} \beta(x,\omega,s)\, ds = 0,$$

$\alpha(R) \equiv \sup_{x \in \mathbb{R}^n} \{ |F(x,y)| : |y| \le R \} < \infty$ for each $R \in (0,\infty)$. At the same time, because

$$\frac{1 + |F(x,R\omega)|^2}{|F(x,R\omega)|^2} \int_{(0,\infty)} \frac{s^2}{1+s^2} \beta(x,\omega,s)\, ds \ge \frac{1}{R},$$

we know that $\alpha(R) \longrightarrow 0$ as $R \searrow 0$. Hence, since (cf. (3.2.6)) $F(x, \cdot)_* M_\mu = M(x, \cdot)$, it follows that

$$\sup_{x \in \mathbb{R}^n} \int_{|y| \le R} |F(x,y)|^2\, M_\mu(dy) \le \sup_{(x,\omega) \in \mathbb{R}^n \times \mathbb{S}^{n-1}} \int_{(0,\alpha(R)]} s^2 \beta(x,\omega,s)\, ds$$

$$\le \left(1 + \alpha(R)^2\right) \sup_{(x,\omega) \in \mathbb{R}^n \times \mathbb{S}^{n-1}} \int_{(0,\alpha(R)]} \frac{s^2}{1+s^2} \beta(x,\omega,s)\, ds$$

is finite for each $R > 0$ and tends to 0 as $R \searrow 0$. Finally, since $F(x,r\omega) = \rho(x,\omega,r)\omega$ and

$$\int_{(\rho(x,\omega,r),\infty)} \beta(x,\omega,s)\, ds = \frac{1}{r},$$

we know that

$$|F(x_1,r\omega) - F(x_0,r\omega)| \le |x_1 - x_0| \int_0^t |\mathrm{grad}_{x_t} \rho(\,\cdot\,,\omega,r)|\, dt,$$

where $x_t = (1-t)x_0 + tx_1$, and that

$$\mathrm{grad}_x \rho(\,\cdot\,,\omega,r) = \frac{\int_{(\rho(x,\omega,r),\infty)} \mathrm{grad}_x \beta(\,\cdot\,,\omega,s)\, ds}{\beta(x,\omega,\rho(x,\omega,r))}.$$

Hence, by once again using $F(x, \cdot)_* M_\mu = M(x, \cdot)$, we see that

$$\int_{|y| \le R} |F(x_1,y) - F(x_0,y)|^2\, M_\mu(dy)$$

$$\le |x_1 - x_0|^2 \sup_{(x,\omega) \in \mathbb{R}^n \times \mathbb{S}^{n-1}} \int_{(0,R]} \frac{\left(\int_{(r,\infty)} |\mathrm{grad}_x \beta(\,\cdot\,,\omega,s)|\, ds\right)^2}{\beta(x,\omega,r)}\, dr. \quad \square$$

COROLLARY 3.2.11. *Let* $\beta : \mathbb{R}^n \times \mathbb{S}^{n-1} \times (0, \infty) \longrightarrow (0, \infty)$ *satisfy the conditions in Theorem 3.2.10,* $\mu \in \mathbf{M}_1(\mathbb{S}^{n-1})$, *and* $n' \geq 2$ *be given. Then there exists a measurable* $F : \mathbb{R}^n \times \mathbb{R}^{n'} \longrightarrow \mathbb{R}^n$ *satisfying the conditions in* **(H2)** *such that* $M(x, \cdot) = F(x, \cdot)_* M'$, *where* $x \rightsquigarrow M(x, \cdot)$ *is given by (3.2.6) and* $M'(dy') = \mathbf{1}_{\mathbb{R}^{n'} \setminus \{0\}}(y)|y|^{-n-1} dy$.

PROOF: Using Lemma 3.2.5, find a measurable $f : \mathbb{R}^{n'} \longrightarrow \mathbb{R}^n$ so that (cf. (3.2.9)) $M_\mu = f_* M'$. Next, using Theorem 3.2.10, construct $F : \mathbb{R}^n \times \mathbb{R}^n \longrightarrow \mathbb{R}^n$ so that **(H2)** is satisfied and $M(x, \cdot) = F(x, \cdot)_* M_\mu$. Finally, replace F by $(x, y') \in \mathbb{R}^n \times \mathbb{R}^{n'} \longmapsto F(x, f(y')) \in \mathbb{R}^n$. □

§3.2.3. Exercises

EXERCISE 3.2.12. In order to develop a feeling for the applicability of Corollary 3.2.11, consider the case when $\beta(x, \omega, r) = \gamma(x, \omega) r^{-\lambda(x,\omega)}$, where $\gamma \in C^{1,0}(\mathbb{R}^n \times \mathbb{S}^{n-1}; (0,1])$ and $\lambda \in C^{1,0}(\mathbb{R}^n \times \mathbb{S}^{n-1}; [\lambda_1, \lambda_2])$ for some $1 < \lambda_1 < \lambda_2 < 3$. Show that the hypotheses of Theorem 3.2.10 hold if

$$(x, \omega) \rightsquigarrow \frac{|\text{grad}_x \gamma(\cdot, \omega)|^2}{\gamma(x, \omega)} + \gamma(x, \omega)|\text{grad}_x \lambda(\cdot, \omega)|^2$$

is uniformly bounded. From an analytic standpoint, especially the perspective of pseudodifferential operators, this example is interesting because the operator L given by (cf. (3.2.6))

$$L\varphi(x) = \int_{\mathbb{R}^n} \left(\varphi(x + y) - \varphi(x) - \mathbf{1}_{(0,1]}(|y|)(y, \text{grad}_x \varphi)_{\mathbb{R}^n} \right) M(x, dy)$$

is a pseudodifferential operator which, in general, will fail to have a well defined *principal part*. Indeed, the singularity of the operator is determined by the exponent λ, which can change from point to point. Thus, it is not clear that the standard theory of pseudodifferential operators would allow one to reach the conclusions (eg. Theorem 4.1.5) which we will draw, on the basis of Itô's construction, in the next chapter. Nonetheless, devotees of pseudodifferential technology have been at work on generators of Markov processes, and a good introduction to their results can be found in [17]

3.3 Some Examples to Keep in Mind

As we said in Remark 3.1.21 above, the meaning of the integral in (3.1.22) is not given by either Riemann or Lebesgue. In order to drive this point home, we give here some examples which, it is hoped, will put the reader on his guard about taking integral equations like (3.1.22) too casually.

Throughout this section, $n = n'$ and $M' = 0$. Remember that the distribution of p' under $\mathbb{P}^0 = \mathbb{P}^{(I,0,0)}$ is that of a Lévy process whose increments over a time interval of length t are centered Gaussian random variables with covariance $tI_{\mathbb{R}^n}$. That is, the $\mathbb{P}^0$-distribution of p is that of Brownian motion. In particular, we may and will think of $\mathbb{P}^0$ being a probability measure on $C\big([0,\infty);\mathbb{R}^n\big)$.

§3.3.1. **The Ornstein–Uhlenbeck Process:** The best understood, nontrivial example of the sort treated in Exercise 3.1.32 was introduced by Ornstein and Uhlenbeck in an attempt to rationalize the physically unacceptable fact the Brownian paths are nowhere differentiable. Namely, they looked at the process, known ever since as the Ornstein–Uhlenbeck process, which is the solution to an integral equation of the form

$$(3.3.1) \qquad X(t,x,p') = x + p'(t) - \beta \int_0^t X(\tau,x,p')\,d\tau.$$

By Exercise 3.1.32, we know that (3.3.1) has a unique solution $X(\,\cdot\,,x,p')$ for each $x \in \mathbb{R}^n$ and $p' \in C\big([0,\infty);\mathbb{R}^n\big)$ and that $(x,p') \rightsquigarrow X(\,\cdot\,,x,p')$ is continuous. Further, if p' is smooth, then, after writing (3.3.1) as a differential equation, it is an easy matter to check that

$$(3.3.2) \qquad X(t,x,p') = e^{-\beta t}x + p'(t) - \beta \int_0^t e^{\beta(\tau-t)}p'(\tau)\,d\tau.$$

Hence, because both sides of (3.3.1) are continuous with respect to p', we now know that (3.3.1) holds for all $p' \in C\big([0,\infty);\mathbb{R}^n\big)$.

The reason why the Ornstein–Uhlenbeck process is so well understood is that an explicit expression for the $\mathbb{P}^0$-distribution of $p' \rightsquigarrow X(t,x,p')$ is readily available. One way to find that distribution is to first integrate the right hand side of (3.3.2) by parts to get

$$(3.3.3) \qquad X(t,x,p') = e^{-\beta t}\big(x + p'(0)\big) + \int_0^t e^{\beta(\tau-t)}\,dp'(\tau),$$

where the integral on the right is taken in the sense of Riemann–Stieltjes. (The justification for both the integration by parts as well as the integral on the right hand side of (3.3.3) can be found in Theorem 1.2.7 of [34].) Given (3.3.3), use a Riemann approximation to evaluate the integral on the right of (3.3.2). That is,

$$\int_0^t e^{\beta\tau}\,dp'(\tau) = \lim_{\ell\to\infty} \sum_{k=0}^{\ell-1} e^{\beta\frac{k}{\ell}}\Big(p'\big(\tfrac{(k+1)t}{\ell}\big) - p'\big(\tfrac{kt}{\ell}\big)\Big),$$

and, under $\mathbb{P}^0$, the random variables $\{p'(\frac{(k+1)t}{\ell}) - p'(\frac{kt}{\ell}) : 0 \le k < \ell\}$ are mutually independent, centered Gaussians with covariance $\frac{t}{\ell}I_{\mathbb{R}^n}$. Hence the preceding sum under $\mathbb{P}^0$ is a centered Gaussian with covariance

$$\left(\ell t \sum_{k=0}^{\ell-1} e^{\beta \frac{2kt}{\ell}} \right) I_{\mathbb{R}^n},$$

and so the $\mathbb{P}^0$-distribution of $p' \rightsquigarrow \int_0^t e^{\beta\tau}\, dp'(\tau)$ is that of a centered Gaussian with covariance

$$\lim_{\ell \to \infty} \left(\ell t \sum_{k=0}^{\ell} e^{\beta \frac{2kt}{\ell}} \right) I_{\mathbb{R}^n} = \left(\int_0^t e^{2\beta\tau}\, d\tau \right) I_{\mathbb{R}^n} = \frac{e^{2\beta t} - 1}{2\beta} I_{\mathbb{R}^n}.$$

Finally, using this together with (3.3.2), we arrive at the conclusion that *the $\mathbb{P}^0$-distribution of $p' \rightsquigarrow X(t, x, p')$ is that of a Gaussian with mean $e^{-\beta t}x$ and covariance $\frac{1-e^{-2\beta t}}{2\beta}I_{\mathbb{R}^n}$.* In particular, if $\beta > 0$, this shows that, as $t \to \infty$ and independent of $x \in \mathbb{R}^n$, the $\mathbb{P}^0$-distribution of $p' \rightsquigarrow X(t, x, p')$ becomes that of a centered Gaussian with covariance $(2\beta)^{-1}$. Equivalently, for the reader who knows what it means, we have just shown that, when $\beta > 0$, the Ornstein–Uhlenbeck process is ergodic and has the centered Gaussian with covariance $(2\beta)^{-1}I$ as its invariant measure.

§3.3.2. Bachelier's Model: The preceding and Exercise 3.1.32 may give the impression that the interpretation of (3.1.22) given by Itô's prescription is always the one obtained by interpreting the $dp'(t)$-integral in (3.1.22) as a Riemann–Stieltjes integral when p' is smooth and then extending the solution by continuity for generic p''s. Thus, it is important to dispel that misconception as soon as possible. Indeed, the situation in Exercise 3.1.32 is the exception and not the rule. In order to drive this point home, we will now discuss an example in which Itô's interpretation is different from Riemann–Stieltjes's even for smooth p''s. Thus, the distinction between his theory and classical theories is inherent and not just an artifact.

The example I have in mind appeared in Bachelier's now famous thesis[3] [1] in which he was attempting to give a mathematical model for price fluctuations on the Paris Bourse. In Bachelier's model, the price of a stock evolves in such a way that its marginal fluctuation over disjoint short time intervals will be mutually independent and that the marginal change over a given short time interval will be a centered Gaussian with variance equal to the length of the interval. That is, if $X(t)$ denotes the price of the stock at

---

[3] His thesis was written in 1900 and directed by no less a figure than H. Poincaré, who apparently had to exercise some of his formidable influence to get it accepted.

time t, then Bachelier's model says that the marginal increment $\frac{dX(t)}{X(t)}$ is the increment of an $\mathbb{R}$-valued Brownian motion.[4] Hence, one way to interpret Bachelier is to think of the stock price as a function of the Brownian motion and to say that, as such, it satisfies

$$\frac{dX(t,x,p')}{X(t,x,p')} = dp'(t) \quad \text{where } x \text{ is the price at time } 0.$$

Alternatively, writing this in integral form, he was saying

$$(3.3.4) \qquad X(t,x,p') = x + \int_0^t X(\tau,x,p')dp'(\tau).$$

The question now is how one should interpret (3.3.4). One way, and this is the one which Bachelier himself seems to have chosen, is to say that Riemann gives the right answer when p' is smooth and $p' \rightsquigarrow X(\,\cdot\,,x,p')$ should be a continuous function of locally uniform convergence. To be precise, when p' is continuously differentiable and one interprets the integral on the right hand side of (3.3.4) *à la* Riemann, it is clear that $X(t,x,p') = xe^{p'(t)}$. Thus, if $p' \rightsquigarrow X(\,\cdot\,,x,p')$ is going to be continuous with respect to uniform convergence on compacts, there is no choice but to take $X(t,x,p') = xe^{p'(t)}$ for $\mathbb{P}^0$-almost every p', and that is what Bachelier did.

Reasonable as the preceding may be, it is *not* the interpretation given by Itô's theory. Namely, take $n = 1 = n'$, $\sigma(x) = x$, $b \equiv 0$, and $F = 0$ in **(H1)**–**(H3)**, and apply the prescription given in **(H4)**. To carry this out in detail, define

$$S_N(t,\epsilon) = \left\{p' : |p'(\tau) - p'(\sigma)| \le \epsilon \text{ for } 0 \le \sigma \le \tau \le t \text{ with } \tau - \sigma \le 2^{-N}\right\}$$

for $t \in [0,\infty)$ and $\epsilon \in \left(0,\frac{1}{2}\right)$. Because our p''s are continuous, $\mathbb{P}^{M'}\big(S_N(t,\epsilon)\big) \longrightarrow 1$ as $N \to \infty$. Moreover, if $p' \in S_N(t,\epsilon)$, then it is an easy matter to check that

$$X^N(t,x,p') = x \prod_{m=0}^{[2^N t]} \left(1 + \Delta_m^N p'(t)\right),$$

where

$$(3.3.5) \qquad \Delta_m^N p'(t) \equiv p'\big(t \wedge (m+1)2^{-N}\big) - p'\big(t \wedge m2^{-N}\big).$$

---

[4] This way of phrasing it is grossly unfair to Bachelier. Indeed, A. Einstein's model of Brownian motion [9] did not appear until 1905.

In order to see what happens when $N \to \infty$, use Taylor's theorem to see that

$$\log(1 + \xi) = \xi - \frac{\xi^2}{2} + R(\xi) \quad \text{where } |R(\xi)| \leq \frac{2|\xi|^3}{3} \text{ for } |\xi| \leq \frac{1}{2}.$$

Hence, for $p' \in S_N(t, \epsilon)$,

$$X^N(t, x, p') = x \exp\left(p'(t) - \frac{1}{2}\sum_{m=0}^{\infty} \Delta_m^N p'(t)^2 + E_N(t, p')\right)$$

$$\text{where } |E_N(t, p')| \leq \frac{2\epsilon}{3} \sum_{m=0}^{\infty} \Delta_m^N p'(t)^2.$$

But (cf. Exercise 4.1.11 in [36]),

$$\sum_{m=0}^{\infty} \Delta_m^N p'(t)^2 \longrightarrow t \text{ for } \mathbb{P}^0\text{-almost every } p',$$

and so the preceding leads to the conclusion that

$$\lim_{N \to \infty} X^N(t, x, p') = x e^{p'(t) - \frac{t}{2}} \quad \text{for } \mathbb{P}^{M'}\text{-almost every } p'.$$

In other words, Itô says that the solution to (3.3.4) is $X(t, x, p') = x e^{p'(t) - \frac{t}{2}}$, which is markedly different from what Bachelier got, whether or not p' is smooth.

Without further input from the model, it is impossible to say which of these two is the "correct" one from an economic standpoint. The reason why Itô's is the one adopted by economists is that, in order to "avoid an arbitrage opportunity," economists want $X(t)$ to be a martingale. Be that as it may, from a purely mathematical standpoint, both interpretations make sense and their difference should be sufficient to give any mathematician pause.

§3.3.3. A Geometric Example: Here is another example which displays the same sort of disturbing characteristics as the preceding.

Take $n = 2 = n'$, b and F both identically 0, and

$$\sigma(x) = (1 \vee |x|^2)^{-1} \begin{pmatrix} x_2^2 & -x_1 x_2 \\ -x_1 x_2 & x_1^2 \end{pmatrix}.$$

When $|x| \neq 0$, $\sigma(x)\xi$ is $1 \wedge |x|^2$ times the orthogonal projection of ξ in the direction perpendicular to x. In particular, if p' is smooth and we interpret

$$(*) \qquad X(t, x, p') = x + \int_0^t \sigma(X^N(\tau, x, p')) \, dp'(\tau)$$

à la Riemann, then $d|X(t,x,p')|^2 = 2(X^N(t,x,p'),\sigma(X^N(t,x,p')dp'(t))_{\mathbb{R}^n}$ $= 0$, and so $|X(t,x,p')| = |x|$ for all $t \geq 0$. Thus, if Itô's procedure were some sort of extension to non-smooth p''s of the classical solution when p' is smooth, then $|X(t,x,p'|$ would have to equal $|x|$ for $\mathbb{P}^0$-almost every p', and this is simply not the case. Indeed, for $m2^{-N} \leq t \leq (m+1)2^{-N}$, because $\sigma(X^N(m2^{-N},x,p'))X^N(m2^{-N},x,p') = 0$ (cf. (3.3.5)),

$$|X^N(t,x,p')|^2 - |X^N(m2^{-N},x,p')|^2$$
$$= \left(X^N(t,x,p') + X^N(m2^{-N},x,p'),\sigma(X^N(m2^{-N},x,p'))\Delta_m^N p'(t)\right)_{\mathbb{R}^n}$$
$$= \left(X^N(t,x,p') - X^N(m2^{-N},x,p'),\sigma(X^N(m2^{-N},x,p'))\Delta_m^N p'(t)\right)_{\mathbb{R}^n}$$
$$= \left|\sigma(X^N(m2^{-N},x,p'))\Delta_m^N p'(t)\right|^2.$$

In particular, $t \rightsquigarrow |X^N(t,x,p')|$ is nondecreasing. Hence, if $|x| = 1$, then $\sigma(X^N([t]_N,x,p'))$ is orthogonal projection onto $X^N([t]_N,x,p')^\perp$ and so the $\mathbb{P}^0$-distribution of $Y^N(t,p') \equiv |\sigma(X^N([t]_N,x,p'))\Delta_m^N p'(t)|^2$ is that of the square of a centered Gaussian whose variance is $t - [t]_N$. Furthermore, if $\bar{Y}^N(t,p') = Y^N(t,p') - (t - [t]_N)$, then $\mathbb{E}^{\mathbb{P}^0}[\bar{Y}^N(t) \mid \mathcal{B}_{[t]_N}] = 0$, and so

$$\mathbb{E}^{\mathbb{P}^0}\left[\left(|X^N(t,x,p')|^2 - t\right)^2\right] = \mathbb{E}^{\mathbb{P}^0}\left[\left(\sum_{m=0}^{[2^N t]+1} \bar{Y}^N(t \wedge m2^{-N})\right)^2\right]$$

$$= \sum_{m=0}^{[2^N t]+1} \mathbb{E}^{\mathbb{P}^0}[\bar{Y}_N(t \wedge m2^{-N})^2] \leq 3(2^N t + 2)2^{-4N} \longrightarrow 0$$

as $N \to \infty$. In other words, for $\mathbb{P}^0$-almost every p', $|X(t,x,p')|^2 = 1 + t$, which is *very* different from what we would get if Itô's interpretation of (*) were an extension of a Riemannian interpretation.

REMARK 3.3.6. The last example, and the one preceding it, highlight one of the pitfalls in Itô's theory. Namely, his whole theory derives from the consideration of Lévy processes, and Lévy processes are inextricably tied to the Euclidean coordinates in which they are described. Anything other than a linear changes of coordinates will render any non-deterministic Lévy process unrecognizable. In Chapter 8, we will introduce a variant of Itô's theory which overcomes, at least partially, this problem.

§3.3.4. Exercises

EXERCISE 3.3.7. Here is a generalization of the Ornstein–Uhlenbeck process. Namely, suppose that $\Sigma \in \mathrm{Hom}(\mathbb{R}^{n'}; \mathbb{R}^n)$ and $B \in \mathrm{Hom}(\mathbb{R}^n; \mathbb{R}^n)$ are given, and take $\sigma(x) = \Sigma$, $b(x) = -Bx$, and $F(x, \cdot) = 0$ for $x \in \mathbb{R}^n$. Now apply the construction in **(H4)** (cf. Exercise 3.1.32) with $M' = 0$, and let $p' \in C\big([0, \infty); \mathbb{R}^{n'}\big) \rightsquigarrow X(\cdot, x, p') \in C\big([0, \infty); \mathbb{R}^n\big)$ be the resulting process described in **(G1)**.

(i) Set $A = \Sigma\Sigma^\top$, and show that, for each $(t, x) \in [0, \infty) \times \mathbb{R}^n$, the $\mathbb{P}^0$-distribution of $p' \rightsquigarrow X(t, x, p')$ is that of a $\mathbb{R}^n$-valued Gaussian with mean $e^{-tB}x$ and covariance

$$C(t) \equiv \int_0^t e^{-\tau B} A e^{-\tau B^\top} \, d\tau.$$

(ii) Assume that there is a $\beta > 0$ such that $(\xi, B\xi)_{\mathbb{R}^n} \geq \beta|\xi|^2$ for all $\xi \in \mathbb{R}^n$. Show that $|X(t, x_1, p') - X(t, x_2, p')| \leq e^{-\beta t}|x_1 - x_2|$ and that, for each $x \in \mathbb{R}^n$, as $t \to \infty$ the $\mathbb{P}^0$-distribution of $p' \rightsquigarrow X(t, x, p')$ tends that of a centered, $\mathbb{R}^n$-valued, Gaussian with covariance $C(\infty) \equiv \lim_{t\to\infty} C(t)$.

(iii) Continuing with the assumption in (ii), let γ denote the centered Gaussian measure on $\mathbb{R}^n$ with covariance $C(\infty)$. Show that, for all $t \geq 0$, the $\gamma \times \mathbb{P}^0$-distribution of $(x, p') \rightsquigarrow X(t, x, p')$ is γ. In other words, the $\gamma \times \mathbb{P}^0$-distribution of $(x, p') \rightsquigarrow X(\cdot, x, p')$ is *stationary* in time.

CHAPTER 4

Further Considerations

In Chapter 3, we explained how Itô's method leads to stochastic processes which are the pathspace analog of the integral curves obtained in §2.2 when we integrated affine vector fields on $\mathbf{M}_1(\mathbb{R}^n)$. In this chapter we will point out a couple of the important properties with which Itô's method endows these processes, and, because it is undoubtedly the most important, we begin with the Markov property.

4.1 Continuity, Measurability, and the Markov Property

Under the condition that we can solve the Kolmogorov's backward equation (3.1.28) for a sufficiently rich class of initial data φ, we showed in Corollary 3.1.27 that Itô's method produces a process which is Markov. In this section we will give an entirely different way of checking the Markov property for the processes he constructs, an approach which gives a stronger result and does not require our knowing that (3.1.28) can be solved.

We will be continuing in the setting described at the beginning of §3.1 and will be using $\mathbb{P}_x = \mathbb{P}_x^{(\sigma,b,F,M)}$ to denote the $\mathbb{P}^{M'}$-distribution of $p' \in D\big([0,\infty);\mathbb{R}^{n'}\big) \longmapsto X(\,\cdot\,,x,p') \in D\big([0,\infty);\mathbb{R}^n\big)$.[1] The first goal of this section is to check that $x \rightsquigarrow \mathbb{P}_x$ is measurable in the sense that if $\Phi : D\big([0,\infty);\mathbb{R}^n\big) \longrightarrow [0,\infty]$ is measurable with respect to $\mathcal{B} \equiv \sigma\big(\{p(\tau) : \tau \geq 0\}\big)$, then $x \rightsquigarrow \langle \Phi, \mathbb{P}_x \rangle$ is measurable. Once we have done this, we show that the family $\{\mathbb{P}_x : x \in \mathbb{R}^n\}$ is Markov.

§4.1.1. Continuity and Measurability: We begin by remarking that, for each $N \geq 0$ and $p' \in D\big([0,\infty);\mathbb{R}^{n'}\big)$, $x \in \mathbb{R}^n \longmapsto X^N(\,\cdot\,,x,p') \in$

---

[1] The reader may justifiably complain that the notation here appears to be inconsistent with that used in Corollary 3.1.27. However, the reason why we are not using $\mathbb{P}_x^L$ here is that, in general, we do not know that the $\mathbb{P}^{M'}$-distribution of $p' \rightsquigarrow X(\,\cdot\,,x,p')$ is uniquely determined by the operator L in (3.1.5). For example, it might have a different distribution if we were to take a different σ such that $a = \sigma\sigma^\top$ or a different F and M' for which $M(x,\,\cdot\,) = F(x,\,\cdot\,)_* M'$. The point made in Corollary 3.1.27 was that, under the hypotheses given there, such changes in σ or F would not lead to a change in the distribution of $p' \rightsquigarrow X(\,\cdot\,,x,p')$. We return to this point in Corollary 4.2.6 and Remark 4.2.7 below.

$D\big([0,\infty);\mathbb{R}^n\big)$ is continuous. Indeed, based on the description given in **(H4)**, one can use induction on $m \in \mathbb{N}$ to see that $x \rightsquigarrow X^N(\,\cdot\,,x,p') \upharpoonright [0, m2^{-N}]$ is continuous. At the same time,

$$\mathbb{P}^{M'}\Big(\big\|X(\,\cdot\,,x_2) - X(\,\cdot\,,x_1)\big\|_{[0,T]} \geq \epsilon\Big)$$

$$\leq \mathbb{P}^{M'}\Big(\big\|X(\,\cdot\,,x_2) - X^N(\,\cdot\,,x_2)\big\|_{[0,T]} \geq \tfrac{\epsilon}{3}\Big)$$

$$+ \mathbb{P}^{M'}\Big(\big\|X^N(\,\cdot\,,x_2) - X^N(\,\cdot\,,x_1)\big\|_{[0,T]} \geq \tfrac{\epsilon}{3}\Big)$$

$$+ \mathbb{P}^{M'}\Big(\big\|X(\,\cdot\,,x_1) - X^N(\,\cdot\,,x_1)\big\|_{[0,T]} \geq \tfrac{\epsilon}{3}\Big).$$

Hence, by Theorem 3.1.20, we find that

$$(4.1.1) \qquad \lim_{\delta \searrow 0} \sup_{\substack{|x_2 - x_1| \leq \delta \\ |x_1| \vee |x_2| \leq R}} \mathbb{P}^{M'}\Big(\big\|X(\,\cdot\,,x_2) - X(\,\cdot\,,x_1)\big\|_{[0,T]} \geq \epsilon\Big) = 0$$

for all $(R,T) \in (0,\infty)^2$ and $\epsilon > 0$. In particular, this proves that, as $y \to x$, $\Phi\big(X(\,\cdot\,,y)\big) \longrightarrow \Phi\big(X(\,\cdot\,,x)\big)$ in $L^1(\mathbb{P}^{M'};\mathbb{R})$ when $\Phi \in C_b\big(D([0,\infty);\mathbb{R}^n);\mathbb{R}\big)$ is (cf. Exercise 2.3.5) $\mathcal{B}$-measurable. Thus (cf. Exercise 3.1.29), we have now proved the first part of the following statement.

THEOREM 4.1.2. *Let $\mathbb{P}_x \in \mathbf{M}_1\big(D([0,\infty);\mathbb{R}^n)\big)$ be the $\mathbb{P}^{M'}$-distribution of the random variable $p' \rightsquigarrow X(\,\cdot\,,x,p')$ described in* **(G1)** *of §3.2.1. Then $x \rightsquigarrow \mathbb{P}_x$ is continuous in the sense that $x \rightsquigarrow \langle \Phi, \mathbb{P}_x \rangle$ is continuous for each $\mathcal{B}$-measurable $\Phi \in C_b\big(D([0,\infty);\mathbb{R}^n);\mathbb{R}\big)$. In particular, $x \rightsquigarrow \langle \Phi, \mathbb{P}_x \rangle$ is measurable if Φ is measurable and either bounded or non-negative. Furthermore, there exists a $C < \infty$ such that, for each $\epsilon \in (0,1]$*

$$\mathbb{P}_x\Big(\|p - x\|_{[0,h]} \geq \epsilon\sqrt{1 + |x|^2}\Big) \leq \frac{Ch}{\epsilon^2}.$$

PROOF: To prove the final assertion, choose $\psi \in C^\infty\big(\mathbb{R}^n;[0,1]\big)$ so that $\psi(0) = 1$ and $\psi \equiv 0$ off of $B_{\mathbb{R}^n}(0,1)$, and, for $\epsilon \in (0,1]$ and $x \in \mathbb{R}^n$, define $\psi_{\epsilon,x} : \mathbb{R}^n \longrightarrow [0,1]$ by

$$\psi_{\epsilon,x}(y) = \psi\left(\frac{y - x}{\epsilon\sqrt{1 + |x|^2}}\right), \qquad y \in \mathbb{R}^n.$$

Proceeding as in the proof of Lemma 3.1.24, one can easily check that $\epsilon^2 \|L\psi_{\epsilon,x}\|_u$ is bounded, independent of $\epsilon \in (0,1]$ and $x \in \mathbb{R}^n$, by a constant $C < \infty$. Next, define $\zeta_\epsilon : D([0,\infty);\mathbb{R}^n) \longrightarrow [0,\infty]$ by

$$\zeta_\epsilon(p) = \inf\big\{t \geq 0 : \|p - p(0)\|_{[0,t]} \geq \epsilon\sqrt{1 + |p(0)|^2}\big\}.$$

Then, by Theorem 3.1.26 and Doob's stopping time theorem,

$$\mathbb{P}_x\Big(\|p-x\|_{[0,h]} \ge \epsilon\sqrt{1+|x|^2}\Big) = \mathbb{P}_x\big(\zeta_\epsilon \le h\big) \le 1 - \mathbb{E}^{\mathbb{P}_x}\big[\psi_{\epsilon,x}\big(p(h\wedge\zeta_\epsilon)\big)\big]$$

$$= -\mathbb{E}^{\mathbb{P}_x}\left[\int_0^{h\wedge\zeta_\epsilon} L\psi_{\epsilon,x}\big(p(\tau)\big)\, d\tau\right] \le \frac{Ch}{\epsilon^2},$$

which is exactly what had to be proved. $\square$

REMARK 4.1.3. It should be admitted that our treatment of these continuity questions yields conclusions which are weaker than those which would have been available if we had worked harder. In particular, it is possible to show that there is a measurable map $(x,p') \in \mathbb{R}^n \times D\big([0,\infty);\mathbb{R}^n\big) \longmapsto \tilde{X}(\,\cdot\,,x,p') \in D\big([0,\infty);\mathbb{R}^n\big)$ with the properties that $x \rightsquigarrow X(\,\cdot\,,x,p')$ is continuous for $\mathbb{P}^{M'}$-almost every $p' \in D\big([0,\infty);\mathbb{R}^{n'}\big)$ and that $\tilde{X}(\,\cdot\,,x) = X(\,\cdot\,,x)$ $\mathbb{P}^{M'}$-almost surely for each $(t,x) \in [0,\infty) \times \mathbb{R}^n$. In fact, $x \rightsquigarrow \tilde{X}(\,\cdot\,,x)$ will be $\mathbb{P}^{M'}$-almost surely Hölder continuous of any order strictly less than 1. The proof of this sort of results can be based on either classical Sobolev embedding theory or on the variant of that theory provided by the multidimensional analog of Kolmogorov's continuity criterion. At least for the case when $M' = 0$, details can be found in H. Kunita's book [20].

§4.1.2. The Markov Property: Our aim in this subsection is to prove that the family $\{\mathbb{P}_x : x \in \mathbb{R}^n\}$ is Markov. However, before stating the version of the Markov property at which we are aiming, we need to discuss the concept of a *stopping time*. For our purposes, a stopping time relative to[2] $\{\mathcal{B}_t : t \ge 0\}$ will be a random variable $\zeta : D\big([0,\infty);\mathbb{R}^n\big) \longrightarrow [0,\infty]$ with the property that $\{p : \zeta(p) < t\} \in \mathcal{B}_t$ for every $t \ge 0$. Actually, this definition is slightly different from the one (cf. §7.1 in [36]) which is usually adopted. The usual definition is more restrictive in that it requires that $\{\zeta \le t\} \in \mathcal{B}_t$ for all $t \ge 0$. As is discussed in Exercise 7.1.31 of [36], the best comparison of the two definitions is provided by the characterization: $\{\zeta < t\} \in \mathcal{B}_t$ for all $t \ge 0$ if and only if $\{\zeta \le t\} \in \mathcal{B}_{t+} \equiv \bigcap_{\epsilon>0} \mathcal{B}_{t+\epsilon}$ for all $t \ge 0$.

Associated with a given stopping time ζ is the σ-algebra $\mathcal{B}_\zeta$ of $A \in \mathcal{B}$ with the property that $A \cap \{\zeta < t\} \in \mathcal{B}_t$ for all $t \ge 0$. Elementary, but vital, facts about stopping times and their associated σ-algebras are (cf. Lemma 3.3.2 in [36])

(4.1.4)
$$p \rightsquigarrow \zeta(p) \ \& \ p \rightsquigarrow p(\zeta) \equiv p\big(\zeta(p)\big) \text{ are } \mathcal{B}_\zeta\text{-measurable on } \{\zeta < \infty\},$$
$$\zeta_1 \le \zeta_2 \implies \mathcal{B}_{\zeta_1} \subseteq \mathcal{B}_{\zeta_2},$$
$$\zeta_1 \vee \zeta_2, \quad \zeta_1 \wedge \zeta_2, \quad \zeta_1 + \zeta_2, \quad \text{and}$$
$$\zeta_F^\zeta \equiv \inf\big\{t \ge \zeta : \overline{\{p(\tau) : \tau \in [0,t]\}} \cap F \ne \emptyset\big\} \text{ are stopping times,}$$

---

[2] Recall that $\mathcal{B}_t = \sigma(\{p(\tau) : \tau \in [0,t]\})$.

where the ζ's stand for various generic stopping times and F is a closed subset of $\mathbb{R}^n$.

Finally, for fixed $s \geq 0$, define the time shift map $\Sigma_s : D\big([0,\infty);\mathbb{R}^n\big) \longrightarrow D\big([0,\infty);\mathbb{R}^n\big)$ so that $\Sigma_s p(t) = p(s+t)$, and for a stopping time ζ set $\Sigma_\zeta p = \Sigma_{\zeta(p)} p$ if $\zeta(p) < \infty$.

THEOREM 4.1.5. *Let ζ be a stopping time and $\Phi : D\big([0,\infty);\mathbb{R}^n\big) \longrightarrow \mathbb{R}$ a $\mathcal{B}$-measurable function which is either bounded or non-negative. Then*

$$\mathbb{E}^{\mathbb{P}_x}\big[\Phi \circ \Sigma_\zeta \,\big|\, \mathcal{B}_\zeta\big] = \mathbb{E}^{\mathbb{P}_{p(\zeta)}}[\Phi] \quad \mathbb{P}_x\text{-almost surely on } \{\zeta < \infty\}.$$

In particular, if $P_t(x, \cdot)$ is the $\mathbb{P}_x$-distribution of $p \rightsquigarrow p(t)$, then $(t,x) \rightsquigarrow P_t(x, \cdot)$ is a transition probability function and $\mathbb{P}_x$ is the Markov process determined by this transition probability function. That is, (1.4.1) holds when $\mathbb{P} = \mathbb{P}_x$, $\mu_0 = \delta_x$, and $(t,x) \rightsquigarrow P_t(x, \cdot)$ is the one here.

PROOF: Without loss in generality, we will assume that $\zeta \leq T$ for some $T \in [0,\infty)$. Further, by the considerations in Exercise 2.3.5, it suffices for us to handle Φ's which are bounded and continuous. Thus, suppose that $\zeta : D\big([0,\infty);\mathbb{R}^n\big) \longrightarrow [0,T]$ is a stopping time, and set $\zeta'(p') = \zeta\big(X(\,\cdot\,,x,p')\big)$. Because $p' \rightsquigarrow X(s,x,p')$ is $\overline{\mathcal{B}'_s}$-measurable for each s, it is easy to check that ζ' is a stopping time relative to $\{\overline{\mathcal{B}'_t} : t \geq 0\}$. Hence it suffices for us show that for any $\zeta' : D\big([0,\infty);\mathbb{R}^{n'}\big) \longrightarrow [0,T]$ which is a stopping time relative to $\{\overline{\mathcal{B}'_t} : t \geq 0\}$, any bounded, continuous, $\mathcal{B}$-measurable $\Phi : D\big([0,\infty);\mathbb{R}^n\big) \longrightarrow \mathbb{R}$, and any $A \in \overline{\mathcal{B}'_{\zeta'}}$

$$(*) \qquad \mathbb{E}^{\mathbb{P}^{M'}}\big[\Phi \circ X(\,\cdot\, + \zeta',x), A\big] = \int_A \mathbb{E}^{\mathbb{P}^L_{X(\zeta'(p'),x,p')}}[\Phi]\, \mathbb{P}^{M'}(dp').$$

The key to our proof of $(*)$ is the observation that, for any $N \in \mathbb{N}$,

$$(4.1.6) \qquad X^N(\,\cdot\, + m2^{-N},x,p') = X^N\big(\,\cdot\,, X^N(m2^{-N},x,p'), \delta_{m2^{-N}}p'\big),$$

where the *increment map* $\delta_s : D\big([0,\infty);\mathbb{R}^{n'}\big) \longrightarrow D\big([0,\infty);\mathbb{R}^{n'}\big)$ is defined so that $\delta_s p'(t) = p'(s+t) - p'(s)$. Hence, because $\mathcal{B}'_s$, and therefore also $\overline{\mathcal{B}'_s}$, is $\mathbb{P}^{M'}$-independent of $\sigma\big(\{\delta_s p'(\tau) : \tau \geq 0\}\big)$ and $\mathbb{P}^{M'}$ is again the distribution of $p' \rightsquigarrow \delta_s p'$,

$$\mathbb{E}^{\mathbb{P}^{M'}}\big[\Phi \circ X^N(\,\cdot\, + m2^{-N},x), A\big]$$
$$= \int_A \left(\int \Phi \circ X^N\big(\,\cdot\,, X^N(m2^{-N},x,p'),q'\big)\, \mathbb{P}^{M'}(dq')\right) \mathbb{P}^{M'}(dp')$$

for any $A \in \overline{\mathcal{B}'_{m2^{-N}}}$.

Now let $\zeta' : D([0,\infty); \mathbb{R}^{n'}) \longrightarrow [0,T]$ be a stopping time relative to $\{\overline{\mathcal{B}}_t' : t \geq 0\}$, and suppose the $\Phi \in C_b(D([0,\infty); \mathbb{R}^n); \mathbb{R})$ is $\mathcal{B}$-measurable. For each $N \in \mathbb{N}$ and $p' \in D([0,\infty); \mathbb{R}^{n'})$, set $\zeta^N(p') = m2^{-N}$ if $(m-1)2^{-N} \leq \zeta'(p') < m2^{-N}$. It is an easy matter to check that $A_m^N \equiv A \cap \{\zeta^N = m2^{-N}\} \in \overline{\mathcal{B}}_{m2^{-N}}'$ for all $m \geq 1$. Hence, by the preceding paragraph,

$$\mathbb{E}^{\mathbb{P}^{M'}}\big[\Phi \circ X^N(\,\cdot\, + \zeta^N, x), A\big] = \sum_{m=1}^{\infty} \mathbb{E}^{\mathbb{P}^{M'}}\big[\Phi \circ X^N(\,\cdot\, + m2^{-N}, x), A_m^N\big]$$

$$= \sum_{m=1}^{\infty} \int_{A_m^N} \left(\int \Phi \circ X^N\big(\,\cdot\,, X^N(m2^{-N}, x, p'), q'\big) \, \mathbb{P}^{M'}(dq') \right) \mathbb{P}^{M'}(dp')$$

$$= \int_A \left(\int \Phi \circ X^N\big(\,\cdot\,, X^N(\zeta^N(p'), x, p'), q'\big) \, \mathbb{P}^{M'}(dq') \right) \mathbb{P}^{M'}(dp').$$

By Theorem 3.1.20 and the fact that $T \geq \zeta^N(p') \searrow \zeta'(p')$ for each p',

$$\lim_{N \to \infty} \mathbb{E}^{\mathbb{P}^{M'}}\big[\Phi \circ X^N(\,\cdot\, + \zeta^N, x), A\big] = \mathbb{E}^{\mathbb{P}^{M'}}\big[\Phi \circ X(\,\cdot\, + \zeta', x), A\big].$$

At the same time, again by Theorem 3.1.20, as $N \to \infty$, $X^N(\zeta^N(p'), x, p')$ $\longrightarrow X(\zeta'(p'), x, p')$ for $\mathbb{P}^{M'}$-almost every p' and

$$\left| \int \Phi \circ X^N\big(\,\cdot\,, X^N(\zeta^N(p'), x, p'), q'\big) \, \mathbb{P}^{M'}(dq') - \mathbb{E}^{\mathbb{P}_{X^N(\zeta^N(p'), x, p')}}[\Phi] \right| \longrightarrow 0$$

if $X^N(\zeta^N(p'), x, p') \longrightarrow X(\zeta'(p'), x, p')$. Finally, because $x \rightsquigarrow \mathbb{E}^{\mathbb{P}_x}[\Phi]$ is continuous, we have now completed the proof of (*).

Given what we have already shown, the final assertion is completely standard. $\square$

REMARK 4.1.7. As we said in the introduction to this section, the argument with which we have just proved the Markov property here is somewhat different from the one which we used in Corollary 3.1.27. Nonetheless, they both turn on uniqueness. There the uniqueness was of the conditional distribution of $p \rightsquigarrow p(s+t)$ given $\mathcal{B}_s$. Here the uniqueness is the almost sure existence of the limit of $\{X^N(\,\cdot\,, x, p')\}_1^{\infty}$.

§4.1.3. Exercises

EXERCISE 4.1.8. Using the Markov property, the last part of Theorem 4.1.2, and part (iii) of Exercise 3.1.31, show that, for each $T \in [0,\infty)$ and $\epsilon > 0$,

$$\lim_{h \to 0} \sup_{\substack{s \in [0,T] \\ x \in \mathbb{R}^n}} \mathbb{P}_x \left(\sup_{t \in [0,h]} |p(s+t) - p(s)| \geq \epsilon \sqrt{1 + |x|^2} \right) = 0.$$

4.2 Differentiability

In the previous section we showed that Itô's method produces maps $x \in \mathbb{R}^n \longmapsto \mathbb{P}_x \in \mathbf{M}_1\big(D\big([0,\infty);\mathbb{R}^n\big)\big)$ which are continuous and so the transition probability function $(t,x) \in [0,\infty) \times \mathbb{R}^n \longmapsto P_t(x,\cdot) \in \mathbf{M}_1(\mathbb{R}^n)$ is continuous. the *Markov semigroup* $\{\mathbf{P}_t : t \geq 0\}$ given by[3]

$$(4.2.1) \quad \mathbf{P}_t\varphi(x) = \int_{\mathbb{R}^n} \varphi(y)\, P_t(x,dy), \quad (t,x) \in [0,\infty) \times \mathbb{R}^n \ \& \ \varphi \in B(\mathbb{R}^n;\mathbb{R}).$$

leaves $C_\mathrm{b}(\mathbb{R}^n;\mathbb{R})$ invariant. Our goal in this section is to find conditions under which $\{\mathbf{P}_t : t \geq 0\}$ will leave $C_\mathrm{b}^k(\mathbb{R}^n;\mathbb{R})$ invariant for k's other than 0. Of particular interest to us will be $k = 2$. Indeed, when $C_\mathrm{b}^2(\mathbb{R}^n;\mathbb{R})$ is $\{\mathbf{P}_t : t \geq 0\}$-invariant we will be able to show (cf. Corollary 4.2.6 below) that the measure $\mathbb{P}_x = \mathbb{P}_x^{(\sigma,b,F,M')}$ is uniquely determined by the fact that it solves the martingale problem for the operator L in (3.1.5) on $C_\mathrm{c}^2(\mathbb{R}^n;\mathbb{R})$.

§4.2.1. First Derivatives: Throughout this subsection, we will be replacing **(H2)** and **(H3)** of §3.1.2 by

(H2)[1] $F(\,\cdot\,,y') \in C^1(\mathbb{R}^n;\mathbb{R})$ for each $y' \in \mathbb{R}^n \setminus \{0\}$ and

(a) $$\lim_{r \searrow 0} \sup_{x\in\mathbb{R}^n} \frac{1}{1+|x|^2} \int_{B_{\mathbb{R}^{n'}}(0,r)} |F(x,y')|^2\, M'(dy') = 0,$$

(b) $$\sup_{x\in\mathbb{R}^n} \int_{\mathbb{R}^n} \left(|F(x,y')|^{2q} + \left\| \frac{\partial F}{\partial x}(x,y') \right\|_{\mathrm{op}}^{2q} \right) M'(dy') < \infty$$

      for each $q \in [1,\infty)$.

(H3)[1] $\sigma : \mathbb{R}^n \longrightarrow \mathrm{Hom}(\mathbb{R}^{n'};\mathbb{R}^n)$ and $b : \mathbb{R}^n \longrightarrow \mathbb{R}^n$ are continuously differentiable functions with bounded first order derivatives,

and under these conditions we want to show that, for each $q \in [1,\infty)$, there is a $\beta_q \in [0,\infty)$ such that

$$\sup_{N\geq 0} \mathbb{E}^{\mathbb{P}^{M'}}\left[\left\| \frac{\partial X^N(t,\,\cdot\,)}{\partial x}\big(X^N(t,x)\big) \right\|_{\mathrm{H.S.}}^{2q} \right] \leq n^{2q} e^{\beta_q t}.$$

Critical for our analysis are the estimates contained in the following lemma.

---

[3] We use $B(E;\mathbb{R})$ to denote the space of bounded measurable functions on a measurable space E.

LEMMA 4.2.2. *For any $q \in [1, \infty)$ and $\xi \in \mathbb{S}^{n-1}$,*

$$\mathbb{E}^{\mathbb{P}^{M'}}\left[\left\|(\xi, w(\,\cdot\,, p'))_{\mathbb{R}^n}\right\|_{[0,T]}^{2q}\right] = T^q \mathbb{E}^{\mathbb{P}^{M'}}\left[\|w(\,\cdot\,, p')\|_{[0,1]}^{2q}\right]$$

$$\leq T^q \left(\frac{2q}{2q-1}\right)^{2q} \frac{1}{\sqrt{2\pi}} \int_{\mathbb{R}} |y|^{2q} e^{-\frac{y^2}{2}} \, dy.$$

In addition, given a measurable $f : \mathbb{R}^n \longrightarrow \mathbb{R}$ which vanishes in a neighborhood of the origin and is M'-integrable, set

$$Y(t, p') = \int_{\mathbb{R}^{n'}} f(y') \, \eta(t, dy', p') \text{ and } \bar{Y}(t, p') = Y(t, p') - t \int_{\mathbb{R}^n} f(y') \, M'(dy')$$

for $(t, p') \in [0, \infty) \times D([0, \infty); \mathbb{R}^{n'})$. Then, for each $q \in [1, \infty)$ there is a universal $C_q \in (0, \infty)$ with the property that

$$\mathbb{E}^{\mathbb{P}^{M'}}\left[\|\bar{Y}\|_{[0,T]}^{2q}\right] \leq C_q \left[\left(T \int_{\mathbb{R}^{n'}} |f(y')|^2 \, M'(dy')\right)^q \right.$$

$$\left. + T \int_{\mathbb{R}^{n'}} |f(y')|^{2q} \, M'(dy')\right].$$

PROOF: The first part is very easy. Namely, the initial equality is an application of Brownian scaling invariance, and the subsequent inequality is an application of Doob's inequality and the fact that the $P^{M'}$-distribution of $p' \rightsquigarrow (\xi, w(1, p'))_{\mathbb{R}^n}$ is that of a standard normal random variable.

Turning to the second part, note that, without loss in generality, we may and will assume that f is bounded, in which case (cf. Lemma 3.1.2) the $\mathbb{P}^{M'}$-distribution of $p' \rightsquigarrow Y(\,\cdot\,, p')$ is that of an $\mathbb{R}$-valued compound Poisson process with intensity $M'(\{y' : f(y') \neq 0\}) < \infty$ and compactly supported jump distribution. In particular, this means that $t \rightsquigarrow \mathbb{E}^{P^{M'}}[|Y(t)|^{2q}]$ is bounded on compact intervals for every $q \in [1, \infty)$. In addition (cf. Exercise 2.4.15), we know that $(\bar{Y}(t), \mathcal{B}'_t, \mathbb{P}^{M'})$ is a martingale and that, for any $\varphi \in C_b^2(\mathbb{R}; \mathbb{R})$,

$$\varphi(\bar{Y}(t)) - \int_0^t \left(\int_{\mathbb{R}^{n'}} \left(\varphi(\bar{Y}(\tau) + f(y')) - \varphi(\bar{Y}(\tau))\right.\right.$$

$$\left.\left. - \left(f(y'), \varphi'(\bar{Y}(\tau))\right)_{\mathbb{R}^n}\right) M'(dy')\right) d\tau$$

is a $\mathbb{P}^{M'}$-martingale relative to $\{\mathcal{B}'_t : t \geq 0\}$. Hence, after an easy cutoff argument, we see that

$$|\bar{Y}(t)|^{2q} - \int_0^t \left(\int_{\mathbb{R}^{n'}} \left(\left|\bar{Y}(\tau) + f(y')\right|^{2q} - |\bar{Y}(\tau)|^{2q}\right.\right.$$

$$\left.\left. - 2q f(y') \bar{Y}(t) |\bar{Y}(\tau)|^{2(q-1)}\right) M'(dy')\right) d\tau$$

is a $\mathbb{P}^{M'}$-martingale relative to $\{\mathcal{B}'_t : t \geq 0\}$. In particular, by Taylor's theorem, $\mathbb{E}^{\mathbb{P}^{M'}}\left[|\bar{Y}(T)|^{2q}\right]$ is dominated by

$$
2q(q-1)\int_0^T \mathbb{E}^{\mathbb{P}^{M'}}\left[\int_{\mathbb{R}^{n'}} |f(y')|^2\left(|\bar{Y}(\tau)| + |f(y')|\right)^{2(q-1)} M'(dy')\right] d\tau
$$

$$
\leq 4^q q(q-1)T\left(\int_{\mathbb{R}^{n'}} |f(y')|^2 M'(dy')\mathbb{E}^{\mathbb{P}^{M'}}\left[\|\bar{Y}\|_{[0,T]}^{2(q-1)}\right]\right.
$$

$$
\left. + \int_{\mathbb{R}^{n'}} |f(y')|^{2q} M'(dy')\right).
$$

At the same time, by Doob's inequality,

$$
\mathbb{E}^{\mathbb{P}^{M'}}\left[\|\bar{Y}\|_{[0,T]}^{2q}\right] \leq \left(\frac{2q}{2q-1}\right)^{2q} \mathbb{E}^{\mathbb{P}^{M'}}\left[|\bar{Y}(T)|^{2q}\right].
$$

Hence, we now know that

$$
\mathbb{E}^{\mathbb{P}^{M'}}\left[\|\bar{Y}\|_{[0,T]}^{2q}\right] \leq B_q T \int_{\mathbb{R}^{n'}} |f(y')|^2 M'(dy')\mathbb{E}^{\mathbb{P}^{M'}}\left[\|\bar{Y}\|_{[0,T]}^{2q}\right]^{1-\frac{1}{q}}
$$

$$
+ B_q T \int_{\mathbb{R}^{n'}} |f(y')|^{2q} M'(dy')
$$

$$
\leq \frac{1}{q}\left(B_q T \int_{\mathbb{R}^{n'}} |f(y')|^2 M'(dy')\right)^q + \left(1 - \tfrac{1}{q}\right)\mathbb{E}^{\mathbb{P}^{M'}}\left[\|\bar{Y}\|_{[0,T]}^{2q}\right]
$$

$$
+ B_q T \int_{\mathbb{R}^{n'}} |f(y')|^{2q} M'(dy'),
$$

where $B_q \equiv \left(\frac{4q}{2q-1}\right)^{2q}\frac{q}{q-1}$ and we have used $ab^{1-\frac{1}{q}} = (a^q)^{\frac{1}{q}}b^{1-\frac{1}{q}} \leq \frac{1}{q}a^q + \left(1 - \frac{1}{q}\right)b$ for all $a, b \geq 0$. Clearly, the desired result follows from this. $\square$

Notice that the hypotheses in $(\mathbf{H2})^1$ and $(\mathbf{H3})^1$ make it obvious that $x \rightsquigarrow X^N(t, x, p')$ is continuously differentiable. In the following, we will be using the notation

$$
(4.2.3) \qquad J^N(t, x, p') \equiv \frac{\partial X^N(t, \cdot, p')}{\partial x}(x) \in \mathrm{Hom}(\mathbb{R}^n; \mathbb{R}^n),
$$

$A^N(t, x, p') \in \mathrm{Hom}(\mathbb{R}^n; \mathbb{R}^n)$ to denote

$$
\frac{\partial \sigma}{\partial x}\left(X^N([t]_N, x, p')\right)\left(w(t, p') - w([t]_N, p')\right)
$$

$$
+ \frac{\partial b}{\partial x}\left(X^N([t]_N, x, p')\right)\left(t - [t]_N\right)
$$

$$
+ \int_{|y'|\geq r_N} \frac{\partial F}{\partial x}\left(X^N([t]_N, x, p'), y'\right)\left(\bar{\eta}(t, dy', p') - \bar{\eta}([t]_N, dy', p')\right),
$$

and

$$\tilde{A}^N(t,x,p') \equiv A^N(t,x,p') + A^N(t,x,p')^\top + A^N(t,x,p')^\top A^N(t,x,p').$$

LEMMA 4.2.4. *For each $q \in [1,\infty)$ there is a $\beta_q \in (0,\infty)$, depending only on the bounds on the quantities appearing in* **(H2)**[1] *and* **(H3)**[1], *such that*

$$\sup_N \mathbb{E}^{\mathbb{P}^{M'}} \left[\|J^N(t,x)\|_{\text{H.S.}}^{2q} \right] \le n^{2q} e^{\beta_q t}.$$

Moreover, for each $(t,x) \in [0,\infty) \times \mathbb{R}^n$, there exists a $\overline{B'_t}$-measurable $p' \in D([0,\infty);\mathbb{R}^{n'}) \longrightarrow J(t,x,p') \in \text{Hom}(\mathbb{R}^{n'}\mathbb{R}^n)$ such that

$$\left\| J^N(t,x) - J(t,x) \right\|_{L^{2q}(\mathbb{P}^{M'};\text{Hom}(\mathbb{R}^n;\mathbb{R}^n))} \longrightarrow 0$$

uniformly for (t,x) in compacts. In particular, $(t,x) \in [0,\infty) \times \mathbb{R}^n \longmapsto J(t,x) \in L^{2q}(\mathbb{P}^{M'};\text{Hom}(\mathbb{R}^n;\mathbb{R}^n))$ is continuous for each $q \in [1,\infty)$. Finally, for each $\varphi \in C_b^1(\mathbb{R}^n;\mathbb{R})$,

$$(\partial_\xi)_x \mathbf{P}_t \varphi = \mathbb{E}^{\mathbb{P}^{M'}} \left[\left(J(t,x)\xi, \text{grad}_{X(t,x)}\varphi \right)_{\mathbb{R}^n} \right],$$

and so $C_b^1(\mathbb{R}^n;\mathbb{R})$ is $\{\mathbf{P}_t : t \ge 0\}$-invariant.

PROOF: We begin with the observation that

$$J^N(t,x,p') = \left(I + A^N(t,x,p') \right) J^N([t]_N,x,p').$$

In particular, this allows us to use an induction argument to prove that $x \in \mathbb{R}^n \longmapsto J^N(t,x,p')$ is continuous for $t \in [m2^{-N}, (m+1)2^{-N}]$ and all $m \in \mathbb{N}$. It also shows that

$$J^N(t,x)^\top J^N(t,x) = J^N([t]_N,x)^\top \left(I + \tilde{A}^N(t,x) \right) J^N([t]_N,x),$$

and so, for any orthonormal basis $(\mathbf{e}_1,\dots,\mathbf{e}_n)$ in $\mathbb{R}^n$,

$$\|J^N(t,x)\|_{\text{H.S}}^{2q} = \left(\sum_{i=1}^n \left(J^N([t]_N,x)\mathbf{e}_i, \left(I + \tilde{A}^N(t,x) \right) J^N([t]_N,x)\mathbf{e}_i \right)_{\mathbb{R}^n} \right)^q$$

$$\le \|J^N([t]_N,x)\|_{\text{H.S.}}^{2(q-1)} \sum_{i=1}^n \left(J^N([t]_N,x)\mathbf{e}_i, \left(I + \tilde{A}^N(t,x) \right) J^N([t]_N,x)\mathbf{e}_i \right)_{\mathbb{R}^n}^q.$$

Hence,

$$\mathbb{E}^{\mathbb{P}^{M'}} \left[\|J^N(t,x)\|_{\text{H.S.}}^{2q} \right] \le \mathbb{E}^{\mathbb{P}^{M'}} \left[B^{N,q}(t,x) \|J^N([t]_N,x)\|_{\text{H.S.}}^{2q} \right],$$

where

$$B^{N,q}(t,x) \equiv \sup_{\xi \in \mathbb{S}^{n-1}} \mathbb{E}^{\mathbb{P}^{M'}} \left[\left(1 + (\xi, \tilde{A}^N(t,x)\xi)_{\mathbb{R}^n} \right)^q \Big| \mathcal{B}'_{[t]_N} \right].$$

Now assume that $q \in \mathbb{Z}^+$, and write

$$\left(1 + (\xi, \tilde{A}^N(t,x)\xi)_{\mathbb{R}^n} \right)^q = 1 + q(\xi, \tilde{A}^N(t,x)\xi)_{\mathbb{R}^n} + \sum_{k=2}^{q} \binom{q}{k} (\xi, \tilde{A}^N(t,x)\xi)^k_{\mathbb{R}^n},$$

the sum on the right being taken equal to 0 if $q = 1$. Since

$$(\xi, \tilde{A}^N(t,x)\xi)_{\mathbb{R}^n} = 2(\xi, A^N(t,x)\xi)_{\mathbb{R}^n} + |A^N(t,x)\xi|^2$$

while

$$\mathbb{E}^{\mathbb{P}^{M'}} \left[A^N(t,x) \Big| \mathcal{B}'_{[t]_N} \right]$$
$$= (t - [t]_N) \left(\frac{\partial b}{\partial x}(X^N([t]_N, x)) + \int_{|y'| \geq 1} \frac{\partial F}{\partial x}(X^N([t]_N, x), y') \, M'(dy') \right),$$

and

$$\mathbb{E}^{\mathbb{P}^{M'}} \left[\|A^N(t,x)\|^2_{\mathrm{op}} \Big| \mathcal{B}'_{[t]_N} \right] \leq 3(t - [t]_N) \left\| \frac{\partial \sigma}{\partial x} \frac{\partial \sigma^\top}{\partial x}(X^N([t]_N, x)) \right\|_{\mathrm{op}}$$
$$+ 3(t - [t]_N)^2 \left\| \frac{\partial b}{\partial x}(X^N([t]_N, x)) + \int_{|y'| \geq 1} \frac{\partial F}{\partial x}(X^N([t]_N, x), y') \, M'(dy') \right\|^2_{\mathrm{op}}$$
$$+ 3(t - [t]_N) \int_{|y'| \geq r_N} \left\| \frac{\partial F}{\partial x}(X^N([t]_N, x), y') \right\|^2_{\mathrm{op}} M'(dy'),$$

we see that

$$(*) \qquad \mathbb{E}^{\mathbb{P}^{M'}} \left[(\xi, \tilde{A}^N(t,x)\xi)_{\mathbb{R}^n} \Big| \mathcal{B}_{[t]_N} \right] \leq C(t - [t]_N).$$

Treatment of the terms $\mathbb{E}^{\mathbb{P}^{M'}} \left[(\xi, \tilde{A}^N(t,x)\xi)^k_{\mathbb{R}^n} \Big| \mathcal{B}_{[t]_N} \right]$ when $k \geq 2$ requires less care. Indeed, from the estimates in Lemma 4.2.4, it is an easy matter to check that each such term is again dominated by a constant times $(t - [t]_N)$, and with this information, together with $(*)$, we arrive first at

$$\mathbb{E}^{\mathbb{P}^{M'}} \left[\|J^N(t,x)\|^{2q}_{\mathrm{H.S.}} \right] \leq (1 + \beta_q(t - [t]_N)) \mathbb{E}^{\mathbb{P}^{M'}} \left[\|J^N([t]_N, x)\|^{2q}_{\mathrm{H.S.}} \right]$$

for some $\beta_q < \infty$ with the required dependence and then, via the discrete version of Gronwall's inequality in Exercise 3.1.30, at the asserted estimated for $q \in \mathbb{Z}^+$. When $q \in [1, \infty) \backslash \mathbb{Z}^+$, the estimate follows by Hölder's inequality.

To prove the asserted convergence result, first note that for $N' > N$

$$
\begin{aligned}
\Delta^{N,N'}(t,x) &\equiv J^{N'}(t,x) - J^N(t,x) \\
&= \left(I + A^{N'}(t,x)\right)\Delta^{N,N'}([t]_{N'},x) + \left(A^{N'}(t,x) - A^N(t,x)\right)J^N(t,x) \\
&\quad + \left(I + A^{N'}(t,x)\right)\left(J^N([t]_{N'},x) - J^N([t]_N,x)\right) \\
&= \left(I + A^{N'}(t,x)\right)\Delta^{N,N'}(t,x) + B^{N,N'}(t,x)J^N([t]_N,x),
\end{aligned}
$$

where

$$
B^{N,N'}(t,x) \equiv A^{N'}(t,x) + A^N([t]_{N'},x) - A^N(t,x) + A^{N'}(t,x)A^N([t]_{N'},x).
$$

Proceeding as we did above (especially, taking advantage of the $\mathbb{P}^{M'}$-independence of future increments) and applying the estimates coming from Theorem 3.1.20, the last part of Theorem 4.1.2 and the first part of this lemma, one finds that

$$
\mathbb{E}^{\mathbb{P}^{M'}}\left[\left\|\Delta^{N,N'}(t,x)\right\|_{\text{H.S.}}^2\right] \le (1 + \beta_2(t - [t]_{N'})\mathbb{E}^{\mathbb{P}^{M'}}\left[\left\|\Delta^{N,N'}([t]_{N'},x)\right\|_{\text{H.S.}}^2\right] \\
+ (t - [t]_{N'})\epsilon_N(x),
$$

where $\epsilon_N(x) \to 0$ uniformly on compacts as $N \to \infty$, and clearly this leads to

$$
\mathbb{E}^{\mathbb{P}^{M'}}\left[\left\|\Delta^{N,N'}(t,x)\right\|_{\text{H.S.}}^2\right] \le \frac{\beta_2}{2^{N'}}\sum_{m=0}\mathbb{E}^{\mathbb{P}^{M'}}\left[\left\|\Delta^{N,N'}(m2^{-N'},x)\right\|_{\text{H.S.}}^2\right] + t\epsilon_N(x).
$$

Finally, by the discrete form of Gronwall's inequality, it follows that

$$
\sup_{N'>N} \mathbb{E}^{\mathbb{P}^{M'}}\left[\left\|\Delta^{N,N'}(t,x)\right\|_{\text{H.S.}}^2\right] \le t\epsilon_N(x)e^{\beta_2 t}.
$$

Obviously, this proves the existence of a measurable $p' \rightsquigarrow J(t,x,p')$ for which $\left\|J^N(t,x) - J(t,x)\right\|_{L^2(\mathbb{P}^{M'})}$ tends to 0 uniformly as (t,x) runs over compacts. Finally, by combining this with the estimates obtained earlier and using Hölder's inequality, one gets the required convergence result in $L^{2q}(\mathbb{P}^{M'})$ for every $q \in [1,\infty)$. In particular, for $\varphi \in C_b^1(\mathbb{R}^n;\mathbb{R})$,

$$
\begin{aligned}
\mathbf{P}_t\varphi(x+\xi) &- \mathbf{P}_t\varphi(x) \\
&= \lim_{N\to\infty}\int_0^1 \mathbb{E}^{\mathbb{P}^{M'}}\left[\left(J^N(t,x+\tau\xi)\xi, \text{grad}_{X^N(t,x+\tau\xi)}\varphi\right)_{\mathbb{R}^n}\right]d\tau \\
&= \int_0^1 \mathbb{E}^{\mathbb{P}^{M'}}\left[\left(J(t,x+\tau\xi)\xi, \text{grad}_{X(t,x+\tau\xi)}\varphi\right)_{\mathbb{R}^n}\right]d\tau. \quad \square
\end{aligned}
$$

§4.2.2. Second Derivatives and Uniqueness: As we said at the beginning of this section, what we really want are conditions which allow us to show that $C_b^2(\mathbb{R}^n; \mathbb{R})$ is $\{\mathbf{P}_t \; : \; t \geq 0\}$-invariant. For this purpose, we strengthen $(\mathbf{H2})^1$ and $(\mathbf{H3})^1$ to

$(\mathbf{H2})^2$ $F \in C^2(\mathbb{R}^n; \mathbb{R})$ satisfies the conditions in $(\mathbf{H2})^1$ and, for each $q \in [1, \infty)$,

$$\sup_{x \in \mathbb{R}^n} \int_{\mathbb{R}^n} \left\| \frac{\partial^2 F}{\partial x^2}(x, y') \right\|_{\mathrm{op}}^{2q} M'(dy') < \infty,$$

where $\frac{\partial^2 F}{\partial x^2}(x, y')$ is the element of $\mathrm{Hom}(\mathbb{R}^n \otimes \mathbb{R}^n; \mathbb{R}^n)$ determined by

$$\frac{\partial^2 F}{\partial x^2}(x, y')(\xi \otimes \eta) = (\partial_\xi)_x \partial_\eta F.$$

$(\mathbf{H3})^2$ $\sigma : \mathbb{R}^n \longmapsto \mathrm{Hom}(\mathbb{R}^{n'}; \mathbb{R}^n)$ and $b : \mathbb{R}^n \longrightarrow \mathbb{R}^n$ are twice continuously differentiable functions with bounded first and second order derivatives.

Under the conditions in $(\mathbf{H2})^2$ and $(\mathbf{H2})^2$, it is clear that $X^N(t, \cdot) \in C^2(\mathbb{R}^n; \mathbb{R}^n)$ and that

$$H^N(t, x) \equiv \frac{\partial^2 X^N(t, \cdot)}{\partial x^2}(x)$$
$$= (I + A^N(t, x)) H^N([t]_N, x) + S^N([t]_N, x) J^N([t]_N, x) \otimes J^N([t]_N, x)$$

where

$$S^N(t, x) \equiv \frac{\partial^2 \sigma}{\partial x^2}(X^N([t]_N, x))(w(t) - w([t]_N))$$
$$+ \frac{\partial^2 b}{\partial x^2}(X^N([t]_N, x))(t - [t]_N)$$
$$+ \int_{|y'| \geq r_N} \frac{\partial^2 F}{\partial x^2}(X^N([t]_N, x))(\bar{\eta}(t, dy') - \bar{\eta}([t]_N, dy')).$$

Thus, by repeating the reasoning used in the proof of Lemma 4.2.4, one can prove all but the final part of the following result.

THEOREM 4.2.5. *Under the conditions in* $(\mathbf{H1})$, $(\mathbf{H2})^2$, *and* $(\mathbf{H3})^2$, *there exists, for each $q \in [1, \infty)$, a $K_q < \infty$, depending only on the bounds on the quantities in* $(\mathbf{H2})^2$, *and* $(\mathbf{H3})^2$, *such that*

$$\mathbb{E}^{\mathbb{P}^{M'}} \left[\left\| H^N(t, x) \right\|_{\mathrm{H.S.}}^{2q} \right] \leq e^{\beta_q t} \left(n^{6q} + K_q \right).$$

Furthermore, for each $(t,x) \in [0,\infty) \times \mathbb{R}^n$, there exists a $\overline{B'_t}$-measurable $p' \in D([0,\infty); \mathbb{R}^{n'}) \longmapsto H(t,x,p') \in \mathrm{Hom}(\mathbb{R}^n \otimes \mathbb{R}^n; \mathbb{R}^n)$ such that

$$\left\| H^N(t,x) - H(t,x) \right\|_{L^{2q}(\mathbb{P}^{M'}; \mathrm{Hom}(\mathbb{R}^n \otimes \mathbb{R}^n; \mathbb{R}^n))} \longrightarrow 0$$

uniformly for (t,x) in compacts, and so $(t,x) \in [0,\infty) \times \mathbb{R}^n \longrightarrow H(t,x) \in L^{2q}(\mathbb{P}^{M'}; \mathrm{Hom}(\mathbb{R}^n \otimes \mathbb{R}^n; \mathbb{R}^n))$ is continuous. In particular, for each $\varphi \in C_b^2(\mathbb{R}^n; \mathbb{R})$ (cf. the notation in Lemma 4.2.4),

$$(\partial_\xi)_x \partial_{\xi'} \mathbf{P}_t \varphi = \mathbb{E}^{\mathbb{P}^{M'}} \left[\left(J(t,x)\xi, \mathrm{Hess}_{X(t,x)} \varphi J(t,x)\xi' \right)_{\mathbb{R}^n} \right]$$
$$+ \mathbb{E}^{\mathbb{P}^{M'}} \left[\left(H(t,x)\xi \otimes \xi', \mathrm{grad}_{X(t,x)} \varphi \right)_{\mathbb{R}^n} \right],$$

and therefore $C_b^2(\mathbb{R}^n; \mathbb{R})$ is $\{\mathbf{P}_t : t \geq 0\}$-invariant. In fact, for each $\varphi \in C_b^2(\mathbb{R}^n; \mathbb{R})$, $t \in [0,\infty) \longmapsto \mathbf{P}_t \varphi \in C_b^2(\mathbb{R}^n; \mathbb{R})$ is continuously differentiable and

$$\partial_t \mathbf{P}_t \varphi = L\mathbf{P}_t \varphi = \mathbf{P}_t L\varphi.$$

PROOF: As we said above, the only the final statement requires comment. To prove it, note that $(\mathbf{H2})^2$ and $(\mathbf{H3})^2$ mean that the conditions of Lemma 3.1.25 are met. Thus, by that lemma, we already know that

$$\mathbf{P}_t \varphi - \varphi = \int_0^t \mathbf{P}_\tau L\varphi \, d\tau$$

for all $\varphi \in C_b^2(\mathbb{R}^n; \mathbb{R})$. Now let $s \in [0,\infty)$ be given, apply the preceding to $\mathbf{P}_s \varphi$, and get

$$\mathbf{P}_{s+t} \varphi - \mathbf{P}_s \varphi = \int_0^t \mathbf{P}_\tau \circ (L\mathbf{P}_s \varphi) \, d\tau.$$

Hence, the desired conclusion follows after one divides by t and lets $t \searrow 0$. $\square$

The reason for our interest in the preceding is contained in following application.

COROLLARY 4.2.6. Assume that σ and b satisfy the conditions in $(\mathbf{H3})^2$ and that, for each $R \in [1,\infty)$, F_R satisfies $(\mathbf{H2})^2$, where

$$F_R(x,y') \equiv \mathbf{1}_{[0,R]}(|y'|)F(x,y').$$

Then, for each $x \in \mathbb{R}^n$, $\mathbb{P}_x^{(\sigma,b,F,M')}$ is the one and only solution $\mathbb{P}_x^L$ to the martingale problem for L on $C_c^2(\mathbb{R}^n; \mathbb{R})$ starting from x.

PROOF: The proof is essentially the same as the proof of Corollary 3.1.27, only here there is one new ingredient. Namely, given $R \geq 1$, let $\{\mathbf{P}_t^R : t \geq 0\}$ be the Markov semigroup corresponding to σ, b, and F_R. By Theorem 4.2.5, for any $\varphi \in C_b^2(\mathbb{R}^n; \mathbb{R})$, $(t, x) \rightsquigarrow \mathbf{P}_t^R \varphi(x)$ satisfies

$$\partial_t \mathbf{P}_t^R \varphi = L \mathbf{P}_t^R \varphi - \mathbf{Q}_t^R \varphi \quad \text{where}$$

$$\mathbf{Q}_t^R \varphi(x) \equiv \int_{|y'|>R} \Big(\mathbf{P}_t \varphi \big(x + F(x, y') \big) - \mathbf{P}_t \varphi(x) \Big) \, M'(dy').$$

Hence, by the argument used in Corollary 3.1.27, we can prove that

$$\mathbb{E}^{\mathbb{P}_x^L} \Big[\varphi \big(p(s+t) \big) \,\Big|\, \mathcal{B}_s \Big] - \mathbf{P}_t \varphi(p(s)) = \mathbb{E}^{\mathbb{P}_x^L} \left[\int_s^{s+t} \mathbf{Q}_\tau^R \varphi \big(p(s+t-\tau) \big) \, d\tau \,\Big|\, \mathcal{B}_s \right].$$

At the same time,

$$\big\| \mathbf{Q}_t^R \varphi \big\|_{\mathrm{u}} \leq 2 \|\varphi\|_{\mathrm{u}} M' \big(\mathbb{R}^{n'} \setminus B_{\mathbb{R}^{n'}}(0, R) \big) \longrightarrow 0$$

and (cf. Lemma 3.1.15)

$$\big\| \mathbf{P}_t \varphi - \mathbf{P}_t^R \varphi \big\|_{\mathrm{u}} \leq 2 \|\varphi\|_{\mathrm{u}} \mathbb{P}^{M'} \big(A(t, R) \complement \big) \longrightarrow 0$$

as $R \to \infty$; and so it follows that

$$\mathbb{E}^{\mathbb{P}_x^L} \Big[\varphi \big(p(s+t) \big) \,\Big|\, \mathcal{B}_s \Big] = \mathbf{P}_t \varphi(p(s)) \quad \mathbb{P}_x^L\text{-almost surely.}$$

But, once we know this, the argument reverts to the one with which we completed the proof of Corollary 3.1.27. □

REMARK 4.2.7. It may not be immediately clear what is the significance of uniqueness statements like those given in Corollaries 3.1.27 and 4.2.6. Indeed, in §4.1.2, we showed that the Markov property does not depend on our knowing such a uniqueness result. On the other hand, the canonical object is the operator L, whereas the choice of σ, F, M', and, for that matter, n' is, in general, quite arbitrary. Thus, if nothing else, it should be comforting to know that the function space measure at which one arrives depends only on L. Moreover, in practice, this sort of uniqueness result does much more than provide comfort. To wit, we have been discussing one procedure, namely, Itô's, for constructing a measure on function space associated with L. However, there are many others. For example, in many applications one wants to have an approximation scheme via discrete Markov processes on rescaled lattices. The importance of uniqueness results like Corollary 4.2.6 is that they free one to use any method one wants as long as it leads to a measure which solves the martingale problem for L. The interested reader might want to consult [41] for a systematic development of these ideas and for proofs of this sort of uniqueness in more challenging circumstances.

CHAPTER 5

Itô's Theory of Stochastic Integration

Up to this point, I have been recognizing but not confronting the challenge posed by integrals of the sort in (3.1.22). There are several reasons for my decision to postpone doing so until now, perhaps the most important of which is my belief that, in spite of, or maybe because of, its elegance, Itô's theory of stochastic integration tends to mask the essential simplicity and beauty of his ideas as we have been developing them heretofore. However, it is high time that I explain his theory of integration, and that is what I will be doing in this chapter. However, we will not deal with the theory in full generality and will restrict our attention to the case when the paths are continuous. In terms of equations like (3.1.22), this means that we will not try to rationalize the "$dp'(t)$" integral except in the case when p' is Brownian motion (i.e., $M' = 0$ and $p' = w(\,\cdot\,, p')$). The general theory is beautiful and has been fully developed by the French school, particularly by C. Dellacherie and P.A. Meyer who have published a detailed account of their findings in [5].

Because it already contains most of the essential ideas, we will devote this chapter to stochastic integration with respect to Brownian motion.

5.1 Brownian Stochastic Integrals

Let $(\Omega, \mathcal{F}, \mathbb{P})$ be a complete probability space. Then $\big(\beta(t), \mathcal{F}_t, \mathbb{P}\big)$ will be called an $\mathbb{R}^n$-valued *Brownian motion* if $\{\mathcal{F}_t : t \geq 0\}$ is a nondecreasing family of $\mathbb{P}$-complete sub σ-algebras of the σ-algebra $\mathcal{F}$ and $\beta : [0, \infty) \times \Omega \longrightarrow \mathbb{R}^n$ is a $\mathcal{B}_{[0,\infty)} \times \mathcal{F}$-measurable map with the properties that[1]

(a) $\beta(0) = 0$ and $t \rightsquigarrow \beta(t, \omega)$ is continuous for $\mathbb{P}$-almost every ω,
(b) $\omega \rightsquigarrow \beta(t, \omega)$ is $\mathcal{F}_t$-measurable for each $t \in [0, \infty)$,
(c) for all $s \in [0, \infty)$ and $t \in (0, \infty)$, $\beta(s + t) - \beta(s)$ is $\mathbb{P}$-independent of $\mathcal{F}_s$ and has the distribution of a centered Gaussian with covariance $t I_{\mathbb{R}^n}$ under $\mathbb{P}$.

---

[1] It should be noticed that our insistence on the completeness of all σ-algebras imposes no restriction. Indeed, if $\mathcal{F}$ or the $\mathcal{F}_t$'s are not complete but (a), (b), and (c) hold for some $\omega \rightsquigarrow \beta(\,\cdot\,, \omega)$, then they will continue to hold after all σ-algebras have been completed.

Notice that the preceding definition is very close to saying that $\beta(\,\cdot\,,\omega)$ is continuous for all ω and that the $\mathbb{P}$-distribution of $\omega \in \Omega \longmapsto \beta(\,\cdot\,,\omega) \in C([0,\infty);\mathbb{R}^n)$ is given by Wiener measure, the measure which would have been denoted by $\mathbb{P}^{(I,0,0)}$ in §2.4 and by $\mathbb{P}^0$ starting in §3.1.1. To be more precise, first observe that, without loss in generality, one can assume that $\beta(\,\cdot\,,\omega) \in C([0,\infty);\mathbb{R}^n)$ for all ω, in which case (a), (b), and (c) guarantee that $\mathbb{P}^{(I,0,0)}$ is the $\mathbb{P}$-distribution of $\omega \rightsquigarrow \beta(\,\cdot\,,\omega)$. Conversely, if $\Omega = C([0,\infty);\mathbb{R}^n)$, $\mathbb{P} = \mathbb{P}^{(I,0,0)}$, and $\mathcal{F}$ and $\mathcal{F}_t$ are, respectively, the $\mathbb{P}$-completions of $\mathcal{B}$ and $\mathcal{B}_t$, then one gets a Brownian motion by taking $\beta(t,p) = p(t)$. Thus, the essential generalization afforded by the preceding definition is that the σ-algebras need not be inextricably tied to the random variables $\beta(t)$. That is, $\mathcal{F}_t$ must contain but need not be the completion of $\sigma(\{\beta(\tau) : \tau \in [0,t]\})$.

§**5.1.1. A Review of the Paley–Wiener Integral:** As an aid to understanding Itô's theory, it may be helpful to recall the theory of stochastic integration which was introduced by Paley and Wiener. Namely, let $(\beta(t), \mathcal{F}_t, \mathbb{P})$ be a Brownian motion on the complete probability space $(\Omega, \mathcal{F}, \mathbb{P})$, and assume $\beta(\,\cdot\,,\omega) \in C([0,\infty);\mathbb{R}^n)$ for all $\omega \in \Omega$. Given a Borel measurable function $\theta : [0,\infty) \longrightarrow \mathbb{R}^n$ which has bounded variation on each finite interval, one can use Riemann–Stieltjes theory[2] to define

$$t \in [0,\infty) \longmapsto I_\theta(t,\omega) = \int_0^t \big(\theta(\tau), d\beta(\tau,\omega)\big)_{\mathbb{R}^n} \in \mathbb{R}$$

Because it is given by a Riemann–Stieltjes integral, we can say that (cf. (3.3.5))

$$(5.1.1) \qquad I_\theta(t,\omega) = \lim_{N\to\infty} \sum_{m=0}^{\infty} \big(\theta(m2^{-N}), \Delta_m^N \beta(t,\omega)\big)_{\mathbb{R}^n},$$

By (c), we know that, for each $N \in \mathbb{N}$,

$$\omega \rightsquigarrow \sum_{m=0}^{\infty} \big(\theta(m2^{-N}), \Delta_m^N \beta(t,\omega)\big)_{\mathbb{R}^n}$$

is a centered Gaussian with variance $\int_0^t |\theta([\tau]_N)|^2 \, d\tau$, and so it is an easy step to the conclusion that $\omega \rightsquigarrow I_\theta(t,\omega)$ is a centered Gaussian with variance equal to $\int_0^t |\theta(\tau)|^2 \, d\tau$. In fact, with only a little more effort, one sees

---

[2] What is needed here is the fact (cf. Theorem 1.2.7 in [34]) that Riemann–Stieltjes theory is completely symmetric: φ is Riemann–Stieltjes integrable with resect to ψ if and only if ψ is with respect to φ. In fact, the integration by parts formula is what allows one to exchange the two.

that $I(s+t) - I(s)$ is $\mathbb{P}$-independent of $\sigma(\{I(\sigma) : \sigma \in [0, s]\})$ and that its $\mathbb{P}$-distribution is that of a centered Gaussian with variance $\int_s^t |\theta(\tau)|^2 \, d\tau$. Moreover, $I_\theta(\cdot, \omega) \in C([0, \infty); \mathbb{R}^n)$, and so we now know that $(I_\theta(t), \mathcal{F}_t, \mathbb{P})$ is a continuous, square integrable martingale. In particular, by Doob's inequality,

$$(5.1.2) \quad \|\theta\|_{L^2([0,\infty);\mathbb{R}^n)}^2 = \lim_{t\to\infty} \mathbb{E}^{\mathbb{P}}\big[I_\theta(t)^2\big] \le \mathbb{E}^{\mathbb{P}}\big[\|I_\theta\|_{[0,\infty)}^2\big] \le 4\|\theta\|_{L^2([0,\infty);\mathbb{R}^n)}^2.$$

The relations in (5.1.2) can be used as the basis on which to extend the definition of $\theta \rightsquigarrow I_\theta$ to square integrable θ's which do not necessarily possess locally bounded variation. Indeed, (5.1.2) says that, as a map taking $\theta \in L^2([0, \infty); \mathbb{R}^n)$ with locally bounded variation into continuous square integrable martingales on $(\Omega, \mathcal{F}_t, \mathbb{P})$, $\theta \rightsquigarrow I_\theta$ is continuous. Hence, because the smooth elements of $L^2([0, \infty); \mathbb{R}^n)$ are dense there, this map admits a unique continuous extension. To be precise, define $\mathcal{M}^2(\mathbb{P}; \mathbb{R})$ to be the space of all $\mathbb{R}$-valued, square integrable, $\mathbb{P}$-almost surely continuous $\mathbb{P}$-martingales M relative to $\{\mathcal{F}_t : t \ge 0\}$ such that

$$(5.1.3) \qquad \|M\|_{\mathcal{M}^2(\mathbb{P};\mathbb{R})} = \sup_{t\in[0,\infty)} \mathbb{E}^{\mathbb{P}}\big[|M(t)|^2\big]^{\frac{1}{2}} < \infty.$$

Although it may not be apparent, $\mathcal{M}^2(\mathbb{P}; \mathbb{R})$ is actually a Hilbert space. In fact, it can be isometrically embedded as a closed subspace of $L^2(\mathbb{P}; \mathbb{R})$. Namely, if $M \in \mathcal{M}^2(\mathbb{P}; \mathbb{R})$, then M is an L^2-bounded martingale and therefore, by the L^2-martingale convergence theorem (cf. Theorem 7.1.16 in [36]), there exists an $M(\infty) \in L^2(\mathbb{P}; \mathbb{R})$ to which $\{M(t) : t \ge 0\}$ converges both $\mathbb{P}$-almost surely and in $L^2(\mathbb{P}; \mathbb{R})$. In particular, this means that, for each $t \ge 0$, $M(t) = \mathbb{E}^{\mathbb{P}}\big[M(\infty)|\mathcal{F}_t\big]$ $\mathbb{P}$-almost surely. Moreover, because $(M(t)^2, \mathcal{F}_t, \mathbb{P})$ is a submartingale,

$$\|M(t)\|_{L^2(\mathbb{P};\mathbb{R})} \nearrow \|M\|_{\mathcal{M}^2(\mathbb{P};\mathbb{R})} = \|M(\infty)\|_{L^2(\mathbb{P};\mathbb{R})}.$$

Hence, the map $M \in \mathcal{M}^2(\mathbb{P}; \mathbb{R}) \longmapsto M(\infty) \in L^2(\mathbb{P}; \mathbb{R})$ is a linear isometry. Finally, to see that $\{M(\infty) : M \in \mathcal{M}^2(\mathbb{P}; \mathbb{R})\}$ is closed in $L^2(\mathbb{P}; \mathbb{R})$, suppose that $\{M_k\}_{k=1}^\infty \subseteq \mathcal{M}^2(\mathbb{P}; \mathbb{R})$ and that $M_k(\infty) \longrightarrow X$ is $L^2(\mathbb{P}; \mathbb{R})$. By Doob's inequality,

$$\sup_{\ell>k} \mathbb{E}^{\mathbb{P}}\Big[\|M_\ell - M_k\|_{[0,\infty)}^2\Big] \le 4\sup_{\ell>k} \|M_\ell(\infty) - M_k(\infty)\|_{L^2(\mathbb{P};\mathbb{R})}^2 \longrightarrow 0$$

as $k \to \infty$. Hence, there exists an $\mathcal{F}$-measurable map $\omega \in \Omega \longmapsto M \in C([0, \infty); \mathbb{R}^n)$ such that $\lim_{k\to\infty} \mathbb{E}^{\mathbb{P}}\big[\|M - M_k\|_{[0,\infty)}^2\big] = 0$. But clearly, for each $t \ge 0$, $M(t) = \mathbb{E}^{\mathbb{P}}[X|\mathcal{F}_t]$ $\mathbb{P}$-almost surely, and so not only is $M \in \mathcal{M}^2(\mathbb{P}; \mathbb{R})$ but also, since X is $\sigma(\bigcup_{t\ge 0} \mathcal{F}_t)$-measurable, $X = M(\infty)$. For future reference, we will collect these observations in a lemma.

LEMMA 5.1.4. *The space $\mathcal{M}^2(\mathbb{P};\mathbb{R})$ with the norm given by (5.1.3) is a Hilbert space. Moreover, for each $M \in \mathcal{M}^2(\mathbb{P};\mathbb{R})$, $M(\infty) \equiv \lim_{t\to\infty} M(t)$ exists both $\mathbb{P}$-almost surely and in $L^2(\mathbb{P};\mathbb{R})$, and the map $M \in \mathcal{M}^2(\mathbb{P};\mathbb{R}) \longmapsto M(\infty) \in L^2(\mathbb{P};\mathbb{R})$ is a linear isometry.*

By combining Lemma 5.1.4 with the remarks which precede it, we arrive at the following statement, which summarizes the Paley–Wiener theory of stochastic integration.

THEOREM 5.1.5. *There is a unique, linear isometry $\theta \in L^2([0,\infty);\mathbb{R}^n) \longmapsto I_\theta \in \mathcal{M}^2(\mathbb{P};\mathbb{R})$ with the property that $I_\theta(t,\omega)$ is given by (5.1.1) when θ has locally bounded variation. In particular, for each $T \geq 0$, $I_\theta(T) = I_{1_{[0,T]}\theta}(\infty)$ $\mathbb{P}$-almost surely. Finally, for each $\theta \in L^2([0,\infty);\mathbb{R}^n)$ and all $0 \leq s < t \leq \infty$, $I_\theta(t) - I_\theta(s)$ is $\mathbb{P}$-independent of $\mathcal{F}_s$ and its $\mathbb{P}$-distribution is that of a centered Gaussian with variance $\int_s^t |\theta(\tau)|^2\,d\tau$.*

§5.1.2. Itô's Extension: Itô's extension of the preceding to θ's which may depend on ω as well as t is completely natural if one keeps in mind the reason for his wanting to make such an extension. Namely, he was trying to make sense out of integrals which appear in his method of constructing Markov processes. Thus, he wanted to find a notion of integration which would allow him to interpret

$$\lim_{N\to\infty} \sum_{m=0}^{\infty} \sigma\big(X^N(m2^{-N},x)\big)\,\Delta_m^N p'(t)$$

as an integral. In particular, he had reason to suppose that it was best to make sure that the integrand is independent of the differential by which it is being multiplied.

With this in mind, we say that a map F on $[0,\infty) \times \Omega$ into a measurable space is *progressively measurable* if $F \upharpoonright [0,T] \times \Omega$ is $\mathcal{B}_{[0,T]} \times \mathcal{F}_T$-measurable for each $T \in [0,\infty)$.[3] The following elementary facts about progressive measurability are proved in Lemma 7.1.2 of [36].

LEMMA 5.1.6. *Let $\mathcal{PM}$ denote the collection of all $A \subseteq [0,\infty) \times \Omega$ such that 1_A is an $\mathbb{R}$-valued, progressively measurable function. Then $\mathcal{PM}$ is a sub σ-algebra of $\mathcal{B}_{[0,\infty)} \times \mathcal{F}$, and a function on $[0,\infty) \times \Omega$ is progressively measurable if and only if it is measurable with respect to $\mathcal{P}(M)$. Furthermore, if F :*

---

[3] So far as I know, the notion of progressive measurability is one of the many contributions which P.A. Meyer made to the subject of stochastic integration. In particular, Itô dealt with *adapted* functions: F's for which $\omega \rightsquigarrow F(T,\omega)$ is $\mathcal{F}_T$-measurable for each T. Even though adaptedness is more intuitively appealing, there are compelling technical reasons for preferring progressive measurability. See Remark 7.1.1 in [36] for further comments on this matter.

$[0, \infty) \times \Omega \longrightarrow E$, where E is a metric space, with the properties that $F(\cdot, \omega)$ is continuous for each ω and $F(T, \cdot)$ is $\mathcal{F}_T$ for each $T \geq 0$, then F is progressively measurable. In fact, if $\mathcal{F}_t$ is $\mathbb{P}$-complete for each $t \geq 0$ and $F(\cdot, \omega)$ is continuous for $\mathbb{P}$-almost every ω, then F is progressively measurable if $F(T, \cdot)$ is $\mathcal{F}_T$-measurable for each $T \geq 0$.

Now let $\Theta^2(\mathbb{P}; \mathbb{R}^n)$ denote the space of all progressively measurable θ : $[0, \infty) \times \Omega \longrightarrow \mathbb{R}^n$ with the property that

$$\|\theta\|_{\Theta^2(\mathbb{P};\mathbb{R}^n)} \equiv \mathbb{E}^{\mathbb{P}}\left[\int_0^\infty |\theta(t)|^2 \, dt\right] < \infty.$$

Since, by Lemma 5.1.6, an equivalent way to describe $\Theta^2(\mathbb{P}; \mathbb{R}^n)$ is as the subspace of progressively measurable elements of $L^2\left(\text{Leb}_{[0,\infty)} \times \mathbb{P}\right)$, we know that $\Theta^2(\mathbb{P}; \mathbb{R}^n)$ is a Hilbert space.

Our goal is to show (cf. Exercise 5.1.20 below) that there is a unique linear isometry $\theta \in \Theta^2(\mathbb{P}; \mathbb{R}^n) \longmapsto I_\theta \in \mathcal{M}^2(\mathbb{P}; \mathbb{R})$ with the property that when θ is an element of $\Theta^2(\mathbb{P}; \mathbb{R}^n)$ such that $\theta(\cdot, \omega)$ has locally bounded variation for each ω

$$I_\theta(t, \omega) = \int_0^t \left(\theta(\tau, \omega), d\beta(\tau, \omega)\right)_{\mathbb{R}^n}$$

(5.1.7)

$$= \left(\theta(t, \omega), \beta(t, \omega)\right)_{\mathbb{R}^n} - \int_0^t \left(\beta(\tau, \omega), d\theta(\tau, \omega)\right)_{\mathbb{R}^n},$$

where the integrals are taken in the sense of Riemann–Stieltjes.

LEMMA 5.1.8. Let[4] $S\Theta^2(\mathbb{P}; \mathbb{R}^n)$ be the subspace of uniformly bounded $\theta \in \Theta^2(\mathbb{P}; \mathbb{R}^n)$ for which there exists an $N \in \mathbb{N}$ with the property that $\theta(t, \omega) = \theta([t]_N, \omega)$ for all $(t, \omega) \in [0, \infty) \times \Omega$. Then $S\Theta^2(\mathbb{P}; \mathbb{R}^n)$ is dense in $\Theta^2(\mathbb{P}; \mathbb{R}^n)$. Moreover, if $\theta \in S\Theta^2(\mathbb{P}; \mathbb{R}^n)$, I_θ is given as in (5.1.7), and

(5.1.9) $$A_\theta(t, \omega) \equiv \int_0^t |\theta(\tau, \omega)|^2 \, d\tau,$$

then $(E_\theta(t), \mathcal{F}_t, \mathbb{P})$ is a martingale when

(5.1.10) $$E_\theta(t, \omega) \equiv \exp\left(I_\theta(t, \omega) - \tfrac{1}{2} A_\theta(t)\right).$$

In particular, both $\left(I_\theta(t), \mathcal{F}_t, \mathbb{P}\right)$ and $\left(I_\theta(t)^2 - A_\theta(t), \mathcal{F}_t, \mathbb{P}\right)$ are martingales.

---

[4] The "S" here stands for "simple."

PROOF: To prove the density statement, first observe that, by any standard truncation procedure, it is easy to check that uniformly bounded elements of $\Theta^2(\mathbb{P}; \mathbb{R}^n)$ which are supported on $[0, T] \times \Omega$ for some $T \in [0, \infty)$ form a dense subspace of $\Theta^2(\mathbb{P}; \mathbb{R}^n)$. Thus, suppose that $\theta \in \Theta^2(\mathbb{P}; \mathbb{R}^n)$ is uniformly bounded and vanishes off $[0, T] \times \Omega$. Choose a $\psi \in C_c^\infty(\mathbb{R}; [0, \infty))$ so that $\psi \equiv 0$ on $(-\infty, 0]$ and $\int_{\mathbb{R}} \psi(t)\, dt = 1$, and set

$$\theta_k(t, \omega) = k \int_0^\infty \psi\big(k(t - \tau)\big)\big)\theta(\tau, \omega)\, d\tau.$$

Then, for each $k \in \mathbb{Z}^+$ and $\omega \in \Omega$, $\theta_k(\,\cdot\,, \omega)$ is smooth and vanishes off of $[0, T+1]$. In addition, because ψ is supported on the right half line, it follows from Lemma 5.1.6 that θ_k is progressively measurable. At the same time, for each ω, $\|\theta_k(\,\cdot\,, \omega)\|_{L^2(\mathbb{P}; \mathbb{R}^n)} \leq \|\theta(\,\cdot\,, \omega)\|_{L^2(\mathbb{P}; \mathbb{R}^n)}$ and

$$\|\theta_k(\,\cdot\,, \omega) - \theta(\,\cdot\,, \omega)\|_{L^2([0,\infty); \mathbb{R}^n)} \longrightarrow 0 \quad \text{as } k \to \infty.$$

Hence, by Lebesgue's dominated convergence theorem, $\|\theta_k - \theta\|_{\Theta^2(\mathbb{P}; \mathbb{R}^n)} \longrightarrow 0$. In other words, we have now proved the density of the uniformly bounded elements of $\Theta^2(\mathbb{P}; \mathbb{R})$ which are supported on $[0, T] \times \Omega$ for some $T \geq 0$ and are smooth as functions of $t \in [0, \infty)$ for each $\omega \in \Theta$. But clearly, if θ is such an element of $\Theta^2(\mathbb{P}; \mathbb{R})$ and $\theta^N(t, \omega) \equiv \theta([t]_N, \omega)$, then $\theta^N \longrightarrow \theta$ in $\Theta^2(\mathbb{P}; \mathbb{R})$.

To prove the second assertion, let θ be a uniformly bounded element of $\Theta^2(\mathbb{P}; \mathbb{R})$ which satisfies $\theta(t, \omega) = \theta([t]_N, \omega)$. Then, for $m2^{-N} \leq s < t \leq (m+1)2^{-N}$,

$$\mathbb{E}^{\mathbb{P}}\big[E_\theta(t) \,|\, \mathcal{F}_s\big]$$

$$= E_\theta(s) e^{-\frac{t-s}{2}|\theta(m2^{-N})|^2} \mathbb{E}^{\mathbb{P}}\Big[\exp\Big(\big(\theta(m2^{-N}), \beta(t) - \beta(s)\big)_{\mathbb{R}^n}\Big) \,\Big|\, \mathcal{F}_s\Big] = E_\theta(s)$$

since $\beta(t) - \beta(s)$ is $\mathbb{P}$-independent of $\mathcal{F}_s$. Clearly, for general $0 \leq s < t$, one gets the same conclusion by iterating the preceding result. Hence, we now know that $\big(E_\theta(t), \mathcal{F}_t, \mathbb{P}\big)$ is a martingale.

To complete the proof from here, first observe that we can replace θ by $\lambda\theta$ for any $\lambda \in \mathbb{R}$. In particular, this means that

$$\mathbb{E}^{\mathbb{P}}\big[e^{|I_\theta(t)|}\big] \leq e^{\frac{1}{2}A^2 t}\Big(\mathbb{E}^{\mathbb{P}}\big[E_\theta(t) + E_{-\theta}(t)\big]\Big) = 2e^{\frac{1}{2}A^2 t},$$

where A is the uniform upper bound on $|\theta(t, \omega)|$. At the same time, by Taylor's theorem,

$$\left| \frac{E_{\lambda\theta}(t) - 1}{\lambda} - I_\theta(t) \right| \leq \frac{\lambda}{2} e^{|I_\theta(t)|}$$

and

$$\left| \frac{E_{\lambda\theta}(t) + E_{-\lambda\theta}(t) - 2}{\lambda^2} - \big(I_\theta(t)^2 - A_\theta(t)\big) \right| \leq \frac{\lambda}{3} e^{|I_\theta(t)|}$$

for $0 < \lambda \leq 1$. Hence, by Lebesgue's dominated convergence Theorem, we get the desired conclusion after letting $\lambda \searrow 0$. $\square$

THEOREM 5.1.11. *There is a unique linear, isometric map $\theta \in \Theta^2(\mathbb{P}; \mathbb{R}^n)$ $\longmapsto I_\theta \in \mathcal{M}^2(\mathbb{P}; \mathbb{R})$ with the properties that I_θ is given by (5.1.7) for each $\theta \in S\Theta^2(\mathbb{P}; \mathbb{R})$. Moreover, given $\theta_1, \theta_2 \in \Theta^2(\mathbb{P}; \mathbb{R})$,*

$$(5.1.12) \qquad \left(I_{\theta_1}(t) I_{\theta_2}(t) - \int_0^t \big(\theta_1(\tau), \theta_2(\tau) \big)_{\mathbb{R}^n}, \mathcal{F}_t, \mathbb{P} \right)$$

is a martingale. Finally, if $\theta \in \Theta^2(\mathbb{P}; \mathbb{R})$ and $E_\theta(t)$ is defined as in (5.1.10), then $\big(E_\theta(t), \mathcal{F}_t, \mathbb{P} \big)$ is always a supermartingale and is a martingale if $A_\theta(T)$ is bounded for each $T \in [0, \infty)$. (See Exercises 5.1.24 and 5.3.4 for more refined information.)

PROOF: The existence and uniqueness of $\theta \rightsquigarrow I_\theta$ is immediate from Lemma 5.1.8. Indeed, from that lemma, we know that this map is linear and isometric on $S\Theta^2(\mathbb{P}; \mathbb{R}^n)$ and that $S\Theta^2(\mathbb{P}; \mathbb{R}^n)$ is dense in $\Theta^2(\mathbb{P}; \mathbb{R}^n)$. Furthermore, Lemma 5.1.8 says that $\big(I_\theta(t)^2 - A_\theta(t), \mathcal{F}_t, \mathbb{P} \big)$ is a martingale when $\theta \in S\Theta^2(\mathbb{P}; \mathbb{R})$, and so the general case follows from the fact that $I_{\theta_k}(t)^2 - A_{\theta_k}(t) \longrightarrow I_\theta(t)^2 - A_\theta(t)$ in $L^1(\mathbb{P}; \mathbb{R})$ when $\theta_k \longrightarrow \theta$ in $\Theta^2(\mathbb{P}; \mathbb{R})$. Knowing that $\big(I_\theta(t)^2 - A_\theta(t), \mathcal{F}_t, \mathbb{P} \big)$ is a martingale for each $\theta \in \Theta^2(\mathbb{P}; \mathbb{R}^n)$, we get (5.1.12) by polarization. That is, one uses the identity

$$I_{\theta_1}(t) I_{\theta_2}(t) - \int_0^t \big(\theta_1(\tau), \theta_2(\tau) \big)_{\mathbb{R}^n} d\tau$$
$$= \frac{1}{4} \left(I_{\theta_1 + \theta_2}(t)^2 - I_{\theta_1 - \theta_2}(t)^2 - A_{\theta_1 + \theta_2}(t) + A_{\theta_1 - \theta_2}(t) \right).$$

Finally, to prove the last assertion, choose $\{\theta_k\}_1^\infty \subseteq S\Theta^2(\mathbb{P}; \mathbb{R})$ so that $\theta_k \longrightarrow \theta$ in $\Theta^2(\mathbb{P}; \mathbb{R}^n)$. Because $\big(E_{\theta_k}(t), \mathcal{F}_t, \mathbb{P} \big)$ is a martingale for each k and $E_{\theta_k}(t) \longrightarrow E_\theta(t)$ in $\mathbb{P}$-probability for each $t \geq 0$, Fatou's lemma implies that $\big(E_\theta(t), \mathcal{F}_t, \mathbb{P} \big)$ is a supermartingale. Next suppose that θ is uniformly bounded by a constant $\Lambda < \infty$, and choose the θ_k's so that they are all uniformly bounded by Λ as well. Then, for each $t \in [0, T]$,

$$\mathbb{E}^{\mathbb{P}} \big[E_{\theta_k}(t)^2 \big] \leq e^{\Lambda t} \mathbb{E}^{\mathbb{P}} \big[E_{2\theta_k}(t) \big] = e^{\Lambda t},$$

and so $E_{\theta_k}(t) \longrightarrow E_\theta(t)$ in $L^1(\mathbb{P}; \mathbb{R})$ for each $t \in [0, T]$. Hence, we now know that $\big(E_\theta(t), \mathcal{F}_t, \mathbb{P} \big)$ is a martingale when θ is bounded. Finally, if $A_\theta(t, \omega) \leq \Lambda(t) < \infty$ for each $t \in [0, \infty)$ and $\theta_m(t, \omega) \equiv \mathbf{1}_{[0,m]} \big(|\theta(t, \omega)| \big) \theta(t, \omega)$, then $E_{\theta_m}(t) \longrightarrow E_\theta(t)$ in $\mathbb{P}$-probability and $\mathbb{E}^{\mathbb{P}} \big[E_{\theta_m}(t)^2 \big] \leq e^{\Lambda(t)}$ for each $t \geq 0$, which again is sufficient to show that $\big(E_\theta(t), \mathcal{F}_t, \mathbb{P} \big)$ is a martingale. $\square$

REMARK 5.1.13. With Theorem 5.1.11, we have completed the basic construction in Itô's theory of Brownian stochastic integration, and, as time goes on, we will increasingly often replace the notation I_θ by the more conventional notation

$$(5.1.14) \qquad \int_0^t \big(\theta(\tau), d\beta(\tau)\big)_{\mathbb{R}^n} = I_\theta(t).$$

Because it recognizes that Itô theory is very like a classical integration theory, (5.1.14) is good notation. On the other hand, it can be misleading. Indeed, one has to keep in mind that, in reality, Itô's "integral" is, like the Fourier transform on $\mathbb{R}$, defined only up to a set of measure 0 and via an L^2-completion procedure. In addition, for the cautionary reasons discussed in §§3.3.2—3.3.3, it is a serious mistake to put too much credence in the notion that an Itô integral behaves like a Riemann–Stieltjes integral.

§5.1.3. Stopping Stochastic Integrals and a Further Extension: The notation $\int_0^t \big(\theta(\tau), d\beta(\tau)\big)_{\mathbb{R}^n}$ for $I_\theta(t)$ should make one wonder to what extent it is true that, for $\zeta_1 \le \zeta_2$,

$$
\begin{aligned}
(5.1.15) \qquad \int_{t\wedge\zeta_1}^{t\wedge\zeta_2} \big(\theta(\tau), d\beta(\tau)\big)_{\mathbb{R}^n} &\equiv I_\theta(t \wedge \zeta_2) - I_\theta(t \wedge \zeta_1) \\
&= \int_0^t \mathbf{1}_{[\zeta_1,\zeta_2)}(t)\big(\theta(\tau), d\beta(\tau)\big)_{\mathbb{R}^n}.
\end{aligned}
$$

Of course, in order for the right hand side of the preceding to even make sense, it is necessary that $\mathbf{1}_{[\zeta_1,\zeta_2)}\theta$ be progressively measurable, which is more or less equivalent to insisting that ζ_1 and ζ_2 be stopping times.

LEMMA 5.1.16. *Given $\theta \in \Theta^2(\mathbb{P}; \mathbb{R}^n)$ and stopping times $\zeta_1 \le \zeta_2$, (5.1.15) holds $\mathbb{P}$-almost surely. In fact, if α is a bounded, $\mathcal{F}_{\zeta_1}$-measurable function and $\alpha\mathbf{1}_{[\zeta_1,\zeta_2)}\theta(t,\omega)$ equals $\alpha(\omega)\theta(t,\omega)$ or 0 depending on whether t is or is not in $[\zeta_1(\omega),\zeta_2(\omega))$, then $\alpha\mathbf{1}_{[\zeta_1,\zeta_2)}\theta \in \Theta^2(\mathbb{P};\mathbb{R}^n)$ and*

$$
\alpha\big(I_\theta(t \wedge \zeta_2) - I_\theta(t \wedge \zeta_1)\big) = \int_0^t \big(\alpha\mathbf{1}_{[\zeta_1,\zeta_2)}\theta(\tau), d\beta(\tau)\big)_{\mathbb{R}^n}.
$$

PROOF: Clearly, to check that $\tilde\theta \equiv \alpha\mathbf{1}_{[\zeta_1,\zeta_2)}\theta$ is in $\Theta^2(\mathbb{P};\mathbb{R}^n)$, it is enough to check that $\tilde\theta$ is progressively measurable, and this is an elementary exercise. Next, set

$$
\Delta(t) = I_\theta(t \wedge \zeta_2) - I_\theta(t \wedge \zeta_1) \quad \text{and} \quad \tilde I(t) = \int_0^t \big(\tilde\theta(\tau), d\beta(\tau)\big)_{\mathbb{R}^n}.
$$

Then, by Doob's stopping time theorem and (5.1.12),

$$\mathbb{E}^{\mathbb{P}}\left[\left|\alpha\Delta(t) - \tilde{I}(t)\right|^2\right] = \mathbb{E}^{\mathbb{P}}\left[\alpha^2\Delta(t)^2\right] - 2\mathbb{E}^{\mathbb{P}}\left[\alpha\Delta(t)\tilde{I}(t)\right] + \mathbb{E}^{\mathbb{P}}\left[\tilde{I}(t)^2\right]$$

$$= \mathbb{E}^{\mathbb{P}}\left[\alpha^2\int_{t\wedge\zeta_1}^{t\wedge\zeta_2}|\theta(\tau)|^2\,d\tau\right] - 2\mathbb{E}^{\mathbb{P}}\left[\alpha\int_{t\wedge\zeta_1}^{t\wedge\zeta_2}\left(\theta(\tau), \tilde{\theta}(\tau)\right)_{\mathbb{R}^n}\,d\tau\right]$$

$$+ \mathbb{E}^{\mathbb{P}}\left[\int_0^t\left|\tilde{\theta}(\tau)\right|^2\,d\tau\right] = 0. \quad\square$$

The preceding makes it possible to introduce the following extension of Itô's theory. Namely, define $\Theta_{\text{loc}}^2(\mathbb{P}; \mathbb{R}^n)$ to be the space of progressively measurable $\theta : [0, \infty) \times \Omega \longrightarrow \mathbb{R}^n$ with the property that, $\mathbb{P}$-almost surely, $A_\theta(t) \equiv \int_0^T |\theta(t)|^2\,dt < \infty$ for all $T \in [0, \infty)$. At the same time, define $\mathcal{M}_{\text{loc}}(\mathbb{P}; \mathbb{R})$ to be the space of continuous *local martingales*. That is, $M \in \mathcal{M}_{\text{loc}}(\mathbb{P}; \mathbb{R})$ if $M : [0, \infty) \times \Omega \longrightarrow \mathbb{R}$ is a progressively measurable function, $M(\,\cdot\,, \omega)$ is continuous for $\mathbb{P}$-almost every ω, and there exists a sequence $\{\zeta_k\}_1^\infty$ of stopping times such that $\zeta_k \nearrow \infty$ $\mathbb{P}$-almost surely and, for each $k \in \mathbb{Z}^+$, $\left(M(t \wedge \zeta_k), \mathcal{F}_t, \mathbb{P}\right)$ is a martingale.

THEOREM 5.1.17. *There is a unique linear map* $\theta \in \Theta_{\text{loc}}^2(\mathbb{P}; \mathbb{R}^n) \longmapsto I_\theta \in \mathcal{M}_{\text{loc}}(\mathbb{P}; \mathbb{R})$ *with the property that for any* $\theta \in \Theta_{\text{loc}}^2(\mathbb{P}; \mathbb{R}^n)$ *and stopping time* ζ:

$$\mathbb{E}^{\mathbb{P}}\left[\int_0^\zeta |\theta(\tau)|^2\,d\tau\right] < \infty \implies I_\theta(t \wedge \zeta) = \int_0^t \mathbf{1}_{[0,\zeta)}(\tau)\left(\theta(\tau), d\beta(\tau)\right)_{\mathbb{R}^n}.$$

Because it is completely consistent to do so, we will continue to use the notation $\int_0^t\left(\theta(\tau), d\beta(\tau)\right)_{\mathbb{R}^n}$ to denote I_θ, even when $\theta \in \Theta_{\text{loc}}^2(\mathbb{P}; \mathbb{R}^n)$.

Another direction in which it is useful to extend Itô's theory is to matrix-valued integrands. Namely, we have the following, which is a more or less trivial corollary of the preceding theorem.

COROLLARY 5.1.18. *Let* $\left(\beta(t), \mathcal{F}_t, \mathbb{P}\right)$ *be an* $\mathbb{R}^{n'}$*-valued Brownian motion, and suppose that* $\sigma : [0, \infty) \times \Omega \longrightarrow \text{Hom}(\mathbb{R}^{n'}; \mathbb{R}^n)$ *is a progressively measurable function with the property that*

$$(5.1.19) \qquad \int_0^T \|\sigma(t)\|_{\text{H.S.}}^2\,dt < \infty \quad \mathbb{P}\text{-almost surely for all } T \in (0, \infty).$$

Then there is a $\mathbb{P}$*-almost surely unique,* $\mathbb{R}^n$*-valued, progressively measurable function* $t \rightsquigarrow \int_0^t \sigma(\tau)\,d\beta(\tau)$ *with the property that,* $\mathbb{P}$*-almost surely,*

$$\left(\xi, \int_0^t \sigma(\tau)\,d\beta(\tau)\right)_{\mathbb{R}^n} = \int_0^t \left(\sigma(\tau)^\top\xi, d\beta(\tau)\right)_{\mathbb{R}^n} \quad \text{for all } (t, \xi) \in [0, \infty) \times \mathbb{R}^n.$$

In particular, if ζ is a stopping time and

$$\mathbb{E}^{\mathbb{P}}\left[\int_0^\zeta \|\sigma(\tau)\|_{\text{H.S.}}^2 \, d\tau\right] < \infty,$$

then

$$\left(\int_0^{t\wedge\zeta} \sigma(\tau) \, d\beta(\tau), \mathcal{F}_t, \mathbb{P}\right)$$

is an $\mathbb{R}^n$-valued martingale, and

$$\left(\left|\int_0^{t\wedge\zeta} \sigma(\tau) \, d\beta(\tau)\right|^2 - \int_0^{t\wedge\zeta} \|\sigma(\tau)\|_{\text{H.S.}}^2 \, d\tau, \mathcal{F}_t, \mathbb{P}\right)$$

is an $\mathbb{R}$-valued martingale.

§5.1.4. Exercises

EXERCISE 5.1.20. We claimed, but did not prove, that Itô's integration theory coincides with Riemann–Stieltjes's when the integrand has locally bounded variation. To be more precise, let $\theta \in \Theta_{\text{loc}}^2(\mathbb{P}; \mathbb{R}^n)$ and assume that $\theta(\,\cdot\,, \omega)$ has locally bounded variation for each $\omega \in \Omega$. Show that one version of the random variable $\omega \rightsquigarrow I_\theta(\,\cdot\,, \omega)$ is given by the indefinite, Riemann–Stieltjes integral of $\theta(\,\cdot\,, \omega)$ with respect to $\beta(\,\cdot\,, \omega)$.

Hint: First show that there is no loss in generality to assume that θ is uniformly bounded and that, for each ω, $\theta(\,\cdot\,, \omega)$ is right continuous at 0, and left continuous on $(0, \infty)$. Second, because $\theta(\,\cdot\,, \omega)$ is Riemann–Stieltjes integrable with respect to $\beta(\,\cdot\,, \omega)$ on each compact interval, verify that the Riemann–Stieltjes integral of $\theta(\,\cdot\,, \omega)$ on $[0, t]$ with respect to $\beta(\,\cdot\,, \omega)$ can be computed as the limit of the Riemann sums

$$\sum_{m=0}^\infty \left(\theta(m2^{-N}, \omega), \Delta_m^N \beta(t, \omega)\right)_{\mathbb{R}^n}.$$

Finally, use the boundedness of θ and the left continuity of $\theta(\,\cdot\,, \omega)$ to see that

$$\lim_{N\to\infty} \mathbb{E}^{\mathbb{P}}\left[\int_0^t |\theta(\tau) - \theta([\tau]_N)|^2 \, d\tau\right] = 0.$$

EXERCISE 5.1.21. One of the most important applications of the Paley–Wiener integral was made by Cameron and Martin. To explain their application, use, as in §3.1.1, $\mathbb{P}^0$ to denote the measure on $C([0, \infty); \mathbb{R}^n)$ corresponding to the Lévy system $(I, 0, 0)$. That is, $\mathbb{P}^0$ is the standard

Wiener measure on $C([0, \infty); \mathbb{R}^n)$, and so $(p(t), \overline{\mathcal{B}_t}, \mathbb{P}^0)$ is a Brownian motion. Next, given $\eta \in L^2([0, \infty); \mathbb{R}^n)$, set $h(t) = \int_0^t \eta(\tau) \, d\tau$, and let $\tau_h :$ $C([0, \infty); \mathbb{R}^n) \longrightarrow C([0, \infty); \mathbb{R}^n)$ denote the translation map given by $\tau_h(p)$ $= p + h$. Then the theorem of Cameron and Martin states that the measure $(\tau_h)_* \mathbb{P}^0$ is equivalent of $\mathbb{P}^0$ and that its Radon–Nikodym derivative R_h is given by the *Cameron–Martin formula*

$$(5.1.22) \qquad R_h = \exp\left(I_\eta(\infty) - \frac{1}{2} \|\eta\|_{L^2([0,\infty);\mathbb{R}^n)}^2 \right).$$

Prove their theorem.

Hint: There are several ways in which to prove their theorem. Perhaps the one best suited to the development here is to first prove that $\mathbb{P}^0$ can be characterized as the unique probability measure $\mathbb{P}$ on $C([0, \infty); \mathbb{R}^n)$ with the property that

$$\mathbb{E}^{\mathbb{P}}\left[e^{I_\theta(\infty)} \right] = \exp\left(\frac{1}{2} \|\theta\|_{L^2([0,\infty);\mathbb{R}^n)}^2 \right)$$

for all piecewise constant, compactly supported $\theta : [0, \infty) \longrightarrow \mathbb{R}^n$. (In this expression, $I_\theta(\infty, p)$ denotes the Riemann–Stieltjes integral of θ with respect to p over taken over the support of θ.) Knowing this, it is easy to check that $R_h^{-1} d(\tau_h)_* \mathbb{P}^0 = d\mathbb{P}^0$ for compactly supported, piecewise constant η's. To complete the proof, one need only construct a sequence $\{\eta_k\}_1^\infty$ of compact supported, piecewise constant functions so that $\eta_k \longrightarrow \eta$ in $L^2([0, \infty); \mathbb{R}^n)$. Since the corresponding functions h_k tend uniformly on compacts to h while the corresponding R_{h_k}'s tend in $L^1(\mathbb{P}^0; \mathbb{R})$ to R_h, the general case follows.

EXERCISE 5.1.23. Because we are dealing with processes which are almost surely continuous, most of the subtlety in the notion of a local martingale is absent. Indeed, given any progressively measurable $M : [0, \infty) \times \Omega \longrightarrow \mathbb{R}$ with the property that $M(\cdot, \omega)$ is continuous for $\mathbb{P}$-almost every ω, show that $(M(t), \mathcal{F}_t, \mathbb{P})$ is a local martingale if and only if, for each $k \in \mathbb{Z}^+$, $(M(t \wedge \zeta_k), \mathcal{F}_t, \mathbb{P})$ is a martingale, where $\zeta_k(\omega) \equiv \inf\{t \geq 0 : |M(t, \omega)| \geq k\}$. In this connection, show that a local martingale $(M(t), \mathcal{F}_t, \mathbb{P})$ is a martingale if $\|M\|_{[0,T]} \in L^1(\mathbb{P})$ for each $T \in [0, \infty)$. In particular, use this to see that an Itô stochastic integral $I_\theta(t)$ is a $\mathbb{P}$-square integrable martingale if and only if

$$\mathbb{E}^{\mathbb{P}}\left[\int_0^T |\theta(t)|^2 \, dt \right] < \infty \quad \text{for all } T \in [0, \infty).$$

In a slightly different direction, show that a local martingale $(M(t), \mathcal{F}_t, \mathbb{P})$ is a supermartingale if it is uniformly bounded from below.

EXERCISE 5.1.24. Assume that $\sigma : [0, \infty) \times \Omega \longrightarrow \mathrm{Hom}(\mathbb{R}^{n'}; \mathbb{R}^n)$ is a progressively measurable function for which (5.1.19) holds, set

$$X(t) = \int_0^t \sigma(\tau) \, d\beta(\tau) \quad \text{and} \quad A(t) = \int_0^t \sigma(\tau)\sigma(\tau)^\top \, d\tau,$$

and assume that, for some $T \in (0, \infty)$ and $\Lambda(T) \in (0, \infty)$, $\big(\xi, A(T)\xi\big)_{\mathbb{R}^n} \leq \Lambda(T)|\xi|^2$, $\xi \in \mathbb{R}^n$. Show that, for each $\epsilon \in (0, 1)$,

(5.1.25) $\qquad \mathbb{E}^{\mathbb{P}}\left[\exp\left(\frac{\epsilon\|X\|_{[0,T]}^2}{2\Lambda(T)}\right)\right] \leq e(1 - \epsilon)^{-\frac{n}{2}}.$

In particular, conclude that

(5.1.26) $\qquad \mathbb{P}\big(\|X\|_{[0,T]} \geq R\big) \leq e(1 - \epsilon)^{-\frac{n}{2}} e^{-\frac{\epsilon R^2}{2\Lambda(T)}} \quad \text{for all } R > 0,$

and

$$\mathbb{E}^{\mathbb{P}}\left[\exp\left(\frac{\|X\|_{[0,T]}^2}{2n\Lambda(T)}\right)\right] \leq e^{\frac{3}{2}} \quad \text{for all } n \geq 1.$$

Hint: Given $\xi \in \mathbb{R}^n$, set

$$X_\xi(t) = \big(\xi, X(t)\big)_{\mathbb{R}^n} \quad \text{and} \quad E(t, \xi) = \exp\left(X_\xi(t) - \tfrac{1}{2}\big(\xi, A(t)\xi\big)_{\mathbb{R}^n}\right),$$

and show, using the last part of Theorem 5.1.11 and Doob's inequality, that, for each $q \in (1, \infty)$,

$$\mathbb{E}^{\mathbb{P}}\left[\sup_{t \in [0,T]} e^{q\big(\xi, X(t)\big)_{\mathbb{R}^n}}\right] \leq e^{\frac{1}{2}q\Lambda(T)|\xi|^2} \mathbb{E}^{\mathbb{P}}\left[\|E(\,\cdot\,, \xi)\|_{[0,T]}^q\right]$$

$$\leq e^{\frac{1}{2}q\Lambda(T)|\xi|^2} \left(\frac{q}{q-1}\right)^q \mathbb{E}^{\mathbb{P}}\left[E(T, \xi)^q\right]$$

$$\leq e^{\frac{1}{2}q^2\Lambda(T)|\xi|^2} \left(\frac{q}{q-1}\right)^q \mathbb{E}^{\mathbb{P}}\left[E(T, q\xi)\right] = e^{\frac{1}{2}q\Lambda(T)|\xi|^2} \left(\frac{q}{q-1}\right)^q.$$

Next, multiply the preceding through by $(2\pi\tau)^{-\frac{n}{2}} e^{-\frac{|\xi|^2}{2\tau}}$ where $\tau = \frac{\epsilon}{q^2\Lambda(T)}$, and integrate with respect to ξ over $\mathbb{R}^n$. Finally, let $q \nearrow \infty$.

EXERCISE 5.1.27. In this exercise we will develop some of the intriguing relations between Itô stochastic integrals and Hermite polynomials. These relations reflect the residual Gaussian characteristics which stochastic integrals inherit from their driving Brownian forebears. Deeper examples of these relations will be investigated in Exercise 6.3.10.

(i) Given $a \geq 0$, define $H_m(x, a)$ for $x \in \mathbb{R}$ so that the identity

$$e^{\lambda x - \frac{\lambda^2}{2} a} = \sum_{m=0}^{\infty} \frac{\lambda^m}{m!} H_m(x, a), \quad \lambda \in \mathbb{R},$$

holds. Show that $H_m(x, 0) = x^m$ and, for $a > 0$, $H_m(x, a) = a^{\frac{m}{2}} H_m\left(a^{-\frac{1}{2}} x\right)$ where $H_m(x) \equiv H_m(x, 1)$. In addition, show that

$$H_m(x) = (-1)^m e^{\frac{x^2}{2}} \partial_x^m e^{-\frac{x^2}{2}},$$

and conclude that the H_m's can be generated inductively from $H_0(x) = 1$ and $H_m = (x - \partial_x) H_{m-1}$. In particular, use this to conclude that H_m is an mth order polynomial with the properties that: the coefficient of x^m is 1, the coefficient of x^ℓ is 0 unless ℓ has the same parity as m, and the constant term in H_{2m} is $(-1)^m \frac{(2m)!}{2^m m!}$.

(ii) Assume that $\theta \in \Theta^2(\mathbb{P}; \mathbb{R})$ with $A_\theta(T)$ is uniformly bounded, and show that $e^{|I_\theta(T)|}$ is $\mathbb{P}$-integrable. Use this along with the last part of Theorem 5.1.11 to justify

$$\mathbb{E}^{\mathbb{P}}\left[H_m\left(I_\theta(T), A_\theta(T) \right) \right] = \frac{d^m}{d\lambda^m} \mathbb{E}^{\mathbb{P}}\left[E_{\lambda\theta}(T) \right] \Big|_{\lambda=0} = 0$$

for all $m \geq 1$.

(iii) By combining (i) and (ii), show that there exist universal constant B_{2m} for $m \in \mathbb{Z}^+$ such that

$$B_{2m}^{-1} \mathbb{E}^{\mathbb{P}}\left[I_\theta(T)^{2m} \right] \leq \mathbb{E}^{\mathbb{P}}\left[A_\theta(T)^m \right] \leq B_{2m} \mathbb{E}^{\mathbb{P}}\left[I_\theta(T)^{2m} \right],$$

which is a primitive version of Burkholder's inequality. Finally, check that these inequalities continue to hold for any $\theta \in \Theta_{\text{loc}}^2(\mathbb{P}; \mathbb{R}^n)$, not just those for which $A_\theta(T)$ is uniformly bounded.

5.2 Itô's Integral Applied to Itô's Construction Method

Because our motivation for introducing stochastic integration was the desire to understand equations like (3.1.22), it seems reasonable to ask whether the theory developed in §5.2 does in fact do anything to increase our understanding. Of course, because we have been dealing with nothing but Brownian stochastic integrals, we will have to restrict our attention to processes for which the Lévy measure is 0.

§5.2.1. **Existence and Uniqueness:** Let $\sigma : [0, \infty) \times \Omega \times \mathbb{R}^{n'} \longrightarrow \text{Hom}(\mathbb{R}^{n'}; \mathbb{R}^n)$ and $b : [0, \infty) \times \Omega \times \mathbb{R}^n \longrightarrow \mathbb{R}^n$ be functions with the properties that, for each $x \in \mathbb{R}^n$, $(t, \omega) \rightsquigarrow \sigma(t, x, \omega)$ and $(t, \omega) \rightsquigarrow b(t, x, \omega)$ are progressively measurable, $(t, \omega) \rightsquigarrow \sigma(t, 0, \omega)$ and $(t, \omega) \rightsquigarrow b(t, 0, \omega)$ are bounded, and

$$
(5.2.1) \quad \sup_{x_1 \neq x_2} \sup_{(t, \omega)} \frac{\|\sigma(t, x_2, \omega) - \sigma(t, x_1, \omega)\|_{\text{H.S.}}}{|x_2 - x_1|}
$$
$$
\vee \frac{|b(t, x_2, \omega) - b(t, x_1, \omega)|}{|x_2 - x_1|} < \infty.
$$

THEOREM 5.2.2. *Given functions σ and b satisfying (5.2.1) and an $\mathbb{R}^{n'}$-valued Brownian motion $(\beta(t), \mathcal{F}_t, \mathbb{P})$, there exists for each $x \in \mathbb{R}^n$ a $\mathbb{P}$-almost surely unique $\mathbb{R}^n$-valued, progressively measurable function $X(\,\cdot\,, x)$ which solves*[5] *(cf. Corollary 5.1.18)*

$$
(5.2.3) \quad X(t, x) = x + \int_0^t \sigma(\tau, X(\tau, x)) \, d\beta(\tau) + \int_0^t b(\tau, X(\tau, x)) \, d\tau, \quad t \geq 0.
$$

Moreover, there exists a $C \in (0, \infty)$, depending only on

$$
\sup_{x \in \mathbb{R}^n} \sup_{(t, \omega)} \frac{\|\sigma(t, x, \omega)\|_{\text{H.S.}} \vee |b(t, x, \omega)|}{1 + |x|},
$$

such that, for all $T \in (0, \infty)$,

$$
\mathbb{E}^{\mathbb{P}}\left[\|X(\,\cdot\,, x)\|_{[0,T]}^2\right] \leq C(1 + |x|^2) e^{CT^2}
$$

and

$$
\mathbb{E}^{\mathbb{P}}\left[|X(t, x) - X(s, x)|^2\right] \leq C(1 + |x|^2) e^{CT^2}(t - s), \quad 0 \leq s < t \leq T.
$$

PROOF: Set $X_0(t, x) \equiv x$. Assuming that $X_N(\,\cdot\,, x)$ has been defined and that $t \rightsquigarrow \sigma(t, X_N(t, x))$ satisfies (5.1.19), set (cf. Corollary 5.1.18)

$$
(*) \quad X_{N+1}(t, x) = x + \int_0^t \sigma(\tau, X_N(\tau, x)) \, d\beta(\tau) + \int_0^t b(\tau, X_N(\tau, x)) \, d\tau,
$$

and observe that $t \rightsquigarrow \sigma(t, X_{N+1}(t, x))$ will again satisfy (5.1.19). Hence, by induction on $N \geq 0$, we can produce the sequence $\{X_N(\,\cdot\,, x) : N \geq 0\}$ so

---

[5] Just in case it is not clear, the condition that $X(\,\cdot\,, x)$ satisfy the equation implicitly contains the condition that $X(\,\cdot\,, x)$ is $\mathbb{P}^0$-almost surely continuous. In particular, this guarantees that the right hand side of (5.2.3) makes sense.

that, for each $N \in \mathbb{N}$, $t \rightsquigarrow \sigma(t, X_N(t, x))$ satisfies (5.1.19) and (*) holds. Moreover, by the last part of Corollary 5.1.18 plus Doob's inequality,

$$\mathbb{E}^{\mathbb{P}}\Big[\big\|X_1(\,\cdot\,,x) - X_0(\,\cdot\,,x)\big\|_{[0,T]}^2\Big] \leq 8\mathbb{E}^{\mathbb{P}}\left[\int_0^T \|\sigma(\tau,x)\|_{\text{H.S.}}^2\, d\tau\right]$$
$$+ 2T\mathbb{E}^{\mathbb{P}}\left[\int_0^T |b(\tau,x)|^2\, d\tau\right] < \infty$$

and, for any $N \geq 1$,

$$\mathbb{E}^{\mathbb{P}}\Big[\big\|X_{N+1}(\,\cdot\,,x) - X_N(\,\cdot\,,x)\big\|_{[0,T]}^2\Big]$$
$$\leq 8\mathbb{E}^{\mathbb{P}}\left[\int_0^T \big\|\sigma(\tau, X_N(\tau,x)) - \sigma(X_{N-1}(\tau,x))\big\|_{\text{H.S.}}^2\right]$$
$$+ 2T\mathbb{E}^{\mathbb{P}}\left[\int_0^T \big|b(\tau, X_N(\tau,x)) - b(X_{N-1}(\tau,x))\big|^2\, d\tau\right].$$

Thus, by (5.2.1), there is a $K < \infty$ such that

$$\mathbb{E}^{\mathbb{P}}\Big[\big\|X_{N+1}(\,\cdot\,,x) - X_N(\,\cdot\,,x)\big\|_{[0,T]}^2\Big]$$
$$\leq K(1+T)\int_0^T \mathbb{E}^{\mathbb{P}}\Big[\big\|X_N(\,\cdot\,,x) - X_{N-1}(\,\cdot\,,x)\big\|_{[0,t]}^2\Big]\, dt$$

for all $N \geq 1$; and so, by induction, we see that, for each $T \in (0,\infty)$,

$$\mathbb{E}^{\mathbb{P}}\Big[\big\|X_{N+1}(\,\cdot\,,x) - X_N(\,\cdot\,,x)\big\|_{[0,T]}^2\Big]$$
$$\leq \frac{K^N(1+T)^N T^N}{N!}\mathbb{E}^{\mathbb{P}}\Big[\big\|X_1(\,\cdot\,,x) - X_0(\,\cdot\,,x)\big\|_{[0,T]}^2\Big].$$

From the preceding it is clear that, for each $T \in (0,\infty)$,

$$\lim_{N\to\infty}\sup_{N'>N} \mathbb{E}^{\mathbb{P}}\Big[\big\|X_{N'}(\,\cdot\,,x) - X_N(\,\cdot\,,x)\big\|_{[0,T]}^2\Big] = 0.$$

Hence, we have now verified the existence statement.

To prove the uniqueness assertion, suppose that $X(\,\cdot\,,x)$ and $X'(\,\cdot\,,x)$ are two solutions, and note that, for each $k \geq 1$,

$$\mathbb{E}^{\mathbb{P}}\Big[\big\|X'(\,\cdot\,,x) - X(\,\cdot\,,x)\big\|_{[0,T\wedge\zeta_k]}^2\Big]$$
$$\leq K(1+T)\int_0^T \mathbb{E}^{\mathbb{P}}\Big[\big\|X'(\,\cdot\,,x) - X(\,\cdot\,,x)\big\|_{[0,t\wedge\zeta_k]}^2\Big]\, dt,$$

where $\zeta_k \equiv \inf\{t \geq 0 : |X(t,x)| \vee |X'(t,x)| \geq k\}$. But (e.g., by Gronwall's inequality) this means that

$$\mathbb{E}^{\mathbb{P}}\left[\left\|X'(\,\cdot\,,x) - X(\,\cdot\,,x)\right\|^2_{[0,T\wedge\zeta_k]}\right] = 0,$$

and so we obtain the desired conclusion after letting $k \nearrow \infty$.

Turning to the asserted estimates, note that

$$\mathbb{E}^{\mathbb{P}}\left[\left\|X(\,\cdot\,,x)\right\|^2_{[0,T]}\right] \leq 3|x|^2 + 12\mathbb{E}^{\mathbb{P}}\left[\int_0^T \left\|\sigma\big(t, X(t,x)\big)\right\|^2_{\text{H.S.}} dt\right]$$
$$+ 3T\mathbb{E}^{\mathbb{P}}\left[\int_0^T \left|b\big(t, X(t,x)\big)\right|^2 dt\right],$$

and so there exists a K, with the required dependence, such that

$$\mathbb{E}^{\mathbb{P}}\left[1 + \left\|X(\,\cdot\,,x)\right\|^2_{[0,T]}\right] \leq 3|x|^2 + K(1+T)\int_0^T \mathbb{E}^{\mathbb{P}}\left[1 + \left\|X(\,\cdot\,,x)\right\|^2_{[0,t]}\right] dt.$$

Hence the first estimate follows from Gronwall's inequality (2.2.12). As for the second estimate, assume that $0 \leq s < t \leq T$, and use

$$\mathbb{E}^{\mathbb{P}}\left[\left|X(t,x) - X(s,x)\right|^2\right] \leq 2\mathbb{E}^{\mathbb{P}}\left[\int_s^t \left\|\sigma\big(\tau, X(\tau,x)\big)\right\|^2_{\text{H.S.}} d\tau\right]$$
$$+ 2T\mathbb{E}^{\mathbb{P}}\left[\int_s^t \left|b\big(\tau, X(\tau,x)\big)\right|^2 d\tau\right]$$

together with the first estimate to arrive at the second estimate. $\square$

COROLLARY 5.2.4. *Assume that* $\sigma : \mathbb{R}^n \longrightarrow \text{Hom}(\mathbb{R}^{n'};\mathbb{R}^n)$ *and* $b : \mathbb{R}^n \longrightarrow \mathbb{R}^n$ *are continuous functions which satisfy*

$$\sup_{x_1 \neq x_2} \frac{\|\sigma(x_2) - \sigma(x_1)\|_{\text{H.S.}} \vee |b(x_2) - b(x_1)|}{|x_2 - x_1|} < \infty.$$

Refer to the setting in §3.1.2, and take $M' = 0$ *and* $F \equiv 0$. *Then* $(p'(t), \overline{\mathcal{B}'_t}, \mathbb{P}^0)$ *is an* $\mathbb{R}^{n'}$-*valued Browning motion and the function* $p' \rightsquigarrow X(\,\cdot\,,x,p') \in C([0,\infty);\mathbb{R}^n)$ *described in* (G1) *is the* $\mathbb{P}^0$-*almost surely unique,* $\{\overline{\mathcal{B}'_t} : t \geq 0\}$-*progressively measurable solution to*

$$(5.2.5) \qquad X(t,x) = x + \int_0^t \sigma\big(X(\tau,x)\big)\, dp'(\tau) + \int_0^t b\big(X(\tau,x)\big)\, d\tau, \ t \geq 0.$$

PROOF: Let $X(\,\cdot\,,x)$ be the solution to (5.2.5), and define $\{X^N(\,\cdot\,,x) : N \geq 0\}$ as in **(H4)** of §3.1.2. Note that

$$X^N(t,x,p') = x + \int_0^t \sigma\big(X^N([\tau]_N,x,p')\big)\, dp'(\tau) + \int_0^t b\big(X^N([\tau]_N,x,p')\big)\, d\tau.$$

Thus,

$$\mathbb{E}^{\mathbb{P}^0}\left[\big\|X(\,\cdot\,,x) - X^N(\,\cdot\,,x)\big\|_{[0,t]}^2\right]$$

$$\leq 16\mathbb{E}^{\mathbb{P}}\left[\int_0^t \big\|\sigma\big(X([\tau]_N,x)\big) - \sigma\big(X^N([\tau]_N,x)\big)\big\|_{\text{H.S.}}^2\, d\tau\right]$$

$$+ 4t\mathbb{E}^{\mathbb{P}}\left[\int_0^t \big|b\big(X([\tau]_N,x)\big) - b\big(X^N([\tau]_N,x)\big)\big|^2\, d\tau\right]$$

$$+ 16\mathbb{E}^{\mathbb{P}}\left[\int_0^t \big\|\sigma\big(X(\tau,x)\big) - \sigma\big(X([\tau]_N,x)\big)\big\|_{\text{H.S.}}^2\, d\tau\right]$$

$$+ 4t\mathbb{E}^{\mathbb{P}}\left[\int_0^t \big|b\big(X(\tau,x)\big) - b\big(X([\tau]_N,x)\big)\big|^2\, d\tau\right];$$

and so, by the last estimates in Theorem 5.2.2 and the Lipschitz estimates on σ and b, we see that, for each $(T,x) \in (0,\infty)\times\mathbb{R}^n$, there is a $K(T,x) < \infty$ such that

$$\mathbb{E}^{\mathbb{P}^0}\left[\big\|X(\,\cdot\,,x) - X^N(\,\cdot\,,x)\big\|_{[0,t]}^2\right]$$

$$\leq K(T,x)2^{-N} + K(T,x)\int_0^t \mathbb{E}^{\mathbb{P}^0}\left[\big\|X(\,\cdot\,,x) - X^N(\,\cdot\,,x)\big\|_{[0,\tau]}^2\right]\, d\tau$$

for all $t \in [0,T]$. Finally, apply Gronwall's inequality to conclude that

$$\lim_{N\to\infty}\mathbb{E}^{\mathbb{P}^0}\left[\big\|X(\,\cdot\,,x) - X^N(\,\cdot\,,x)\big\|_{[0,t]}^2\right] = 0. \quad \square$$

REMARK 5.2.6. In the literature, (5.2.5) is called the *stochastic integral equation* for a diffusion processes with *diffusion coefficient* $\sigma\sigma^\top$ and *drift coefficient* b. Often such equations are written in differential notation:

$$(5.2.7) \quad dX(t,x) = \sigma\big(X(t,x)\big)\, dp'(t) + b\big(X(t,x)\big)\, dt \quad \text{with } X(0,x) = x,$$

in which case they are called *stochastic differential equations*. It is important to recognize that the joint distribution of $p' \rightsquigarrow (p', X(\,\cdot\,,x,p'))$ under $\mathbb{P}^0$ would be the same were we to replace the canonical Brownian motion $(p'(t),\overline{\mathcal{B}_t'},\mathbb{P}^0)$ by any other $\mathbb{R}^{n'}$-valued Brownian motion $(\beta(t),\mathcal{F}_t,\mathbb{P})$. Indeed, as both the construction in Theorem 5.2.2 and the one in §3.1.2 make clear, this joint distribution depends only on the distribution of $\omega \rightsquigarrow \beta(\,\cdot\,,\omega)$, which is the same no matter which realization of Brownian motion is used.

§5.2.2. Subordination: When solving a system of ordinary differential equations one can sometimes take advantage of an inherent lower triangularity in the system. That is, one may be able to arrange the equations in such a way that the system contains a subsystem which is autonomous onto itself. In this case, the entire system can be solved by first solving the autonomous subsystem, plugging the solution into the remaining equations, and then solving resulting, now time dependent, equations.

The same device applies when solving stochastic differential equations, and is particularly interesting in the following situation. Let $\sigma : \mathbb{R}^n \longrightarrow \mathrm{Hom}(\mathbb{R}^{n'}; \mathbb{R}^n)$ and $b : \mathbb{R}^n \longrightarrow \mathbb{R}^n$ be as in Corollary 5.2.4. Suppose that $n = k + \ell$, $n' = k' + \ell'$, and

$$\sigma(z) = \begin{pmatrix} \alpha(x) & 0 \\ 0 & \beta(x,y) \end{pmatrix} \ \& \ b(z) = \begin{pmatrix} v(x) \\ w(x,y) \end{pmatrix} \quad \text{for } z = (x, y) \in \mathbb{R}^k \times \mathbb{R}^\ell,$$

where $v : \mathbb{R}^k \longrightarrow \mathbb{R}^k$, $w : \mathbb{R}^n \longrightarrow \mathbb{R}^\ell$, $\alpha : \mathbb{R}^k \longrightarrow \mathrm{Hom}(\mathbb{R}^{k'}; \mathbb{R}^k)$ and $\beta : \mathbb{R}^n \longrightarrow \mathrm{Hom}(\mathbb{R}^{\ell'}; \mathbb{R}^\ell)$. Next, define $\Sigma : [0, \infty) \times \mathbb{R}^\ell \times C([0, \infty); \mathbb{R}^k) \longrightarrow \mathrm{Hom}(\mathbb{R}^{\ell'}; \mathbb{R}^\ell)$ and $B : [0, \infty) \times \mathbb{R}^\ell \times C([0, \infty); \mathbb{R}^\ell) \longrightarrow \mathbb{R}^\ell$ so that

$$\Sigma(t, y; p) = \beta(p(t), y) \text{ and } B(t, y; p) \equiv w(p(t), y).$$

Finally, write $r'(t) = (p'(t), q'(t))$, where $p'(t)$ and $q'(t)$ are, respectively, the orthogonal projections of $r'(t)$ onto $\mathbb{R}^{k'}$ and $\mathbb{R}^{\ell'}$ thought of as subspaces of $\mathbb{R}^{n'} = \mathbb{R}^{k'} \oplus \mathbb{R}^{\ell'}$, and note that, when $C([0, \infty); \mathbb{R}^{n'})$ is identified with $C([0, \infty); \mathbb{R}^{k'}) \times C([0, \infty); \mathbb{R}^{\ell'})$, $\mathbb{P}^0$ can be identified with $\mathbb{P} \times \mathbb{Q}$, where $\mathbb{P}$ and $\mathbb{Q}$ are the standard Wiener measures on $C([0, \infty); \mathbb{R}^{k'})$ and $C([0, \infty); \mathbb{R}^{\ell'})$, respectively.

Under these conditions, we want go about solving (5.2.5) by first solving the equation

$$X(t, x, p') = x + \int_0^t \alpha(X(\tau, x, p')) \, dp'(\tau)$$

$$+ \int_0^t v(X(\tau, x, p')) \, d\tau,$$

relative to the Wiener measure $\mathbb{P}$, then, for each $p \in C([0, \infty); \mathbb{R}^k)$ and $y \in \mathbb{R}^\ell$, solving

$$Y(t, y, q'; p) = y + \int_0^t \Sigma(\tau, Y(\tau, y, q'; p); p) \, dq'(\tau)$$

$$+ \int_0^t B(\tau, Y(\tau, y, q'; p); p) \, d\tau$$

relative to $\mathbb{Q}$, and finally showing that

(5.2.8) $t \rightsquigarrow Z\big(t, z, (p', q')\big) = \begin{pmatrix} X(t, x, p') \\ Y\big(t, y, q'; X(\,\cdot\,, x, p')\big) \end{pmatrix}$

is a solution to original stochastic integral equation whose coefficients were σ and b.

The following theorem says that this subordination procedure works.

THEOREM 5.2.9. *Let $r' = (p', q') \rightsquigarrow Z(\,\cdot\,, z, r')$ be defined for $z = (x, y)$ as in (5.2.8). Then $r' \rightsquigarrow Z(\,\cdot\,, z, r')$ is the $\mathbb{P}^0$-almost surely unique solution to the stochastic integral equation*

$$\tilde{Z}(t, z, r') = z + \int_0^t \sigma\big(\tilde{Z}(\tau, z, r')\big)\, dr'(\tau) + \int_0^t b\big(\tilde{Z}(\tau, z, r')\big)\, d\tau, \quad t \geq 0.$$

Hence, if $\Phi : C\big([0, \infty); \mathbb{R}^k\big) \times C\big([0, \infty); \mathbb{R}^\ell\big) \longrightarrow \mathbb{R}$ is measurable and either bounded or non-negative, then

$$\mathbb{E}^{\mathbb{P}^0}\Big[\Phi\big(\tilde{Z}(\,\cdot\,, z, r')\big)\Big]$$

$$= \int \left(\int \Phi\big(X(\,\cdot\,, x, p'); Y(\,\cdot\,, y, q'; X(\,\cdot\,, x, p'))\big)\, \mathbb{Q}(dq') \right) \mathbb{P}(dp').$$

PROOF: The proof of this result is really just an exercise in the use of Fubini's theorem. Namely, let $r' \rightsquigarrow \tilde{Z}(\,\cdot\,, z, r')$ be the $\mathbb{P}^0$-almost surely unique solution to the stochastic integral equation

$$\tilde{Z}(t, z, r') = z + \int_0^t \sigma\big(\tilde{Z}(\tau, z, r')\big)\, dr'(\tau) + \int_0^t b\big(\tilde{Z}(\tau, z, r')\big)\, d\tau, \quad t \geq 0.$$

What we have to show is that $Z(\,\cdot\,, z, r') = \tilde{Z}(\,\cdot\,, z, r')$ for $\mathbb{P}^0$-almost every $r' \in C\big([0, \infty); \mathbb{R}^n\big)$.

Let $\tilde{X}(t, z, r')$ and $\tilde{Y}(t, z, r')$ be the orthogonal projections of $\tilde{Z}(t, z, r')$ onto $\mathbb{R}^k$ and $\mathbb{R}^\ell$. By Fubini's theorem, we know that, for $\mathbb{Q}$-almost every q', the function $p' \rightsquigarrow \tilde{X}\big(\,\cdot\,, z, (p', q')\big)$ satisfies

$$\tilde{X}\big(t, z, (p', q')\big) = x + \int_0^t \alpha\big(\tilde{X}(\tau, z, (p', q'))\big)\, dp'(\tau)$$

$$+ \int_0^t v\big(\tilde{X}(\tau, z, (p', q'))\big)\, d\tau, \quad t \geq 0,$$

$\mathbb{P}$-almost surely. Thus, by uniqueness, for $\mathbb{Q}$-almost every q', $\tilde{X}\big(\,\cdot\,, z, (p', q')\big) = X(\,\cdot\,, x, p')$ for $\mathbb{P}$-almost every p'.

Starting from the preceding and making another application of Fubini's theorem, we see that, for $\mathbb{P}$-almost every p', the function $q' \rightsquigarrow \tilde{Y}(\cdot, z, (p', q'))$ satisfies

$$\tilde{Y}(t, z, (p', q')) = y + \int_0^t \beta\Big(X(\tau, x, p'), \tilde{Y}(\tau, z, (p', q'))\Big) dq'(\tau)$$
$$+ \int_0^t w\Big(X(\tau, x, p'), \tilde{Y}(\tau, z, (p', q'))\Big) d\tau, \quad t \geq 0,$$

$\mathbb{Q}$-almost surely. Hence, by uniqueness, for $\mathbb{P}$-almost every p', $\tilde{Y}(\cdot, z, (p', q')) = Y(\cdot, y, q'; X(\cdot, x, p'))$ $\mathbb{Q}$-almost surely. After combining these two and making yet another application of Fubini's theorem, we have now shown that $\tilde{Z}(\cdot, z, r') = Z(\cdot, z, r')$ for $\mathbb{P}^0$-almost every r'. $\quad\square$

§5.2.3. Exercises

EXERCISE 5.2.10. Here is an example which indicates how Theorem 5.2.9 gets applied. Namely, referring to the setting in that theorem, suppose that β and w are independent of y. That is, $\beta(x, y) = \beta(x)$ and $w(x, y) = w(x)$. Next, define $V : [0, \infty) \times C([0, \infty); \mathbb{R}^k) \longrightarrow \mathrm{Hom}(\mathbb{R}^\ell; \mathbb{R}^\ell)$ and $B : [0, \infty) \times C([0, \infty); \mathbb{R}^k) \longrightarrow \mathbb{R}^\ell$ so that

$$V(t, p) = \int_0^t \beta\beta^\top(p(\tau)) \, d\tau \text{ and } B(t, p) = \int_0^t w(p(\tau)) \, d\tau,$$

and let $\Gamma_t(p, d\eta)$ denote the Gaussian measure on $\mathbb{R}^\ell$ with mean $B(t, p)$ and covariance $V(t, p)$. Given a stopping time $\zeta : C([0, \infty); \mathbb{R}^{k'}) \longmapsto [0, \infty]$ and a bounded measurable $f : \mathbb{R}^n \longrightarrow \mathbb{R}$ show that

$$\mathbb{E}^{\mathbb{P}^0}\Big[f\big(\tilde{Z}(\zeta(p'), z, r')\big), \, \zeta(p') < \infty\Big] = \mathbb{E}^{\mathbb{P}}[\Phi, \, \zeta < \infty] \quad \text{where}$$

$$\Phi(p') = \int_{\mathbb{R}^\ell} f\big(X(\cdot, x, p'), y + \eta\big) \Gamma_{\zeta(p')}\big(X(\cdot, x, p'), d\eta\big) \text{ for } \zeta(p') < \infty.$$

5.3 Itô's Formula

The jewel in the crown of Itô's stochastic integration theory is Itô's formula. Depending on one's point of view, his formula can be seen as the solution to any one of a variety of problems. From the point of view which comes out of the considerations in Chapters 3 and 4, especially **(G2)** in §3.1.2, it gives us a representation for the martingales being discussed there. From a more general standpoint, Itô's formula is the fundamental theorem of the calculus for which his integration theory is the integral, and, as such, it is the identity on which nearly everything else relies.

In the following statements, $(\beta(t), \mathcal{F}_t, \mathbb{P})$ is an $\mathbb{R}^n$-valued Brownian motion, and

(**X**) $X : [0, \infty) \times \Omega \longrightarrow \mathbb{R}^k$ is progressively measurable and, for $\mathbb{P}$-almost every ω, $X(\cdot, \omega)$ is a continuous function of locally bounded variation;

(**Y**) $Y : [0, \infty) \times \Omega \longrightarrow \mathbb{R}^\ell$ is a progressively measurable function and, for each $1 \leq j \leq \ell$, the jth component Y_j of $Y - Y(0)$ is the $d\beta(t)$-Itô stochastic integral I_{θ_j} of some $\theta_j \in \Theta^2_{\text{loc}}(\mathbb{P}; \mathbb{R}^n)$;

(**Z**) $Z(t) = \big(X(t), Y(t)\big)$.

With this notation, *Itô's formula* is the formula given in the next theorem. Our proof is based on the technique introduced by Kunita and Watanabe in their now famous article [21].

THEOREM 5.3.1. *Given any* $F \in C^{1,2}(\mathbb{R}^k \times \mathbb{R}^\ell; \mathbb{R})$,

$$
F\big(Z(t)\big) - F\big(Z(0)\big)
$$

$$
= \sum_{i=1}^{k} \int_0^t \partial_{x_i} F\big(Z(\tau)\big) \, dX_i(\tau) + \sum_{j=1}^{\ell} \int_0^t \partial_{y_j} F\big(Z(\tau)\big) \big(\theta_j(\tau), d\beta(\tau)\big)_{\mathbb{R}^n}
$$

$$
+ \frac{1}{2} \sum_{j,j'=1}^{\ell} \big(\theta_j(\tau), \theta_{j'}(\tau)\big)_{\mathbb{R}^n} \partial_{y_j} \partial_{y_{j'}} F\big(Z(\tau)\big) \, d\tau, \quad t \geq 0,
$$

$\mathbb{P}$-almost surely. Here, dX_i-integrals are taken in the sense of Riemann–Stieltjes and $d\beta$-integrals are taken in the sense of Itô.

PROOF: We begin by making several reductions. In the first place, without loss in generality, we may assume that $Z(\cdot, \omega)$ is continuous for all $\omega \in \Omega$. Secondly, we may assume that all the X_i's have uniformly bounded variation and that all the θ_j's are elements of $\Theta^2(\mathbb{P}; \mathbb{R}^n)$. Indeed, if this is not the case already, then we can introduce the stopping times

$$
\zeta_R = \inf \left\{ \sum_{i=1}^{k} \text{var}_{[0,t]}(X_i) + \sum_{j=1}^{\ell} \int_0^t \big|\theta_j(\tau)\big|^2 \, d\tau \geq R \right\},
$$

replace $Z(t)$ by $Z(t \wedge \zeta_R)$, and, at the end, let $R \nearrow \infty$. Similarly, we may assume that F has compact support. Finally, under these assumptions, a standard mollification procedure makes it clear that we need only handle F's which are both smooth and compactly supported. Thus, from now on, we will be assuming0 that $Z(\cdot, \omega)$ is continuous for all ω, the X_i's have uniformly bounded variation, the θ_j's are elements of $\Theta^2(\mathbb{P}; \mathbb{R}^n)$, and F is smooth and compactly supported. Finally, by continuity, it suffices to prove that the identity holds $\mathbb{P}$-almost surely for each $t \geq 0$.

Given $N \in \mathbb{N}$ and $t \in (0, \infty)$, define stopping times $\{\zeta_m^N : m \geq 0\}$ so that $\zeta_0^N \equiv 0$ and

$$\zeta_{m+1}^N = \inf\left\{\tau \geq \zeta_m^N : \left|Z(\tau) - Z(\zeta_m^N)\right| \vee \max_{1 \leq j \leq \ell} \int_{\zeta_m^N}^\tau |\theta_j(\sigma)|^2 \, d\sigma \geq 2^{-N}\right\} \wedge t.$$

Next, set $Z_m^N = (X_m^N, Y_m^N) = Z(\zeta_m^N)$, and $\Delta_{m,j}^N = Y_j(\zeta_{m+1}^N) - Y_j(\zeta_m^N)$. By continuity, we know that, for each ω, $\zeta_m^N(\omega) = t$ for all but a finite number of m's. Hence,

$$F\big(Z(t)\big) - F\big(Z(0)\big) = \sum_{m=0}^\infty \big(F(Z_{m+1}^N) - F(Z_m^N)\big).$$

Furthermore, by the fundamental theorem of calculus,

$$F(Z_{m+1}^N) - F(Z_m^N) = F(X_m^N, Y_{m+1}^N) - F(Z_m^N)$$
$$+ \sum_{i=1}^k \int_{\zeta_m^N}^{\zeta_{m+1}^N} \partial_{x_i} F\big(X(\tau), Y_{m+1}^N\big) \, dX_i(\tau),$$

and, by Taylor's theorem,

$$F(X_m^N, Y_{m+1}^N) - F(Z_m^N) = \sum_{j=1}^\ell \partial_{y_j} F(Z_m^N) \Delta_{m,j}^N$$
$$+ \frac{1}{2} \sum_{j,j'=1}^\ell \partial_{y_j} \partial_{y_{j'}} F(Z_m^N) \Delta_{m,j}^N \Delta_{m,j'}^N + E_m^N,$$

where there exists a $C_3 < \infty$, depending only on the bound on the third order derivatives of F, such that

$$|E_m^N| \leq C_3 2^{-N} \sum_{j=1}^\ell (\Delta_{m,j}^N)^2.$$

Next, we use Lemma 5.1.16 to first write

$$\partial_{y_j} F(Z_m^N) \Delta_{m,j}^N = \int_0^t \big(\partial_{y_j} F(Z_m^N) \mathbf{1}_{[\zeta_m^N, \zeta_{m+1}^N)} \theta_j(\tau), d\beta(\tau)\big)_{\mathbb{R}^n},$$

and then conclude that

$$F\big(Z(t)\big) - F\big(Z(0)\big) = \sum_{i=1}^k \int_0^t F_i^N(\tau) \, dX_i(\tau) + \sum_{j=1}^\ell \int_0^t \partial_{y_j} \big(\tilde{\theta}_j^N(\tau), d\beta(\tau)\big)_{\mathbb{R}^n}$$
$$+ \frac{1}{2} \sum_{m=0}^\infty \sum_{j,j'=1}^\ell \partial_{y_j} \partial_{y_{j'}} F(Z_m^N) \Delta_{m,j}^N \Delta_{m,j'}^N + \sum_{m=0}^\infty E_m^N,$$

where

$$F_i^N(\tau) \equiv \partial_{x_i} F\big(X(\tau), Y_{m+1}^N\big) \ \& \ \tilde{\theta}_j^N(\tau) \equiv \partial_{y_j} F(Z_m^N)\theta_j(\tau) \text{ for } \tau \in [\zeta_m^N, \zeta_{m+1}^N).$$

Because $|E_m^N| \le C_3 2^{-N} \sum_{j=1}^\ell (\Delta_{m,j}^N)^2$ and

$$\mathbb{E}^\mathbb{P} \left[\sum_{m=0}^\infty (\Delta_{m,j}^N)^2 \right] = \mathbb{E}^\mathbb{P} \left[\left(\sum_{m=0}^\infty (\Delta_{m,j}^N) \right)^2 \right] = \mathbb{E}^\mathbb{P} \big[(Y_j(t) - Y_j(0))^2 \big],$$

we know that $\sum_{m=0}^\infty E_m^N$ tends to 0 in $L^2(\mathbb{P}; \mathbb{R})$ as $N \to \infty$. At the same time, it is clearly that, if C_2 is a bound on the second derivatives of F, then

$$\left| \int_0^t F_i^N(\tau)\, dX_i(\tau) - \int_0^t \partial_{x_i} F\big(Z(\tau)\big)\, dX_i(\tau) \right| \le C_2 2^{-N} \mathrm{var}_{[0,t]}(X_i)$$

and

$$\mathbb{E}^\mathbb{P} \left[\left| \int_0^t \big(\tilde{\theta}_j^N(\tau), d\beta(\tau) \big)_{\mathbb{R}^n} - \int_0^t \big(\partial_{y_j} FZ(\tau) \big)\big(\theta_j(\tau), d\beta(\tau) \big)_{\mathbb{R}^n} \right|^2 \right]$$

$$= \mathbb{E}^\mathbb{P} \left[\int_0^t \big| \tilde{\theta}_j^N(\tau) - \partial_{y_j} F(Z(\tau)) \theta_j(\tau) \big|^2 d\tau \right] \le C_2^2 4^{-N} \|\theta_j\|_{\Theta^2(\mathbb{P};\mathbb{R}^n)}^2.$$

Hence, since it is obvious that

$$\sum_{m=0}^\infty \int_{\zeta_m^N}^{\zeta_{m+1}^N} \big(\theta_j(\tau), \theta_{j'}(\tau) \big)_{\mathbb{R}^n} \partial_{y_j} \partial_{y_{j'}} F\big(Z(\zeta_m^N) \big)\, d\tau$$

$$\longrightarrow \int_0^t \big(\theta_j(\tau), \theta_{j'}(\tau) \big)_{\mathbb{R}^n} \partial_{y_j} \partial_{y_{j'}} F\big(Z(\tau) \big)\, d\tau,$$

all that remains is to show that, for all $1 \le j, j' \le \ell$,

$$\sum_{m=0}^\infty \partial_{y_j} \partial_{y_{j'}} F(Z_m^N) \left(\Delta_{m,j}^N \Delta_{m,j'}^N - \int_{\zeta_m^N}^{\zeta_{m+1}^N} \big(\theta_j(\tau), \theta_{j'}(\tau) \big)_{\mathbb{R}^n}\, d\tau \right) \longrightarrow 0$$

in $\mathbb{P}$-measure. To this end, remark that

$$\mathbb{E}^\mathbb{P} \big[\Delta_{m,j}^N \Delta_{m,j'}^N \mid \mathcal{F}_{\zeta_t^N} \big] = \mathbb{E}^\mathbb{P} \big[Y_j(\zeta_{m+1}^N) Y_{j'}(\zeta_{m+1}^N) - Y_j(\zeta_m^N) Y_{j'}(\zeta_m^N) \mid \mathcal{F}_{\zeta_m^N} \big]$$

$$= \mathbb{E}^\mathbb{P} \left[\int_{\zeta_m^N}^{\zeta_{m+1}^N} \big(\theta_j(\tau), \theta_{j'}(\tau) \big)_{\mathbb{R}^n}\, d\tau \mid \mathcal{F}_{\zeta_m^N} \right],$$

and conclude that the terms

$$S^N_{m,j,j'} \equiv \partial_{y_j} \partial_{y_{j'}} F(Z^N_m) \Delta^N_{m,j} \Delta^N_{m,j'}$$

$$- \int_{\zeta^N_m}^{\zeta^N_{m+1}} \left(\theta_j(\tau), \theta_{j'}(\tau) \right)_{\mathbb{R}^n} \partial_{y_j} \partial_{y_{j'}} F(Z^N_m) \, d\tau$$

are orthogonal for different m's. Hence,

$$\mathbb{E}^{\mathbb{P}} \left[\left(\sum_{m=0}^{\infty} S^N_{m,j,j'} \right)^2 \right] = \sum_{m=0}^{\infty} \mathbb{E}^{\mathbb{P}} \left[(S^N_{m,j,j'})^2 \right].$$

At the same time,

$$(S^N_{m,j,j'})^2 \leq C_2^2 4^{-N} \left((\Delta^N_{m,j})^2 + (\Delta^N_{m,j'})^2 + \int_{\theta_j}^{\zeta^N_{m+1}} \left(|\theta_j(\tau)|^2 + +|\theta_{j'}(\tau)|^2 \right) d\tau \right).$$

Hence,

$$\mathbb{E}^{\mathbb{P}} \left[\left(\sum_{m=0}^{\infty} S^N_{m,j,j'} \right)^2 \right]$$

$$\leq C_2 4^{-N} \mathbb{E}^{\mathbb{P}} \left[\left(Y_j(t) - Y_j(0) \right)^2 + \left(Y_{j'}(t) - Y_{j'}(0) \right)^2 \right.$$

$$\left. + \int_0^t \left(|\theta_j(\tau)|^2 + \theta_{j'}(\tau)|^2 \right) d\tau \right]$$

$$\leq 2C_2 4^{-N} \mathbb{E}^{\mathbb{P}} \left[|Y(t) - Y(0)|^2 \right]. \quad \square$$

REMARK 5.3.2. If one likes to think about the fundamental theorem of calculus in terms of differentials, then the following is an appealing way to think about Itô's formula. In the first place, given a progressively measurable function $\alpha : [0, \infty) \times \Omega \longrightarrow \mathbb{R}$ which satisfies

$$\mathbb{P} \left(\int_0^t \alpha(\tau)^2 |\theta_j(\tau)|^2 \, d\tau < \infty \right) = 1 \quad \text{for all } t \geq 0,$$

one should introduce the notation

$$\int_0^t \alpha(\tau) dY_j(\tau) \equiv \int_0^t \left(\alpha(\tau)\theta_j(\tau), d\beta(\tau) \right)_{\mathbb{R}^n}.$$

That is, in terms of "differentials," "$dY_j(t) \equiv \big(\theta_j(t), d\beta(t)\big)_{\mathbb{R}^N}$." At the same time, one should define

$$\int_0^t \alpha(\tau)\, dY_j(\tau) dY_{j'}(\tau) \equiv \int_0^t \alpha(\tau)\big(\theta_j(\tau), \theta_{j'}(\tau)\big)_{\mathbb{R}^n} d\tau$$

for progressively measurable α's satisfying

$$\mathbb{P}\left(\int_0^t |\alpha(\tau)|\big(|\theta_j(\tau)|^2 + |\theta_{j'}(\tau)|^2\big)\, d\tau < \infty\right) = 1 \quad \text{for all } t \geq 0.$$

In terms of "differentials," this means that we are taking "$dY_j(t)dY_{j'}(t) = \big(\theta_j(t), \theta_{j'}(t)\big)_{\mathbb{R}^n} dt$." Finally, define "$dX_i(t)dX_{i'}(t) = 0 = dX_i(t)dY_j(t)$" for all $1 \leq i, i' \leq k$ and $1 \leq j \leq \ell$. With this notation, a differential version of Itô's formula becomes

$$(5.3.3) \quad dF\big(Z(t)\big) = \big(\mathrm{grad}_{Z(t)}, dZ(t)\big)_{\mathbb{R}^{k+\ell}} + \frac{1}{2}\Big(dZ(t), \mathrm{Hess}_{Z(t)} F\, dZ(t)\Big)_{\mathbb{R}^{k+\ell}}.$$

Although it looks a little questionable, this way of writing Itô's formula has more than mnemonic value. In fact, it highlights the essential fact on which his formula rests. Namely, "$d\beta_m(t)$" may be a "differential," but it is *not*, in a classical sense, an "infinitesimal." Indeed, there is abundant evidence that "$|d\beta_m(t)|$" is of order "$\sqrt{dt}$."[6] Thus, it is only reasonable that, when dealing with Brownian paths, one will not get a differential relation in which the left hand side differs from the right hand side by $o(dt)$ unless one uses a two-place Taylor's expansion. Furthermore, it is clear why $\big(d\beta_m(t)\big)^2 = dt$. In order to figure out how "$d\beta_m d\beta_{m'}$" should be interpreted when $m \neq m'$, remember that $\frac{\beta_m + \beta_{m'}}{\sqrt{2}}$ and $\frac{\beta_m - \beta_{m'}}{\sqrt{2}}$ are both Brownian motions. Hence,

$$\text{"}2d\beta_m(t)d\beta_{m'}(t) = \left(d\frac{\beta_m + \beta_{m'}}{\sqrt{2}}(t)\right)^2 - \left(d\frac{\beta_m - \beta_{m'}}{\sqrt{2}}(t)\right)^2 = dt - dt = 0\, dt.\text{"}$$

In other words, "$d\beta_m(t)d\beta_{m'}(t) = \delta_{m,m'} dt$," which now "explains" why the preceding works. Namely,

$$\text{"}dX_i(t)dX_{i'}(t) = 0\, dt \text{ and } dX_i(t)dY_j(t) = dX_i(t)\big(\theta_j(t), d\beta(t)\big)_{\mathbb{R}^n} = 0\, dt\text{"}$$

because "$dX_i(t)dX_{i'}(t)$" and "$d\beta(t)dY_j(t)$" are "$o(dt)$," and

$$\text{"}dY_j(t)dY_{j'}(t) = \sum_{m,m'} \theta_j(t)_m \theta_{j'}(t)_{m'} d\beta_m(t)d\beta_{m'}(t) = \big(\theta_j(t), \theta_{j'}(t)\big)_{\mathbb{R}^n} dt\text{"}$$

because "$d\beta_m(t)d\beta_{m'}(t) = \delta_{m,m'} dt$."

---

[6] Lévy understood this point thoroughly and in fact made systematic use of the notation $\sqrt{dt}$.

§5.3.1. Exercises

EXERCISE 5.3.4. Given an $\mathbb{R}^n$-valued Brownian motion $(\beta(t), \mathcal{F}_t, \mathbb{P})$ and a $\theta \in \Theta_{\mathrm{loc}}^2(\mathbb{R}^n)$, define E_θ as in (5.1.10).

(i) Show that E_θ is $\mathbb{P}$-uniquely determined by the fact that it is progressively measurable and

$$(5.3.5) \qquad dE_\theta(t) = E_\theta(t)\big(\theta(t), d\beta(t)\big)_{\mathbb{R}^n} \quad \text{and } E_\theta(0) = 1.$$

For this reason, E_θ is sometimes called the *Itô exponential*.

Hint: Checking that E_θ is a solution to (5.1.10) is an elementary application of Itô's formula. To see that it is the only solution, suppose that X is a second solution, and apply Itô's formula to see that $d\frac{X(t)}{E_\theta(t)} = 0$.

(ii) Show that $\big(E_\theta(t), \mathcal{F}_t, \mathbb{P}\big)$ is always a supermartingale and that it is a martingale if and only if $\mathbb{E}^{\mathbb{P}}\big[E_\theta(t)\big] = 1$ for all $t \geq 0$.

(iii) From now on, assume that

$$(5.3.6) \qquad \mathbb{E}^{\mathbb{P}}\Big[e^{\frac{1}{2}A_\theta(T)}\Big] < \infty \quad \text{for each } T \geq 0.$$

First observe that $\big(I_\theta(t), \mathcal{F}_t, \mathbb{P}\big)$ is a square integrable martingale, and conclude that, for every $\alpha \geq 0$, $\big(e^{\alpha I_\theta(t)}, \mathcal{F}_t, \mathbb{P}\big)$ is a submartingale. In addition, show that

$$\mathbb{E}^{\mathbb{P}}\big[e^{\frac{1}{2}I_\theta(T)}\big] \leq \mathbb{E}^{\mathbb{P}}\big[e^{\frac{1}{2}A_\theta(T)}\big]^{\frac{1}{2}} < \infty \quad \text{for all } T \geq 0.$$

Hint: Write $e^{\frac{1}{2}I_\theta} = E_\theta^{\frac{1}{2}} e^{\frac{1}{4}A_\theta}$, and remember that $\mathbb{E}^{\mathbb{P}}\big[E_\theta(T)\big] \leq 1$.

(iv) Given $\lambda \in (0,1)$, determine $p_\lambda \in (1, \infty)$ so that $\frac{\lambda p_\lambda(p_\lambda - \lambda)}{1 - \lambda^2} = \frac{1}{2}$. For $p \in [1, p_\lambda]$, set $\gamma(\lambda, p) = \frac{\lambda p(p - \lambda)}{1 - \lambda^2}$, note that $E_{\lambda\theta}^{p^2} = \big(e^{\gamma(\lambda, p)I_\theta}\big)^{1 - \lambda^2} E_{p\theta}^{\lambda^2}$, and conclude that, for any stopping time ζ,

$$\mathbb{E}^{\mathbb{P}}\big[E_{\lambda\theta}(T \wedge \zeta)^{p^2}\big] \leq \mathbb{E}^{\mathbb{P}}\big[e^{\frac{1}{2}I_\theta(T)}\big]^{2\lambda p(p - \lambda)}.$$

By taking $p = p_\lambda$ and applying part (iii), see that this provides sufficient uniform integrability to check that $\big(E_{\lambda\theta}(t), \mathcal{F}_t, \mathbb{P}\big)$ is a martingale for each $\lambda \in (0,1)$.

(v) By taking $p = 1$ in the preceding, justify

$$1 = \mathbb{E}^{\mathbb{P}}\big[E_{\lambda\theta}(T)\big] \leq \mathbb{E}^{\mathbb{P}}\big[e^{\frac{1}{2}I_\theta(T)}\big]^{2\lambda(1 - \lambda)} \mathbb{E}^{\mathbb{P}}\big[E_\theta(T)\big]^{\lambda^2}$$

for all $\lambda \in (0,1)$. After letting $\lambda \nearrow 1$, conclude first that $\mathbb{E}^{\mathbb{P}}\big[E_\theta(T)\big] = 1$ and then that $\big(E_\theta(t), \mathcal{F}_t, \mathbb{P}\big)$ is a martingale. The fact that (5.3.6) implies $\big(E_\theta(t), \mathcal{F}_t, \mathbb{P}\big)$ is a martingale is known as *Novikov's criterion*.

CHAPTER 6

Applications of Stochastic Integration

to Brownian Motion

This chapter contains a highly incomplete selection of ways in which Itô's theory of stochastic integration, especially his formula, has contributed to our understanding of Brownian motion. For a much more complete selection, see Revuz and Yor's book [27].

6.1 Tanaka's Formula for Local Time

Perhaps the single most beautiful application of Itô's formula was made by H. Tanaka when (as reported in [22]) he applied it to prove the existence of local time for one-dimensional Brownian motion.

Before one can understand Tanaka's idea, it is necessary to know what the preceding terminology means. Thus, let $(\beta(t), \mathcal{F}_t, \mathbb{P})$ be a one-dimensional Brownian motion. What we are seeking is a function $\ell : [0, \infty) \times \mathbb{R} \times \Omega \longmapsto [0, \infty)$ with the properties that

(a) for each $y \in \mathbb{R}$, $(t, \omega) \rightsquigarrow \ell(t, y, \omega)$ is progressively measurable;
(b) for each $(y, \omega) \in \mathbb{R} \times \Omega$, $\ell(0, y, \omega) = 0$ and $\ell(\cdot, y, \omega)$ is nondecreasing;
(c) for each $\omega \in \Omega$, $(t, y) \rightsquigarrow \ell(t, y, \omega)$ is continuous;
(d) for $\mathbb{P}$-almost every $\omega \in \Omega$ and all $t \in [0, \infty)$,

$$\int_0^t \mathbf{1}_{[a,b)} \big(\beta(\tau, \omega) \big) \, d\tau = \int_a^b \ell(t, y, \omega) \, dy, \quad \text{for all } -\infty < a < b < \infty.$$

Alternatively, for each $(t, \omega) \in (0, \infty) \times \Omega$, define the Brownian motion, occupation time distribution $\mu_{t,\omega}$ to be the Borel measure on $\mathbb{R}$ determined by the rule that $\mu_{t,\omega}(\Gamma)$ be the amount of time $\beta(\cdot, \omega) \restriction [0, t]$ spends in Γ. Then $\ell(t, \cdot, \omega)$ is to be the Radon–Nikodym derivative of $\mu_{t,\omega}$ with respect to Lebesgue measure.

For each $y \in \mathbb{R}$, $(t, \omega) \rightsquigarrow \ell(t, y, \omega)$ is known as the *local time functional at y of the Brownian path* $\beta(\cdot, \omega)$. P. Lévy was the first person to discuss this functional, although H. Trotter is usually credited for providing the first rigorous proof of its existence.

§6.1.1. Tanaka's Construction: The idea underlying Tanaka's proof is to first write the somewhat fanciful expression

$$``\ell(t,y) = \int_0^t \delta\big(y - \beta(\tau)\big)\,d\tau,"$$

where δ is Dirac's delta function. Although it seems doubtful that such an expression should be taken seriously, Tanaka realized that Itô's formula can be used to rationalize it. Namely, apply Itô's formula to the function $F(x) = x \vee y$, remember that $F'(x) = \mathbf{1}_{[y,\infty)}(x)$ and $F''(x) = \delta(y - x)$, and come to the conclusion that

$$``\frac{1}{2}\int_0^t \delta\big(y - \beta(\tau)\big)\,d\tau = \beta(t) \vee y - 0 \vee y - \int_0^t \mathbf{1}_{[y,\infty)}\big(\beta(\tau)\big)\,d\beta(\tau)."$$

With the preceding heuristic argument in mind, we now define

$$(6.1.1) \qquad \tilde{\ell}(t,y) = 2\left(\beta(t) \vee y - 0 \vee y - \int_0^t \mathbf{1}_{[y,\infty)}\big(\beta(\tau)\big)\,d\beta(\tau)\right).$$

When reading (6.1.1), it is important to keep in mind that it gives a definition of $(t,\omega) \rightsquigarrow \tilde{\ell}(t,y,\omega)$ for each $y \in \mathbb{R}$ separately. However, because, for each y, the definition has an ambiguity on a $\mathbb{P}$-null set, it does not define $(t,\omega) \rightsquigarrow \tilde{\ell}(t,y,\omega)$ simultaneously for all $y \in \mathbb{R}$, and that is one of the reasons why we have used the notation $\tilde{\ell}$ instead of ℓ. In any case, we already know that, for each y and $\mathbb{P}$-almost every ω, $\tilde{\ell}(\,\cdot\,,y,\omega)$ is a continuous function which is 0 at time 0. We next want to show that it is, $\mathbb{P}$-almost surely, a nondecreasing function. For this purpose, choose a $\psi \in C_c^\infty(\mathbb{R};[0,\infty))$ which is supported in $[-1,1]$ and has total integral 1, and take $F_\epsilon(x) = \psi_\epsilon \star F$, where $\psi_\epsilon(x) = \epsilon^{-1}\psi(\epsilon^{-1}x)$. By Itô's formula,

$$F_\epsilon\big(\beta(t)\big) - F_\epsilon(0) - \int_0^t F_\epsilon'\big(\beta(\tau)\big)\,d\beta(\tau) = \frac{1}{2}\int_0^t F_\epsilon''\big(\beta(\tau)\big)\,d\tau.$$

Hence, since $F_\epsilon'' \geq 0$, the left hand side is $\mathbb{P}$-almost surely nondecreasing as a function of $t \geq 0$. Furthermore, $F_\epsilon \longrightarrow F$ uniformly on compacts, while $\big|F_\epsilon' - \mathbf{1}_{[y,\infty)}\big| \leq \mathbf{1}_{[y-\epsilon,y+\epsilon]}$ and therefore

$$\mathbb{E}^{\mathbb{P}}\left[\left\|\int_0^\cdot F_\epsilon'\big(\beta(\tau)\big)\,d\beta(\tau) - \int_0^\cdot \mathbf{1}_{[y,\infty)}\big(\beta(\tau)\big)\,d\beta(\tau)\right\|_{[0,t]}^2\right]$$

$$\leq 4\mathbb{E}^{\mathbb{P}}\left[\int_0^t \big|F_\epsilon'\big(\beta(\tau)\big) - \mathbf{1}_{[y,\infty)}\big(\beta(\tau)\big)\big|^2\,d\tau\right]$$

$$\leq 4\int_0^t \frac{1}{\sqrt{2\pi\tau}}\left(\int_{y-\epsilon}^{y+\epsilon} e^{-\frac{\xi^2}{2\tau}}\,d\xi\right)d\tau \leq 8\epsilon\sqrt{\frac{t}{\pi}}.$$

Hence, as $\epsilon \searrow 0$, $\left\|\int_0^\cdot F_\epsilon''(\beta(\tau)) \, d\tau - \tilde{\ell}(\,\cdot\,,y)\right\|_{[0,t]} \longrightarrow 0$ in $\mathbb{P}$-measure, and so we now know that, for each $y \in \mathbb{R}$, $\tilde{\ell}(\,\cdot\,,y)$ is $\mathbb{P}$-almost surely a continuous, nondecreasing function.

The next step is to use Kolmogorov's continuity criterion (cf. Exercise 2.4.17) to prove that, for each y, $\tilde{\ell}(\,\cdot\,,y)$ can be modified on a $\mathbb{P}$-null set in such a way that the resulting function $\ell(\,\cdot\,,y)$ is, $\mathbb{P}$-almost surely, a continuous function of y. Namely, we will show that, for each $T \in (0,\infty)$, there exists a $C(T) < \infty$ for which

$$\mathbb{E}^{\mathbb{P}}\left[\left\|\tilde{\ell}(\,\cdot\,,y_2) - \tilde{\ell}(\,\cdot\,,y_1)\right\|_{[0,T]}^4\right] \leq C(T)(y_2 - y_1)^2 \quad \text{when } y_1 \leq y_2.$$

To check this, first observe that $|\beta(t) \vee y_2 - \beta(t) \vee y_1| \leq |y_2 - y_1|$. Thus, what still must be checked is that

$$\mathbb{E}^{\mathbb{P}}\left[\left\|\int_0^\cdot \mathbf{1}_{[y_2,\infty)}(\beta(\tau)) \, d\beta(\tau) - \int_0^\cdot \mathbf{1}_{[y_1,\infty)}(\beta(\tau)) \, d\beta(\tau)\right\|_{[0,t]}^4\right]$$
$$= \mathbb{E}^{\mathbb{P}}\left[\left\|\int_0^\cdot \mathbf{1}_{[y_1,y_2)}(\beta(\tau)) \, d\beta(\tau)\right\|_{[0,T]}^4\right] \leq C(T)(y_2 - y_1)^2.$$

But, by Doob's inequality,

$$\mathbb{E}^{\mathbb{P}}\left[\left\|\int_0^\cdot \mathbf{1}_{[y_1,y_2)}(\beta(\tau)) \, d\beta(\tau)\right\|_{[0,T]}^4\right]$$
$$\leq \left(\frac{4}{3}\right)^4 \mathbb{E}^{\mathbb{P}}\left[\left(\int_0^T \mathbf{1}_{[y_1,y_2)}(\beta(\tau)) \, d\beta(\tau)\right)^4\right],$$

and, by Itô's formula,

$$\left(\int_0^T \mathbf{1}_{[y_1,y_2)}(\beta(\tau)) \, d\beta(\tau)\right)^4$$
$$= 4\int_0^T \mathbf{1}_{[y_1,y_2)}(\beta(t)) \left(\int_0^t \mathbf{1}_{[y_1,y_2)}(\beta(\tau)) \, d\beta(\tau)\right)^3 d\beta(t)$$
$$+ 6\int_0^T \mathbf{1}_{[y_1,y_2)}(\beta(t)) \left(\int_0^t \mathbf{1}_{[y_1,y_2)}(\beta(\tau)) \, d\beta(\tau)\right)^2 dt.$$

Hence,

$$
\mathbb{E}^{\mathbb{P}} \left[\left\| \int_0^{\cdot} \mathbf{1}_{[y_1,y_2)}\big(\beta(\tau)\big)\, d\beta(\tau) \right\|_{[0,T]}^4 \right]
$$

$$
\leq 6 \left(\frac{4}{3}\right)^4 \mathbb{E}^{\mathbb{P}} \left[\int_0^T \mathbf{1}_{[y_1,y_2)}\big(\beta(t)\big) \left(\int_0^t \mathbf{1}_{[y_1,y_2)}\big(\beta(\tau)\big)\, d\beta(\tau) \right)^2 dt \right]
$$

$$
\leq 6 \left(\frac{4}{3}\right)^4 \mathbb{E}^{\mathbb{P}} \left[\int_0^T \mathbf{1}_{[y_1,y_2)}\big(\beta(t)\big)\, dt \left\| \int_0^{\cdot} \mathbf{1}_{[y_1,y_2)}\big(\beta(\tau)\big)\, d\beta(\tau) \right\|_{[0,T]}^2 \right]
$$

$$
\leq 6 \left(\frac{4}{3}\right)^4 \mathbb{E}^{\mathbb{P}} \left[\left(\int_0^T \mathbf{1}_{[y_1,y_2)}\big(\beta(t)\big)\, dt \right)^2 \right]^{\frac{1}{2}}
$$

$$
\times \mathbb{E}^{\mathbb{P}} \left[\left\| \int_0^{\cdot} \mathbf{1}_{[y_1,y_2)}\big(\beta(\tau)\big)\, d\beta(\tau) \right\|_{[0,T]}^4 \right]^{\frac{1}{2}},
$$

and so

$$
\mathbb{E}^{\mathbb{P}} \left[\left\| \int_0^{\cdot} \mathbf{1}_{[y_1,y_2)}\big(\beta(\tau)\big)\, d\beta(\tau) \right\|_{[0,T]}^4 \right] \leq K \mathbb{E}^{\mathbb{P}} \left[\left(\int_0^T \mathbf{1}_{[y_1,y_2)}\big(\beta(t)\big)\, dt \right)^2 \right],
$$

where $K = 36 \left(\frac{4}{3}\right)^8$. Finally, write the square of the integral over $[0,T]$ as twice the integral over the simplex $0 \leq \tau \leq t \leq T$, and note that

$$
\mathbb{E}^{\mathbb{P}} \left[\int_0^T \mathbf{1}_{[y_1,y_2)}\big(\beta(t)\big) \left(\int_0^t \mathbf{1}_{[y_1,y_2)}\big(\beta(\tau)\big)\, d\tau \right) dt \right]
$$

$$
= \frac{1}{2\pi} \iint\limits_{0<\tau<t\leq T} \frac{1}{\sqrt{\tau(t-\tau)}} \left(\iint\limits_{[y_1,y_2)^2} e^{-\frac{\xi_1^2}{2\tau}} e^{-\frac{(\xi_2-\xi_1)^2}{2(t-\tau)}}\, d\xi_1 d\xi_2 \right) d\tau dt
$$

$$
\leq \frac{(y_2-y_1)^2}{2\pi} \iint\limits_{0<\tau<t\leq T} \frac{1}{\sqrt{\tau(t-\tau)}}\, d\tau dt.
$$

Hence, by Kolmogorov's criterion, there exists each $\omega \in \Omega$ a continuous $y \in \mathbb{R} \longmapsto \ell(\,\cdot\,,y,\omega) \in C\big([0,\infty);\mathbb{R}\big)$ such that $\ell(\,\cdot\,,y) = \tilde{\ell}(\,\cdot\,,y)$ $\mathbb{P}$-almost surely for each $y \in \mathbb{R}$. In fact, after further adjustment on a $\mathbb{P}$-null set, we may and will assume that, for all $\omega \in \Omega$, $(t,y) \rightsquigarrow \ell(t,y,\omega)$ is continuous, and, for all (y,ω), $\ell(0,y,\omega) = 0$ and $\ell(\,\cdot\,,y,\omega)$ is nondecreasing.

It remains to show that ℓ satisfies **(d)**. To this end, let $\varphi \in C_c^\infty(\mathbb{R};\mathbb{R})$ be given, and set $\Phi(x) = \int_{\mathbb{R}} x \vee y\varphi(y)\, dy$. Then $\Phi'' = \varphi$, and so, by the mean value theorem, we can choose $\xi_m^N \in \left(m2^{-N}, (m+1)2^{-N}\right)$ so that $\Phi'\left((m+1)2^{-N}\right) - \Phi'(m2^{-N}) = 2^{-N}\varphi(\xi_m^N)$. Thus

$$\frac{1}{2}\int_{\mathbb{R}} \varphi(y)\ell(t,y)\,dy = \lim_{N\to\infty} 2^{-N}\sum_{m\in\mathbb{Z}} \varphi(\xi_m^N)\ell(t,m2^{-N})$$

$$= \Phi\big(\beta(t)\big) - \Phi(0) - \lim_{N\to\infty}\sum_{m\in\mathbb{Z}}\int_0^t \Phi'(m2^{-N})\mathbf{1}_{[(m-1)2^{-N},m2^{-N})}\big(\beta(\tau)\big)\,d\beta(\tau)$$

$$= \Phi\big(\beta(t)\big) - \Phi(0) - \int_0^t \Phi'\big(\beta(\tau)\big)\,d\beta(\tau)$$

since

$$\int_0^t \Phi'\big(\beta(\tau)\big)\,d\beta(\tau) - \sum_{m\in\mathbb{Z}}\int_0^t \Phi'(m2^{-N})\mathbf{1}_{[(m-1)2^{-N},m2^{-N})}\big(\beta(\tau)\big)\,d\beta(\tau)$$

$$= \int_0^t \left(\Phi'\big(\beta(\tau)\big) - \Phi'\big([\beta(\tau)]_N\big)\right)d\beta(\tau)$$

and

$$\mathbb{E}^{\mathbb{P}}\left[\left(\int_0^t \left(\Phi'\big(\beta(\tau)\big) - \Phi'\big([\beta(\tau)]_N\big)\right)d\beta(\tau)\right)^2\right]$$

$$= \mathbb{E}^{\mathbb{P}}\left[\int_0^t \left(\Phi'\big(\beta(\tau)\big) - \Phi'\big([\beta(\tau)]_N\big)\right)^2 d\tau\right] \longrightarrow 0$$

as $N \to \infty$. Thus, we now know that

$$\frac{1}{2}\int_{\mathbb{R}} \varphi(y)\ell(t,y)\,dy = \Phi\big(\beta(t)\big) - \Phi(0) - \int_0^t \Phi'\big(\beta(\tau)\big)\,d\beta(\tau)$$

$\mathbb{P}$-almost surely. At the same time, by Itô's formula, the right hand side of the preceding is also $\mathbb{P}$-almost surely equal to $\frac{1}{2}\int_0^t \varphi\big(\beta(\tau)\big)\,d\tau$, and so we also know that, for each $\varphi \in C_c^\infty(\mathbb{R};\mathbb{R})$

$$\int_{\mathbb{R}} \varphi(y)\ell(t,y,\omega)\,dy = \frac{1}{2}\int_0^t \varphi\big(\beta(\tau,\omega)\big)\,d\tau = \frac{1}{2}\int_{\mathbb{R}} \varphi(y)\,\mu_{t,\omega}(dy), \quad t \geq 0,$$

for all $\mathbb{P}$-almost every ω. Starting from this, it is an elementary exercise in measure theory to see that ℓ has the property in **(d)**.

We summarize our progress in the following theorem, in which the concluding equality is *Tanaka's formula*.

THEOREM 6.1.2. *Given a one dimensional Brownian motion* $(\beta(t), \mathcal{F}_t, \mathbb{P})$ *there exists a* $\mathbb{P}$-*almost surely unique function* $\ell : [0, \infty) \times \mathbb{R} \times \Omega \longrightarrow [0, \infty)$ *with the properties described in* (a)–(d) *above. Moreover, for each* $y \in \mathbb{R}$,

$$(6.1.3) \quad \frac{\ell(t, y)}{2} = \beta(t) \vee y - 0 \vee y - \int_0^t \mathbf{1}_{[y, \infty)} \big(\beta(\tau)\big) \, d\beta(\tau), \quad t \in [0, \infty)$$

$\mathbb{P}$-*almost surely.*

REMARK 6.1.4. It is important to realize that the very existence of $\ell(t, \cdot)$ is a reflection of just how *fuzzy* the graph of a Brownian path must be. Indeed, if $p : [0, t] \longrightarrow \mathbb{R}$ is a smooth path, then its occupation time distribution μ_t will usually have a nontrivial part which is singular to Lebesgue measure. For example, if the derivative $\dot{p}$ of p vanishes at some $\tau \in [0, t]$, then it is easy to see that this prevents μ_t from being absolutely continuous with respect to Lebesgue measure. Moreover, even if $\dot{p}$ never vanishes on $[0, t]$, and therefore μ_t is absolutely continuous, the Radon–Nikodym derative of μ_t will have discontinuities at both $\min_{[0,t]} p(\tau)$ and $\max_{[0,t]} p(\tau)$. The way that Brownian paths avoid these problems is that they are changing directions so fast that they, as distinguished from a well behaved path, never slow down. Of course, one might think that their speed could cause other problems. Namely, if they are moving so fast, how do they spend enough time anywhere to have their occupancy time recordable on a Lebesgue scale? The explanation is in the *fuzz* alluded to above. That is, although a Brownian path exits small intervals too fast to have their presence recorded on a Lebesgue scale, it does not depart when it exits but, instead, dithers back and forth.

From a more analytic point of view, the existence of the Brownian local time functional is a stochastic manifestation of the fact that points have positive capacity in one dimension. This connection is an inherent ingredient in Tanaka's derivation. In fact, his argument works only because the the fundamental solution, $x \vee y$, is locally bounded, which is also the reason why points have positive capacity in one dimension. In two and more dimensions, the fundamental solution is too singular for points to have positive capacity.

§**6.1.2. Some Properties of Local Time:** Having constructed ℓ, we now want to derive a couple of its elementary properties. To begin with, suppose that $f : [0, \infty) \times \mathbb{R} \times \Omega \longrightarrow \mathbb{R}$ is a bounded, measurable function with the properties that $(t, y) \rightsquigarrow f(t, y, \omega)$ is continuous for $\mathbb{P}$-almost every ω. Using Riemann sum approximations, one can easily see that, for $\mathbb{P}$-almost every ω,

$$(6.1.5) \quad \int_0^t f\big(\tau, \beta(\tau, \omega), \omega\big) \, d\tau = \int_{\mathbb{R}} \left(\int_0^t f(\tau, y, \omega) \ell(d\tau, y, \omega) \right) dy, \quad t \geq 0,$$

where $\ell(d\tau, y, \omega)$ denotes integration with respect to the measure on $[0, \infty)$ determined by the nondecreasing function $\ell(\cdot, y, \omega)$. In particular, if $\eta \in C^{\infty}(\mathbb{R}; [0, 1])$ vanishes on $[-1, 1]$ and equals 1 off $(2, 2)$, $\psi \in C_c^{\infty}(\mathbb{R}; [0, \infty))$ vanishes off $[-1, 1]$ and has total integral 1, and

$$f_{\epsilon, R}(t, y, \omega) = \epsilon^{-1}\eta\Big(R\big(\beta(t, \omega) - y_0\big)\Big)\psi\big(\epsilon^{-1}(y - y_0)\big),$$

then, by (6.1.5), $\mathbb{P}$-almost surely,

$$\epsilon^{-1}\int_{\mathbb{R}} \psi\big(\epsilon^{-1}(y - y_0)\big)\left(\int_0^t \eta\Big(R\big(\beta(\tau) - y_0\big)\Big)\,\ell(d\tau, y)\right)dy = 0$$

whenever $0 < \epsilon < R^{-1}$. Hence, after first letting $\epsilon \searrow 0$ and then $R \nearrow \infty$, we see that, for each $y \in \mathbb{R}$,

(6.1.6) $\ell\big(\{\tau \geq 0 : \beta(\tau) \neq y\}, y\big) = 0$ $\mathbb{P}$-almost surely.

An interesting consequence of (6.1.6) is that, for each y, $\ell(\cdot, y)$ is $\mathbb{P}$-almost surely a singular, continuous, nondecreasing function. Indeed, $\ell(\cdot, y)$ is $\mathbb{P}$-almost surely supported on $\{\tau : \beta(\tau) = y\}$, whereas

$$\mathbb{E}^{\mathbb{P}}\Big[\mathrm{Leb}\big(\{\tau : \beta(\tau) = y\}\big)\Big] = \mathbb{E}^{\mathbb{P}}\left[\int_0^{\infty} \mathbf{1}_{\{y\}}\big(\beta(\tau)\big)\,d\tau\right]$$
$$= \int_0^{\infty} \mathbb{P}\big(\beta(\tau) = y\big)\,d\tau = 0.$$

In order to take the next step, notice that, from (6.1.3),

$$\beta(t) \wedge 0 = \beta(t) - \beta(t) \vee 0 = \int_0^t \mathbf{1}_{(-\infty, 0)}\big(\beta(\tau)\big)\,d\beta(\tau) - \frac{\ell(t, 0)}{2}.$$

Hence, after subtracting this from (6.1.3), we find that

(6.1.7) $$|\beta(t)| = \int_0^t \mathrm{sgn}\big(\beta(\tau)\big)\,d\beta(\tau) + \ell(t, 0),$$

where, for the sake of definiteness[1] we will take $\mathrm{sgn}(0) = 1$.

Among the many interesting consequences of (6.1.7) are the two discussed in the next statement, both of which were discovered by P. Lévy. For much more information about local time, the reader should consult the books [15] and [27].

---

[1] It really does not matter how one takes sgn at 0 since, $\mathbb{P}$-almost surely, β spends no time there.

THEOREM 6.1.8. *With* $\mathbb{P}$*-probability* 1, $\ell(t,0) > 0$ *for all* $t > 0$. *Furthermore, if, for* $t > 0$ *and* $\epsilon > 0$, $N_\epsilon(t)$ *denotes the number of times* $|\beta| \upharpoonright [0,t]$ *downcrosses[2] the interval* $[0,\epsilon]$, *then*

(6.1.9) $\lim_{\epsilon \searrow 0} \|\epsilon N_\epsilon(\cdot) - \ell(\cdot,0)\|_{[0,t]} = 0$ $\mathbb{P}$*-almost surely for each* $t > 0$.

PROOF: To prove the first statement, set $B(t) = \int_0^t \mathrm{sgn}(\beta(\tau)) \, d\beta(\tau)$. In view of (6.1.7), it suffices for us to prove that

(*) $\mathbb{P}(\forall t > 0 \, \exists \tau \in (0,t] \; B(\tau) < 0) = 1$.

For this purpose, we will first show that $(B(t), \mathcal{F}_t, \mathbb{P})$ is a Brownian motion. To do this, apply Itô's formula to see that

$$\left(\exp\left(\lambda B(t) - \frac{\lambda^2 t}{2}\right), \mathcal{F}_t, \mathbb{P}\right)$$

is a martingale for every $\lambda \in \mathbb{C}$. In particular, this means that, for all $0 \le s < t$,

$$\mathbb{E}^{\mathbb{P}}\left[e^{\sqrt{-1}\,\xi(B(t)-B(s))} \mid \mathcal{F}_s\right] = e^{-\frac{\xi^2(t-s)}{2}} \quad \mathbb{P}\text{-almost surely,}$$

which, together with $B(0) = 0$, is enough to identify the distribution of $\omega \rightsquigarrow B(\cdot, \omega)$ as that of a Brownian motion.

Returning to (*), there are several ways in which it can be approached. One way is to apply Blumenthal's 0–1 law (cf. Theorem 8.1.20 in [36]) which says that, because B is a Brownian motion under $\mathbb{P}$, any set from $\bigcap_{t>0} \sigma(\{B(\tau) : \tau \in [0,t]\})$ has $\mathbb{P}$-probability 0 or 1. In particular, this means that

$$\mathbb{P}(\forall \, t > 0 \, \exists \tau \in [0,t] \; B(\tau) < 0) \in \{0,1\}$$

Hence, since

$$\mathbb{P}(\forall \, t > 0 \, \exists \tau \in [0,t] \; B(\tau) < 0) = \lim_{t \searrow 0} \mathbb{P}(\exists \tau \in [0,t] \; B(\tau) < 0)$$

$$\ge \overline{\lim}_{t \searrow 0} \mathbb{P}(B(t) < 0) = \frac{1}{2},$$

(*) follows. Another, less abstract approach is to use

$$\mathbb{P}(\exists \tau \in [0,t] \; B(\tau) < 0) \ge \lim_{\delta \searrow 0} \mathbb{P}(\exists \tau \in [\delta,t] \; B(\tau) < 0)$$

$$= \lim_{\delta \searrow 0} \frac{1}{\sqrt{2\pi\delta}} \int_{\mathbb{R}} e^{-\frac{x^2}{2\delta}} \mathbb{P}(\exists \, \tau \in [0, t-\delta] \; B(\tau) < -x) \, dx$$

$$= \frac{1}{2} + \lim_{\delta \searrow 0} \frac{1}{\sqrt{2\pi\delta}} \int_0^\infty e^{-\frac{x^2}{2\delta}} 2\mathbb{P}(B(t-\delta) \le -x) \, dx = 1,$$

---

[2] That is, $N_\epsilon(t) \ge m$ if and only if there exist $0 \le \tau_1 < \tau_2 < \cdots < \tau_{2m} \le t$ such that $|\beta(\tau_{2\ell-1})| \ge \epsilon$ and $\beta(\tau_{2\ell}) = 0$ for $1 \le \ell \le m$.

where, in the passage to the last line, we used the reflection principle (cf. (4.3.5) in [36]) when we replaced $\mathbb{P}\big(\exists\, \tau \in [0, t - \delta]\ B(\tau) \leq -x\big)$ by $2\mathbb{P}\big(B(t - \delta) \leq -x\big)$ for $x < 0$.

In order to prove (6.1.9), we introduce the stopping times $\{\zeta_\ell : \ell \geq 0\}$ so that $\zeta_0 \equiv 0$, $\zeta_{2\ell+1} = \inf\{t \geq \zeta_{2\ell} : |\beta(t)| \geq \epsilon\}$ for $\ell \geq 0$, and $\zeta_{2\ell} = \inf\{t \geq \zeta_{2\ell-1} : \beta(t) = 0\}$ for $\ell \geq 1$. Next, set $I_\ell = [\zeta_{2\ell-1}, \zeta_{2\ell})$, and observe that

$$\sum_{\ell=1}^{\infty} \Big(\big|\beta(t \wedge \zeta_{2\ell})\big| - \big|\beta(t \wedge \zeta_{2\ell-1})\big| \Big) = -\epsilon N_\epsilon(t) + \big(|\beta(t)| - \epsilon\big) \sum_{\ell=1}^{\infty} \mathbf{1}_{I_\ell}(t).$$

At the same time, because $\beta(\tau) \neq 0$ when $\tau \in I_\ell$ and therefore $\ell(t \wedge \zeta_{2\ell}) = \ell(t \wedge \zeta_{2\ell-1})$, (6.1.7) tells us that

$$\sum_{\ell=1}^{\infty} \Big(\big|\beta(t \wedge \zeta_{2\ell})\big| - \big|\beta(t \wedge \zeta_{2\ell-1})\big| \Big) = \sum_{\ell=1}^{\infty} \int_0^t \mathbf{1}_{I_\ell}(\tau)\mathrm{sgn}\big(\beta(\tau)\big)\, d\beta(\tau)$$

$$= |\beta(t)| - \sum_{\ell=0}^{\infty} \int_0^t \mathbf{1}_{J_\ell}(\tau)\mathrm{sgn}\big(\beta(\tau)\big)\, d\beta(\tau) - \ell(t, 0),$$

where $J_\ell \equiv [\zeta_{2\ell}, \zeta_{2\ell+1})$. Hence, we have now shown that

$$\epsilon N_\epsilon(t) - \ell(t, 0) = -|\beta(t)| \sum_{\ell=0}^{\infty} \mathbf{1}_{J_\ell}(t) - \epsilon \sum_{\ell=0}^{\infty} \mathbf{1}_{I_\ell}(t) + \int_0^t \theta_\epsilon(\tau)\, d\beta(\tau),$$

where $\theta_\epsilon(\tau) \equiv \sum_{\ell=0}^{\infty} \mathbf{1}_{J_\ell}(\tau)\mathrm{sgn}\big(\beta(\tau)\big)$. Finally, notice that $t \in J_\ell \implies |\beta(t)| \leq \epsilon$, and so the absolute values of the first two terms on the right are each dominated by ϵ. As for the stochastic integral term on the right, observe that $|\theta_\epsilon(\tau)| \leq \mathbf{1}_{[-\epsilon, \epsilon]}\big(\beta(\tau)\big)$, and therefore

$$\mathbb{E}^{\mathbb{P}}\left[\left\| \int_0^{\cdot} \theta_\epsilon(\tau)\, d\beta(\tau) \right\|_{[0,t]}^2 \right] \leq 8\epsilon \int_0^t \frac{1}{\sqrt{2\pi\tau}}\, d\tau.$$

Thus, by taking $\epsilon = \frac{1}{m^2}$, we conclude that

$$\lim_{m \to \infty} \big\| m^{-2} N_{1/m^2} - \ell(\,\cdot\,, 0) \big\|_{[0,t]} = 0 \quad \mathbb{P}\text{-almost surely for each } t > 0.$$

But

$$(m+1)^{-2} \leq \epsilon < m^{-2} \implies (m+1)^{-2} N_{1/m^2} \leq \epsilon N_\epsilon \leq m^{-2} N_{1/(m+1)^2},$$

and so the asserted result follows. $\square$

It should be clear that the results in Theorem 6.1.8 are further evidence of the "fuzziness" alluded to in Remark 6.1.4.

§6.1.3. Exercises

EXERCISE 6.1.10. Let $(\beta(t), \mathcal{F}_t, \mathbb{P})$ be a one-dimensional Brownian motion, and let $\ell(\,\cdot\,, 0)$ be its local time at 0.

(i) Show that $\ell(t, 0)$ is measurable with respect to the $\mathbb{P}$-completion $\mathcal{G}_t$ of $\sigma(\{|\beta(\tau)| : \tau \in [0, t]\})$ and that $\sigma(\{\text{sgn}(\beta(\tau)) : \tau \geq 0\})$ is $\mathbb{P}$-independent of $\mathbb{P}$-completion $\mathcal{G}$ of $\sigma(\{|\beta(\tau)| : \tau \geq 0\})$.

(ii) Set $B(t) = \int_0^t \text{sgn}(\beta(\tau))\, d\beta(\tau)$, and show that $(B(t), \mathcal{G}_t, \mathbb{P})$ is a one-dimensional Brownian motion.

(iii) Refer to part (ii), but think of $t \rightsquigarrow B(t)$ as being a Brownian motion relative of the σ-algebras $t \rightsquigarrow \mathcal{F}_t$, instead of $t \rightsquigarrow \mathcal{G}_t$. Show that

$$\beta(t) = \int_0^t \text{sgn}(\beta(\tau))\, dB(\tau), \quad t \geq 0,$$

but that $\beta(t)$ *is not* measurable with respect to the $\mathbb{P}$-completion of $\sigma(\{B(\tau) : \tau \geq 0\})$.

The conclusion drawn in (iii) demonstrates that it is important to make a distinction between the notion of *strong* and *weak* solutions to a stochastic differential equation. Without going into details, suffice it to say that strong solutions are those which are measurable functions of the driving Brownian motion, whereas weak solutions are those which simply solve the stochastic differential equation relative to some family of σ-algebras with respect to which the driving process happens to be a Brownian motion. For more details, see [27].

6.2 An Extension of the Cameron–Martin Formula

In Exercise 5.1.20 we discussed the famous formula of Cameron and Martin for the Radon–Nikodym derivative R_h of the $(\tau_h)_* \mathbb{P}^0$ with respect to $\mathbb{P}^0$ when $h(t) = \int_0^t \eta(\tau)\, d\tau$ for some $\eta \in L^2([0, \infty); \mathbb{R}^n)$. Because, on average, Brownian motion has no preferred direction, it is reasonable to think of the addition of h to a Brownian path p as the introduction of a *drift*. That is, because h does have a velocity, it forces $p + h$ to have a net drift which p by itself does not possess. Thus, the Cameron–Martin formula can be interpreted as saying that the introduction of a drift into a Brownian motion produces a path whose distribution is equivalent to that of the unperturbed Brownian motion as long as the perturbation has a square integrable derivative. Further, as I. Segal (cf. Exercise 5.2.38 in [36]) showed, only such perturbations leave the measure class unchanged. In fact, $\tau_h \mathbb{P}^0$ is singular to $\mathbb{P}^0$ if h fails to have a square integrable derivative. The intuitive reason underlying these results has to do with the "fuzz" alluded to in Remark 6.1.4. Namely, because the "fuzz" is manifested in almost sure properties, the perturbation must be smooth enough to not disturb the "fuzz." For example, the drift must not alter $\mathbb{P}^0$-almost sure properties like

$$\lim_{N\to\infty} \sum_{m=1}^{2^N} \left| p(m2^{-N})) - p((m-1)2^{-N} \right|^2 = n.$$

In this section, we will give an extension of their formula, one which will allow us to introduce Radon-Nikodym derivatives to produce a random "drift".

§6.2.1. Introduction of a Random Drift: The key result here is the following application of Itô's formula.

THEOREM 6.2.1. *Suppose that* $\beta : [0, \infty) \times C\big([0, \infty); \mathbb{R}^n\big) \longrightarrow \mathbb{R}^n$ *is a* $\{\mathcal{B}_t : t \geq 0\}$*-progressively measurable function, and assume that* $\mathbb{P}$ *is a probability measure on* $C\big([0, \infty); \mathbb{R}^n\big)$ *for which* $\big(\beta(t), \overline{\mathcal{B}_t}^{\mathbb{P}}, \mathbb{P}\big)$[3] *is a Brownian motion. If* $\theta \in \Theta^2_{\mathrm{loc}}(\mathbb{P}; \mathbb{R}^n)$ *and (cf. (5.1.10))* $\mathbb{E}^{\mathbb{P}}[E_\theta(T)] = 1$ *for all* $T \geq 0$, *then there exists a unique* $\mathbb{Q} \in \mathbf{M}_1\big(C\big([0, \infty); \mathbb{R}^n\big)\big)$ *such that*

$$\mathbb{Q}(A) = \mathbb{E}^{\mathbb{P}}\big[E_\theta(T), A\big] \quad \text{for all } T \geq 0 \text{ and } A \in \overline{\mathcal{B}_T}^{\mathbb{P}}.$$

Furthermore, if η *is a* $\{\mathcal{B}_t : t \geq 0\}$*-progressively measurable satisfying* $\int_0^T |\eta(\tau)|\, d\tau < \infty$ $\mathbb{P}$*-almost surely for all* $T \in [0, \infty)$ *and if* $X(t) \equiv \int_0^t \eta(\tau)\, d\tau$, *then, for each* $\varphi \in C^{1,2}\big(\mathbb{R}^m \times \mathbb{R}^n; \mathbb{C}\big)$,[4]

$$\left(\varphi\big(X(t), \beta(t)\big) - \int_0^t H_\varphi(\tau)\, d\tau, \overline{\mathcal{B}_t}^{\mathbb{P}}, \mathbb{Q} \right) \quad \text{with}$$

$$H_\varphi(t) \equiv \int_0^t \left(\big(\eta(\tau), \mathrm{grad}_x\varphi\big)_{\mathbb{R}^m} + \tfrac{1}{2}\Delta_y\varphi + \big(\theta(\tau), \mathrm{grad}_y\varphi\big)_{\mathbb{R}^n} \right)\big(X(\tau), \beta(\tau)\big)\, d\tau$$

is a local martingale.

PROOF: In order to prove the first assertion, note that, by part **(ii)** of Exercise 5.3.4, $\big(E_\theta(t), \overline{\mathcal{B}_t}^{\mathbb{P}}, \mathbb{P}\big)$ is a martingale. Hence, if the Borel probability measure $\mathbb{Q}^T$ is determined on $\overline{\mathcal{B}_T}^{\mathbb{P}}$ by $\mathbb{Q}(A) = \mathbb{E}^{\mathbb{P}}\big[E_\theta(T), \beta \in A\big]$ for all $A \in \overline{\mathcal{B}_T}^{\mathbb{P}}$, then the family $\{\mathbb{Q}^T : T \geq 0\}$ is consistent in the sense that, for each $0 \leq T_1 < T_2$, $\mathbb{Q}^{T_1}$ is the restriction of $\mathbb{Q}^{T_2}$ to $\mathcal{B}_{T_1}$. Thus, by a minor

---

[3] Given a probability measure $\mathbb{P}$ defined on a σ-algebra $\mathcal{F}$, $\overline{\mathcal{F}}^{\mathbb{P}}$ is used to denote the $\mathbb{P}$-completion of $\mathcal{F}$.

[4] In the following, the subscripts "x" and "y" are used to indicate to which variables the operation is being applied. The "x" refers to the first m coordinates, and the "y" refers to the last n coordinates. Also, Δ denotes the standard Euclidean Laplacian for $\mathbb{R}^n$.

variant (cf. Theorem 1.3.5 in [41]) of Kolmogorov's extension theorem, there is a unique Borel probability measure $\mathbb{Q}$ whose restriction to $\mathcal{B}_T$ is $\mathbb{Q}^T$ for each $T \geq 0$.

To prove the second part, define

$$\zeta_R(p) = \inf \left\{ t \geq 0 : |\beta(t)| \vee \int_0^t |\eta(\tau)| \vee |\theta(\tau)|^2 \, d\tau \geq R \right\}.$$

Because

$$\int_0^T |\eta(\tau)| \, d\tau \vee \int_0^T |\theta(\tau)|^2 \, d\tau < \infty \quad \mathbb{P}\text{-almost surely}$$

and $\mathbb{Q} \upharpoonright \overline{\mathcal{B}_T}^\mathbb{P} \ll \mathbb{P} \upharpoonright \overline{\mathcal{B}_T}^\mathbb{P}$ for each $T \geq 0$, we know that $\zeta_R \nearrow \infty$ $\mathbb{Q}$-almost surely as $R \nearrow \infty$. Thus, it is enough for us to show that, for each $\varphi \in C^{1,2}(\mathbb{R}^m \times \mathbb{R}^n; \mathbb{C})$, $0 \leq t_1 < t_2$, $A \in \overline{\mathcal{B}_{t_1}}^\mathbb{P}$, and $R > 0$,

(*) $\mathbb{E}^\mathbb{Q}\big[M(t_2 \wedge \zeta_R), \, A\big] = \mathbb{E}^\mathbb{Q}\big[M(t_1 \wedge \zeta_R), \, A\big],$

where

$$M(t) \equiv \varphi\big(X(t), \beta(t)\big) - \int_0^t H_\varphi(\tau) \, d\tau.$$

By the definition of $\mathbb{Q}$, (*) is equivalent to

$$\mathbb{E}^\mathbb{P}\big[E_\theta(t_2) M(t_2 \wedge \zeta_R), \, A\big] = \mathbb{E}^\mathbb{P}\big[E_\theta(t_2) M(t_1 \wedge \zeta_R), \, A\big],$$

which, because $\big(E_\theta(t), \overline{\mathcal{B}_t}^\mathbb{P}, \mathbb{P}\big)$ is a martingale, is equivalent to

(**) $\mathbb{E}^\mathbb{P}\big[\tilde{M}(t_2 \wedge \zeta_R), \, A\big] = \mathbb{E}^\mathbb{P}\big[\tilde{M}(t_1 \wedge \zeta_R), \, A\big],$

where

$$\tilde{M}(t) \equiv E_\theta(t) \varphi\big(X(t), \beta(t)\big) - \int_0^t E_\theta(\tau) H_\varphi(\tau) \, d\tau.$$

But, by Itô's formula and (5.3.5),

$$E_\theta(t) \varphi\big(X(t), \beta(t)\big) - \varphi(0,0) - \int_0^t E_\theta(\tau) H_\varphi(\tau) \, d\tau$$

$$= \int_0^t E_\theta(\tau) \Big(\varphi\big(X(\tau), \beta(\tau)\big) \theta(\tau) + \mathrm{grad}_y \varphi\big(X(\tau), \beta(\tau)\big), d\beta(\tau)\Big)_{\mathbb{R}^n}.$$

Finally, from the definition of ζ_R and the estimate in (5.1.25), it is an easy matter to check that

$$\mathbb{E}^{\mathbb{P}}\left[\int_0^{T\wedge\zeta_R} E_\theta(\tau)^2 \Big|\varphi\big(X(\tau),\beta(\tau)\big)\theta(\tau) + \mathrm{grad}_y\varphi\big(X(\tau),\beta(\tau)\big)\Big|^2 d\tau\right]$$

$$\leq \mathbb{E}^{\mathbb{P}}\left[E_\theta(T\wedge\zeta_R)^2 \int_0^{T\wedge\zeta_R} \Big|\varphi\big(X(\tau),\beta(\tau)\big)\theta(\tau)\right.$$

$$\left. + \mathrm{grad}_y\varphi\big(X(\tau),\beta(\tau)\big)\Big|^2 d\tau\right] < \infty,$$

where, in passing to the second line, we have used the fact that $\big(E_\theta(t)^2, \overline{\mathcal{B}}_t^{\mathbb{P}},$ $\mathbb{P}\big)$ is a submartingale. But (cf. Exercise 5.1.20) this means that $\big(\tilde{M}(t\wedge\zeta_R),$ $\mathcal{B}_t, \mathbb{P}\big)$ is a square-integrable martingale, and so (**) holds. $\square$

The preceding leaves open the problem of determining when $\mathbb{E}^{\mathbb{P}}\big[E_\theta(T)\big] = 1$ for all $T \geq 0$. Of course, Novikov's condition (cf. part **(iv)** of Exercise 5.3.4) gives one such condition. In particular, it tells us that there is no problem when θ is bounded. However, when θ is not bounded, Novikov's condition is often too crude, and one needs to use the sort of considerations given in our next result, which is a variant on the Cameron–Martin theme usually attributed to I.V. Girsanov.

COROLLARY 6.2.2. *Let* $\theta : [0,\infty) \times C\big([0,\infty);\mathbb{R}^n\big) \longrightarrow \mathbb{R}^n$ *be a* $\{\mathcal{B}_t : t \geq 0\}$*-progressively measurable function with the property that* $\int_0^T |\theta(\tau,p)|^2 \, d\tau < \infty$ *for each* $(T,p) \in [0,\infty) \times C\big([0,\infty);\mathbb{R}^n\big)$*. Then, for each* $x \in \mathbb{R}^n$*, there is at most one* $\mathbb{Q}_x \in \mathbf{M}_1\big(C\big([0,\infty);\mathbb{R}^n\big)\big)$ *with the property that*

$$p \in C\big([0,\infty);\mathbb{R}^n\big) \longmapsto \beta(t,p) \equiv p(t) - x - \int_0^t \theta(\tau,p)\,d\tau \in C\big([0,\infty);\mathbb{R}^n\big)$$

is a $\mathbb{Q}_x$*-Brownian motion relative to* $\{\overline{\mathcal{B}}_t^{\mathbb{Q}_x} : t \geq 0\}$*. Moreover, if* $\mathbb{P}_x^0 \equiv (\tau_x)_*\mathbb{P}^0$ *and* $(t,p) \rightsquigarrow E_\theta(t,p)$ *is defined relative to the Brownian motion* $\big(p(t) - x, \overline{\mathcal{B}}_t^{\mathbb{P}_x^0}, \mathbb{P}_x^0\big)$*, then* $\mathbb{Q}_x$ *exists if and only if*

$$\mathbb{E}^{\mathbb{P}_x^0}\big[E_\theta(T)\big] = 1 \quad \text{for all } T \geq 0,$$

in which case $\mathbb{Q}_x \restriction \mathcal{B}_T$ *is equivalent to* $\mathbb{P}_x^0 \restriction \mathcal{B}_T$ *and* $d\mathbb{Q}_x = E_\theta(T)\,d\mathbb{P}_x^0$ *on* $\mathcal{B}_T$ *for each* $T \geq 0$*.*

PROOF: First suppose that $\mathbb{E}^{\mathbb{P}_x^0}\big[E_\theta(T)\big] = 1$ for all $T \geq 0$, and apply Theorem 6.2.1 to produce $\mathbb{Q}_x$ so that $d\mathbb{Q}_x \restriction \overline{\mathcal{B}}_T^{\mathbb{P}} = E_\theta(T)\,d\mathbb{P}_x^0 \restriction \overline{\mathcal{B}}_T^{\mathbb{P}}$ for each $T \geq 0$. Next, apply the second part of that same theorem to see that

$$\left(\exp\Big(\sqrt{-1}\big(\xi,\beta(t)\big)_{\mathbb{R}^n} + \tfrac{t}{2}|\xi|^2\Big), \mathcal{B}_t, \mathbb{Q}_x\right)$$

is a local martingale, which, because it is bounded on $[0, T] \times C([0, \infty); \mathbb{R}^n)$ for each $T \geq 0$, must therefore be a martingale. In particular, this proves that $\beta(s + t) - \beta(s)$ is $\mathbb{Q}_x$-independent of $\mathcal{B}_s$ and its $\mathbb{Q}_x$-distribution is that of a centered Gaussian with covariance tI. Equivalently, $(\beta(t), \overline{\mathcal{B}}_t^{\mathbb{Q}_x}, \mathbb{Q}_x)$ is a Brownian motion.

Now suppose that $\mathbb{Q}$ has the property that $(\beta(t), \overline{\mathcal{B}}_t^{\mathbb{Q}}, \mathbb{Q})$ is a Brownian motion. We must show that $d\mathbb{Q} \upharpoonright \mathcal{B}_T = E_\theta(T) \, d\mathbb{P}_x^0 \upharpoonright \mathcal{B}_T$ for each $T \geq 0$. For this purpose, set

$$S(t) = \exp\left(-\int_0^t (\theta(\tau), d\beta(\tau))_{\mathbb{R}^n} - \frac{1}{2}\int_0^t |\theta(\tau)|^2 \, d\tau\right).$$

The key step in our argument is to show that $\mathbb{E}^{\mathbb{Q}}[S(T)] = 1$ for all $T \geq 0$. To this end, define

$$\zeta_R(p) = \inf\left\{t \geq 0 : \int_0^t |\theta(\tau, p)|^2 \, d\tau \geq R\right\}.$$

Then

$$S(t \wedge \zeta_R) = S^R(t) \equiv \exp\left(\int_0^t (\theta^R(\tau), d\beta(\tau))_{\mathbb{R}^n} - \frac{1}{2}\int_0^t |\theta^R(\tau)|^2 \, d\tau\right),$$

where $\theta^R(t) \equiv \mathbf{1}_{[0,t]}(\zeta_R)\theta(t)$, and so

$$\mathbb{E}^{\mathbb{Q}}[S(T), \, \zeta_R > T] = \mathbb{E}^{\mathbb{Q}}[S^R(T)] - \mathbb{E}^{\mathbb{Q}}[S^R(T), \, \zeta_R \leq T].$$

To take the next step, note that $\mathbb{E}^{\mathbb{Q}}[S^R(T)] = 1$ for each $(T, R) \in [0, \infty)^2$, and, for a fixed $R > 0$, use Theorem 6.2.1 to determine $\mathbb{P}^R$ so that $d\mathbb{P}^R \upharpoonright \mathcal{B}_T = S^R(T) \, d\mathbb{Q} \upharpoonright \mathcal{B}_T$ for all $T \geq 0$. By repeating the argument given in the first part of the present proof, one can apply the second part of Theorem 6.2.1 to see that $\left(p^R(t) - x, \overline{\mathcal{B}}_t^{\mathbb{P}^R}, \mathbb{P}^R\right)$ is a Brownian motion when

$$p^R(t) \equiv p(t) - \int_0^t (\theta(\tau) - \theta^R(\tau)) \, d\tau.$$

At the same time, because (cf. Exercise 4.3.45 in [36]) ζ_R is a $\{\mathcal{B}_t : t \geq 0\}$-stopping time and $p \upharpoonright [0, \zeta_R(p)) = p^R \upharpoonright [0, \zeta_R(p))$, one can check that $\zeta_R(p) = \zeta_R(p^R)$ for all $p \in C([0, \infty); \mathbb{R}^n)$. Hence,

$$\mathbb{E}^{\mathbb{Q}}[S^R(T), \, \zeta_R \leq T] = \mathbb{P}^R(\zeta_R \circ p^R \leq T) = \mathbb{P}_x^0(\zeta_R \leq T) \longrightarrow 0$$

as $R \to \infty$. That is, $\mathbb{E}^{\mathbb{Q}}[S(T)] = 1$.

Knowing that $\mathbb{E}^{\mathbb{Q}}[S(T)] = 1$ for all $T \geq 0$, we can again apply Theorem 6.2.1 to see that $d\mathbb{P}_x^0 \restriction \mathcal{B}_T = S(T)\, d\mathbb{Q} \restriction \mathcal{B}_T$ for all $T \geq 0$. Hence, all that remains is to show that

$$S(T,p)^{-1} = \exp\left(\int_0^T \big(\theta(\tau,p), dp(\tau)\big)_{\mathbb{R}^n} - \frac{1}{2}\int_0^T |\theta(\tau,p)|^2\, d\tau \right)$$

$\mathbb{P}_x^0$-almost surely, which is almost obvious. Indeed, because $p(t) = x + \beta(t,p) + \int_0^t \theta(\tau)\, d\tau$, it is only reasonable that

$$\int_0^t \big(\theta(\tau), dp(\tau)\big)_{\mathbb{R}^n} = \int_0^t \big(\theta(\tau), d\beta(\tau)\big)_{\mathbb{R}^n} + \int_0^t |\theta(\tau)|^2\, d\tau$$

$\mathbb{P}_x^0$-almost surely. Unfortunately, we have not yet (cf. Exercise 7.2.12) developed the machinery with which to justify this line of reasoning directly, and so we will have to take the following circuitous route. Let $R > 0$ be given, and choose (cf. Lemma 5.1.8) $\{\eta^N : N \geq 0\} \subseteq S\Theta^2(\mathbb{P}_x^0; \mathbb{R}^n)$ so that $\eta^N(t) = \eta^N([t]_N)$ and

$$\mathbb{E}^{\mathbb{Q}}\left[\int_0^{\zeta_R} |\theta(\tau) - \eta^N(\tau)|^2\, d\tau \right] + \mathbb{E}^{\mathbb{P}_x^0}\left[\int_0^{\zeta_R} |\theta(\tau) - \eta^N(\tau)|^2\, d\tau \right] \longrightarrow 0.$$

By taking

$$\zeta_R^N = \inf\left\{ t \geq 0 : \int_0^t |\eta^N(\tau)|^2\, d\tau \geq R \right\}$$

and replacing $\eta^N(t)$ by $\theta^N(t) \equiv \mathbf{1}_{[0,\zeta_R^N)}(t)\eta^N(t)$, we preserve the L^2-convergence, still have $\theta^N(t) = \theta^N([t]_N)$ for $t \in [0, \zeta_R^N)$, and achieve the bound

$$\int_0^{\zeta_R^N} |\theta^N(\tau)|^2\, d\tau \leq R \quad \text{for all } N \geq 0.$$

Now let $T > 0$ be given, set $V^N(p)$ equal to

$$\exp\bigg(\sum_{m=0}^{\infty} \Big(\theta^N(m2^{-N}, p), \beta\big(T \wedge \zeta_R^N(p) \wedge (m+1)2^{-N}, p\big)$$

$$-\, \beta\big(T \wedge \zeta_R^N(p) \wedge m2^{-N}, p\big)\Big)_{\mathbb{R}^n}$$

$$+\, \frac{1}{2}\int_0^{T \wedge \zeta_R^N(p)} \big(\theta(\tau,p), \theta^N(\tau,p)\big)_{\mathbb{R}^n}\, d\tau \bigg),$$

and observe that another expression for $V^N(p)$ is

$$
\exp\left(\sum_{m=0}^{\infty}\left(\theta^N(m2^{-N},p),p\big(T\wedge\zeta_R^N(p)\wedge(m+1)2^{-N}\big)\right.\right.
$$

$$
\left.\left.-\,p\big(T\wedge\zeta_R^N(p)\wedge m2^{-N}\big)\right)_{\mathbb{R}^n}
$$

$$
-\frac{1}{2}\int_0^{T\wedge\zeta_R^N(p)}\big(\theta(\tau),\theta_N(\tau,p)\big)_{\mathbb{R}^n}\,d\tau\right),
$$

Moreover, for any bounded, $\mathcal{B}_T$-measurable F,

$$
\mathbb{E}^{\mathbb{P}_x^0}\big[V^N F,\,\zeta_R>T\big]=\mathbb{E}^{\mathbb{Q}}\big[S(T)V^N F,\,\zeta_R>T\big]
$$

$$
=\mathbb{E}^{\mathbb{Q}}\big[S(T\wedge\zeta_R)V^N F,\,\zeta_R>T\big].
$$

Using the second expression for V^N, we see that, as $N\to\infty$, $V^N\longrightarrow$ $E_\theta(T\wedge\zeta_R)$ in $\mathbb{P}_x^0$-probability. In fact, by taking into account the estimates in (5.1.25), one sees that this convergence is happening in $L^1(\mathbb{P}_x^0;\mathbb{R})$. At the same time, by using the first expression for V^N, the same line of reasoning shows that $V^N S(T\wedge\zeta_R)\longrightarrow 1$ in $L^1(\mathbb{Q};\mathbb{R})$. Thus, we have proved that

$$
\mathbb{E}^{\mathbb{Q}}\big[F,\,\zeta_R>T\big]=\mathbb{E}^{\mathbb{P}_x^0}\big[E_\theta(T)F,\,\zeta_R>T\big],
$$

and, after letting $R\nearrow\infty$, we have proved that $\mathbb{E}^{\mathbb{P}_x^0}[E_\theta(T)]=1$ and that $d\mathbb{Q}\upharpoonright\mathcal{B}_T=E_\theta(T)\,d\mathbb{P}_x^0\upharpoonright\mathcal{B}_T$. $\square$

COROLLARY 6.2.3. *Let $b:[0,\infty)\times\mathbb{R}^n\longrightarrow\mathbb{R}^n$ be a locally bounded, measurable function. Then there exists a $\mathbb{Q}_x^b\in\mathbf{M}_1\big(C\big([0,\infty);\mathbb{R}^n\big)\big)$ with the property that*

$$
\left(p(t)-x-\int_0^t b\big(\tau,p(\tau)\big)\,d\tau,\,\overline{\mathcal{B}}_t^{\mathbb{Q}_x^b},\,\mathbb{Q}_x^b\right)
$$

is a Brownian motion if and only if (cf. the notation in Corollary 6.2.2)

$$
\mathbb{E}^{\mathbb{P}_x^0}\big[R^b(T)\big]=1\quad\text{for all }T\ge 0,
$$

where

$$
R^b(T,p)\equiv\exp\left(\int_0^T\big(b(\tau,p(\tau)),dp(\tau)\big)_{\mathbb{R}^n}-\frac{1}{2}\int_0^T\big|b(\tau,p(\tau))\big|^2\,d\tau\right),
$$

in which case $d\mathbb{Q}_x^b\upharpoonright\mathcal{B}_T=R^b(T)\,d\mathbb{P}_x^0\upharpoonright\mathcal{B}_T$ for each $T\ge 0$. In particular, there is at most one such $\mathbb{Q}_x^b$.

REMARK 6.2.4. There is a subtlety about the result in Corollary 6.2.3. Namely, it might lead one to believe that, for any bounded, measurable $b : \mathbb{R}^n \longrightarrow \mathbb{R}^n$ and $\mathbb{P}^0$ almost every $p \in C([0, \infty); \mathbb{R}^n)$, one can solve the equation

$$(*) \qquad X(t, p) = p(t) + \int_0^t b\big(X(\tau, p)\big)\, d\tau, \quad t \geq 0,$$

and that the solution is unique. However, this is not what it says! Instead, it says that there is a unique $\mathbb{Q}_0^b \in \mathbf{M}_1\big(C([0, \infty); \mathbb{R}^n)\big)$ with the property that $\mathbb{P}^0$ is the $\mathbb{Q}_0^b$-distribution of

$$p \rightsquigarrow \beta(\,\cdot\,, p) \equiv p(\,\cdot\,) - \int_0^{\cdot} b\big(p(\tau)\big)\, d\tau.$$

There is no implication that p can be recovered from $\beta(\,\cdot\,, p)$. In other words, we are, in general, dealing here with the kind of weak solutions alluded to at the end of Exercise 6.1.10. Of course, if b is locally Lipschitz continuous and therefore the solution $X(\,\cdot\,, p)$ to $(*)$ can be constructed (e.g., by Picard iteration) "p by p," then the $\mathbb{P}^0$-distribution of $p \rightsquigarrow X(\,\cdot\,, p)$ is $\mathbb{Q}_0^b$.

§6.2.2. **An Application to Pinned Brownian Motion:** A remarkable property of Wiener measure is that if $T \in (0, \infty)$, $x, y \in \mathbb{R}^n$, and

$$(6.2.5) \qquad p_{T,y}(t) \equiv p(t \wedge T) + \frac{t \wedge T}{T}\big(y - p(T)\big) \quad \text{for } p \in C([0, \infty); \mathbb{R}^n),$$

then *the $\mathbb{P}_x^0$-distribution of $p \rightsquigarrow p_{T,y} \upharpoonright [0, T]$ is the $\mathbb{P}_x^0$-distribution of $p \rightsquigarrow p \upharpoonright [0, T]$ conditioned on the event $p(T) = y$*. To be more precise, if $F : C([0, \infty); \mathbb{R}^n) \longrightarrow \mathbb{R}$ is a bounded, $\mathcal{B}_T$-measurable function and $\Gamma \in \mathcal{B}_{\mathbb{R}^n}$, then (cf. Theorem 4.2.18 in [36] or Lemma 8.3.6 below)

$$\mathbb{E}^{\mathbb{P}_x^0}\big[F, p(T) \in \Gamma\big] = \frac{1}{(2\pi T)^{\frac{n}{2}}} \int_{\mathbb{R}^n} \left(\int_{C([0,\infty);\mathbb{R}^n)} F\big(p_{T,y}\big)\, \mathbb{P}^0(dp) \right) e^{-\frac{|y-x|^2}{2T}}\, dy.$$

For this reason, $p \rightsquigarrow p_{T,y} \upharpoonright [0, T]$ is sometimes called *pinned Brownian motion*, or, in greater detail, Brownian motion pinned to y at time T.

The fact that (6.2.5) gives a representation of Brownian motion conditioned to be at y at time T is one of the many peculiarities (especially its Gaussian nature) of Brownian motion which sets it apart from all other diffusions. For this reason, it is interesting to find another representation, one that has a better chance of admitting a generalization (cf. Remark 6.2.8 below). With this in mind, let $t \in [0, T)$ be given, notice that, by definition, for any bounded, $\mathcal{B}_t$-measurable F,

$$\mathbb{E}^{\mathbb{P}^0_x}\big[F \,\big|\, p(T) = y\big] = \mathbb{E}^{\mathbb{P}^0}\left[F \frac{\gamma_{T-t}\big(y - p(t)\big)}{\gamma_T(y - x)}\right],$$

where $\gamma_\tau(\xi) \equiv (2\pi\tau)^{-\frac{n}{2}} e^{-\frac{|\xi|^2}{2\tau}}$ denotes the centered Gauss kernel on $\mathbb{R}^n$ with covariance $\tau I_{\mathbb{R}^n}$. Next, recall that

$$\partial_\tau \gamma_{T-\tau}(y - \cdot) + \tfrac{1}{2}\Delta\gamma_{T-\tau}(y - \cdot) = 0 \quad \text{on } [0, T) \times \mathbb{R}^n,$$

where Δ denotes the Euclidean Laplacian for $\mathbb{R}^n$; and apply Itô's formula to conclude that

$$\frac{\gamma_{T-t}\big(y - p(t)\big)}{\gamma_T(y - x)} = 1 - \int_0^t \frac{\gamma_{T-\tau}\big(y - p(\tau)\big)}{\gamma_T(y - x)}\left(\frac{p(\tau) - y}{T - \tau}, dp(\tau)\right)_{\mathbb{R}^n}$$

for $t \in [0, T)$; and so, by Exercise 5.3.4,

$$(6.2.6) \qquad
\begin{aligned}
&\frac{\gamma_{T-t}\big(y - p(t)\big)}{\gamma_T(y - x)} \\
&= \exp\left(-\int_0^t \left(\frac{p(\tau) - y}{T - \tau}, dp(\tau)\right)_{\mathbb{R}^n} - \frac{1}{2}\int_0^t \left|\frac{p(\tau) - y}{T - \tau}\right|^2 d\tau\right)
\end{aligned}$$

for $t \in [0, T)$. Furthermore, by the Chapman–Kolmogorov equation,

$$\mathbb{E}^{\mathbb{P}^0_x}\left[\frac{\gamma_{T-t}\big(y - p(t)\big)}{\gamma_T(y - x)}\right] = \int_{\mathbb{R}^n} \gamma_t(\xi - x)\frac{\gamma_{T-t}(y - \xi)}{\gamma_T(y - x)}\, d\xi = 1.$$

Hence, by Corollary 6.2.3, we have already proved a good deal of the following statement.

THEOREM 6.2.7. *Given* $T > 0$, *there a continuous map* $(y, p) \in \mathbb{R}^n \times C\big([0, \infty); \mathbb{R}^n\big) \longmapsto X_y(\,\cdot\,, p) \in C\big([0, T); \mathbb{R}^n\big)$ *such that*

$$X_y(t, p) = p(t) - \int_0^t \frac{X_y(\tau, p) - y}{T - \tau}\, d\tau \quad \text{for } t \in [0, T).$$

Furthermore, for each $x \in \mathbb{R}^n$ *and* $\mathbb{P}^0_x$-*almost every* p, $\lim_{t \nearrow T} X_y(t, p) = y$, *and so there exists a unique* $\mathbb{P}^0_{T,x,y} \in \mathbf{M}_1\big(C([0, T]; \mathbb{R}^n)\big)$ *with the property that the* $\mathbb{P}^0_{T,x,y}$-*distribution of* $p \rightsquigarrow p \upharpoonright [0, T)$ *is the same as the* $\mathbb{P}^0_x$-*distribution of* $p \rightsquigarrow X_y(\,\cdot\,, p) \upharpoonright [0, T)$. *In fact,* $\mathbb{P}^0_{T,x,y}\big(p(T) = y\big) = 1$,

$\mathbb{P}^0_{T,x,y} = (\tau_x)_* \mathbb{P}^0_{T,0,y-x}$, $(x,y) \rightsquigarrow \mathbb{P}^0_{T,x,y}$ *is continuous, and, if* $\check{p}(t) = p(T-t)$ *for* $p \in C([0,T]; \mathbb{R}^n)$ *and* $t \in [0,T]$, *then* $\mathbb{P}^0_{T,y,x}$ *is the* $\mathbb{P}^0_{T,x,y}$-*distribution of* $p \rightsquigarrow \check{p}$. *Finally, for any* $\mathcal{B}_T$-*measurable* F *which is either bounded or non-negative,* $\mathbb{E}^{\mathbb{P}^0_{T,x,y}}[F]$ *is the* $\mathbb{P}^0_x$-*conditional expectation value* $\mathbb{E}^{\mathbb{P}^0_x}[F \,|\, p(T) = y]$ *of* F *given* $p(T) = y$ *in the sense that*

$$\mathbb{E}^{\mathbb{P}^0_x}[F \,|\, \sigma(p(T))] = \mathbb{E}^{\mathbb{P}^0_{T,x,p(T)}}[F] \quad \mathbb{P}^0_x\text{-almost surely.}$$

PROOF: The existence and continuity of $(y,p) \rightsquigarrow X_y(\,\cdot\,,p) \upharpoonright [0,T)$ follow easily from the standard Picard iteration procedure. Moreover, by uniqueness, it is easy to see that $X_y(\,\cdot\,,\tau_x p) = x + X_{y-x}(\,\cdot\,,p)$, and so the distribution of $p \rightsquigarrow X_y(\,\cdot\,,p)$ under $\mathbb{P}_x$ coincides with the $\mathbb{P}^0$-distribution of $p \rightsquigarrow \tau_x X_{y-x}(\,\cdot\,,p)$. Thus, if they exist, then $\mathbb{P}^0_{T,x,y} = (\tau_x)_* \mathbb{P}^0_{T,0,y-x}$.

By Corollary 6.2.3, (6.2.6), and the preceding discussion, we know that, for any $t \in [0,T)$ and $A \in \mathcal{B}_t$,

$$(*) \qquad \mathbb{E}^{\mathbb{P}^0_x}\left[\frac{\gamma_{T-t}(p(t) - y)}{\gamma_T(y-x)}, A\right] = \mathbb{P}^0_x(X_y(\,\cdot\,) \in A).$$

Our first application of (*) will be to show that the $\mathbb{P}^0_x$ distribution of $p \rightsquigarrow X_y(\,\cdot\,,p) \upharpoonright (0,T)$ coincides with the $\mathbb{P}^0_y$-distribution of $p \rightsquigarrow X_x(T - \,\cdot\,,p) \upharpoonright (0,T)$; and for this purpose, it suffices to observe that, by repeated application of (*), for any $m \geq 1$ and $0 < t_1 < \cdots < t_m$,

$$\mathbb{P}^0_x\left(X_y(t_1) \in d\xi_1, \ldots X_y(t_m) \in d\xi_m\right)$$
$$= \frac{\gamma_{t_1}(\xi_1 - x) \cdots \gamma_{t_m - t_{m-1}}(\xi_m - \xi_{m-1})\gamma(y - \xi_m)}{\gamma_T(y - x)} d\xi_1 \cdots d\xi_m$$
$$= \mathbb{P}^0_y\left(X_x(T - t_m) \in d\xi_m, \ldots X_x(T - t_1) \in d\xi_1\right).$$

In particular, we have shown that

$$\mathbb{P}^0_x\left(\lim_{t \nearrow T} X_y(t) = y\right) = \mathbb{P}^0_y\left(\lim_{t \searrow 0} X_x(t) = y\right) = 1,$$

and so we now know that $\mathbb{P}^0_{T,x,y}$ exists, $\mathbb{P}^0_{T,x,y}(p(T) = y) = 1$, and that $\mathbb{P}^0_{T,y,x}$ is the $\mathbb{P}^0_{T,x,y}$-distribution of $p \rightsquigarrow \check{p}$.

We next want to check the continuity of $(x,y) \rightsquigarrow \mathbb{P}^0_{T,x,y}$. To this end, remember (cf. the end of the first paragraph of this proof) that we know $\mathbb{P}^0_{T,x,y} = (\tau_x)_* \mathbb{P}^0_{T,0,y-x}$. Hence, it is enough for us to show that $y \rightsquigarrow \mathbb{P}^0_{T,0,y}$ is continuous. Next, observe that, for any $\delta \in (0,1)$,

$$\left\|X_{y'}(\,\cdot\,,p) - X_y(\,\cdot\,,p)\right\|_{[0,T)}$$
$$\leq \left(|y' - y| + \left\|X_{y'}(\,\cdot\,,p) - X_y(\,\cdot\,,p)\right\|_{[0,(1-\delta)T]}\right)\log\frac{1}{\delta}$$
$$+ \int_{(1-\delta)T}^T \frac{|X_{y'}(\tau,p) - y'|}{T - \tau} d\tau + \int_{(1-\delta)T}^T \frac{|X_y(\tau,p) - y|}{T - \tau} d\tau.$$

Since, for any $\delta \in (0,1)$ and $\epsilon > 0$, we know that

$$\lim_{y' \to y} \left\| X_{y'}(\,\cdot\,,p) - X_y(\,\cdot\,,p) \right\|_{[0,(1-\delta)T]} = 0 \quad \text{for all } p,$$

we will know that

(**)
$$\lim_{y' \to y} \mathbb{P}^0 \left(\left\| X_{y'}(\,\cdot\,) - X_y(\,\cdot\,) \right\|_{[0,T]} \geq \epsilon \right) = 0,$$

once we show that, for each $R > 0$,

$$\lim_{\delta \searrow 0} \sup_{|y| \leq R} \int_{(1-\delta)T}^{T} \frac{\mathbb{E}^{\mathbb{P}^0} \left[|X_y(\tau) - y| \right]}{T - \tau} \, d\tau = 0.$$

But

$$\int_{(1-\delta)T}^{T} \frac{\mathbb{E}^{\mathbb{P}^0} \left[|X_y(\tau) - y| \right]}{T - \tau} \, d\tau = \int_{0}^{\delta T} \frac{\mathbb{E}^{\mathbb{P}^0_y} \left[|X_0(\tau) - y| \right]}{\tau} \, d\tau,$$

and it is an easy matter to check that

$$\sup_{|y| \leq R} \sup_{\tau \in (0, \frac{T}{2}]} \frac{\mathbb{E}^{\mathbb{P}^0_y} \left[|X_0(\tau) - y| \right]|}{\sqrt{\tau}} < \infty.$$

After combining these, we know that (**) holds and therefore that $(x,y) \rightsquigarrow \mathbb{P}^0_{T,x,y}$ is continuous.

All that remains is to identify $\mathbb{P}^0_{T,x,y}$ as the conditional distribution of $\mathbb{P}^0_x \restriction \mathcal{B}_T$ given $p(T) = y$. However, from (*), we know that, for any $\Gamma \in \mathcal{B}_{\mathbb{R}^n}$,

$$\mathbb{E}^{\mathbb{P}^0_x} \left[A \cap \{ p(T) \in \Gamma \} \right] = \int_{\Gamma} \mathbb{P}_{T,x,y}(A) \, \gamma_T(y - x) \, dy$$

for all $A \in \mathcal{B}_t$ when $t \in [0, T]$. Since the set of $A \in \mathcal{B}_T$ for which this relation holds is a σ-algebra, it follows that it holds for all $A \in \mathcal{B}_T$. $\square$

REMARK 6.2.8. Certainly the most intriguing aspect of the preceding result is the conclusion that $\lim_{t \nearrow T} X_y(t,p) = y$ for $\mathbb{P}^0_x$-almost every p. Intuitively, one knows that the drift term $-\frac{X(t,p)-y}{T-t}$ is "punishing" $X(t,p)$ for not being close to y, and that, as $t \nearrow T$, the strength of this penalty becomes infinite. On the other hand, this intuition reveals far less than the whole story. Namely, it completely fails to explain why the penalization takes this particular form instead of, for example, $-2\frac{X(t,p)-y}{T-t}$ or $-\frac{(X(t,p)-y)^3}{T-t}$. As a careful examination of the preceding reveals, the reason why $-\frac{X(t,p)-y}{T-t}$ is precisely the correct drift is that it is equal to $\text{grad}_{X(t,p)} \log g_{T-t}(\,\cdot\,,y)$ where $g_\tau(\,\cdot\,,y)$ is the fundamental solution (in this case $\gamma_\tau(\,\cdot\, - y)$) to the heat equation. It is this observation which allows one to generalize the considerations to diffusions other than Brownian motion.

§6.2.3. Exercises

EXERCISE 6.2.9. The purpose of this exercise is to interpret the meaning of (cf. Corollary 6.2.3) $\mathbb{E}^{\mathbb{P}^0}\big[R^b(T)\big]$ failing to be 1.

Let $b : \mathbb{R}^n \longrightarrow \mathbb{R}^n$ be a locally Lipschitz continuous function. Choose $\psi \in C^\infty(\mathbb{R}^n; [0,1])$ so that $\psi \equiv 1$ on $\overline{B_{\mathbb{R}^n}(0,1)}$ and $\psi \equiv 0$ off $B_{\mathbb{R}^n}(0,2)$, and set $b_R(x) = \psi(R^{-1}x)b(x)$ for $R > 0$ and $x \in \mathbb{R}^n$. Given $R > 0$ and $p \in C([0,\infty); \mathbb{R}^n)$, define $\zeta_R(p) = \inf\{t \geq 0 : |p(t)| \geq R\}$ and determine $X_R(\,\cdot\,, p) \in C([0,\infty); \mathbb{R}^n)$ by the equation

$$X_R(t,p) = p(t) + \int_0^t b_R\big(X(\tau, p)\big)\, d\tau \quad t \in [0,\infty).$$

(i) Given $0 < R_1 < R_2$, show that $X_{R_2}(t,p) = X_{R_1}(t,p)$ for $0 \leq t \leq \zeta_{R_1} \circ X_{R_1}(\,\cdot\,, p)$, and conclude that

$$\zeta_{R_1} \circ X_{R_1}(\,\cdot\,, p) = \zeta_{R_1} \circ X_{R_2}(\,\cdot\,, p) \leq \zeta_{R_2} \circ X_{R_2}(\,\cdot\,, p).$$

Next, define the *explosion time* $\mathfrak{e}(p) = \lim_{R \nearrow \infty} \zeta_R \circ X_R(\,\cdot\,, p)$, and determine $t \in [0, \mathfrak{e}(p)) \longmapsto X(t,p) \in \mathbb{R}^n$ so that $X(t,p) = X_R(t,p)$ for $0 \leq t \leq \zeta_R(p)$.

(ii) Define R^b as in Corollary 6.2.3, show that

$$\mathbb{E}^{\mathbb{P}^0}\big[R^b(T),\, \zeta_R > T\big] = \mathbb{P}^0\big(\zeta_R \circ X_R(\,\cdot\,) > T\big),$$

and conclude that $\mathbb{E}^{\mathbb{P}^0}\big[R^b(T)\big] = \mathbb{P}^0(\mathfrak{e} > T)$. That is, $\mathbb{E}^{\mathbb{P}^0}\big[R^b(T)\big]$ is equal to the probability that $X(\,\cdot\,, p)$ has not exploded by time T.

EXERCISE 6.2.10. When the drift b in Corollary 6.2.3 is a gradient, the Radon–Nikodym $R^b(T)$ becomes much more tractable. Namely, apply Itô's formula to show that if $b(t,x) = \operatorname{grad}_x U(t, \cdot\,)$, then

(6.2.11)
$$R^b(T,p) = \exp\left(U\big(T, p(T)\big) - U\big(0, p(0)\big) - \frac{1}{2}\int_0^T V\big(\tau, p(\tau)\big)\, d\tau\right)$$

where $V(t,x) \equiv \big|\operatorname{grad}_x U(t, \cdot\,)\big|^2 + \big(\partial_t + \tfrac{1}{2}\Delta\big)U(t,x)$.

(i) Recall the Ornstein–Uhlenbeck process $p' \rightsquigarrow X(\,\cdot\,, x, p')$ given by (3.3.1), and let $\mathbb{Q}_x$ denote the distribution of $p \rightsquigarrow X(\,\cdot\,, x,, p)$ under $\mathbb{P}^0$. Using the preceding, show that

$$\mathbb{E}^{\mathbb{Q}_x}\big[\varphi\big(p(T)\big)\big]$$

$$= \mathbb{E}^{\mathbb{P}^0}\left[\varphi\big(p(T)\big)\exp\left(\beta\frac{nT - |p(T)|^2 + |x|^2}{2} - \frac{\beta^2}{2}\int_0^T |p(\tau)|^2\, d\tau\right)\right]$$

for any measurable $\varphi : \mathbb{R}^n \longrightarrow [0,\infty)$.

(ii) After combining (i) with the considerations at the end of §3.3.1, come to the conclusion that

$$\mathbb{E}^{\mathbb{P}^0_x}\left[\varphi\big(p(T)\big)\exp\left(-\frac{1}{2}\int_0^T |p(\tau)|^2\,d\tau\right)\right]$$
$$= e^{-\frac{nT+|x|^2}{2}}\int_{\mathbb{R}^n}\varphi(y)e^{\frac{|y|^2}{2}}\gamma_{1/2(1-e^{-2T})}\big(y-e^{-T}x\big)\,dy.$$

In particular, conclude from this that

$$\mathbb{E}^{\mathbb{P}^0_x}\left[\exp\left(-\frac{1}{2}\int_0^T |p(\tau)|^2\,d\tau\right)\right] = \big(\sqrt{\cosh T}\big)^{-n}\exp\left(-\frac{|x|^2}{2}\tanh T\right).$$

(iii) By applying the Brownian scale invariance property, note that the $\mathbb{P}^0$-distributions of

$$p \rightsquigarrow \exp\left(-\frac{1}{2}\int_0^{\lambda T} |p(\tau)|^2\,d\tau\right) \quad\text{and}\quad p \rightsquigarrow \exp\left(-\frac{1}{2}\lambda^2\int_0^T |p(\tau)|^2\,d\tau\right)$$

are the same for every choice of $\lambda > 0$. Hence,

$$\mathbb{E}^{\mathbb{P}^0}\left[\exp\left(-\frac{\lambda}{2}\int_0^T |p(\tau)|^2\,d\tau\right)\right] = \left(\sqrt{\cosh\sqrt{\lambda}T}\right)^{-n}.$$

(iv) Suppose that, for some $A > 0$,

$$\mathbb{E}^{\mathbb{P}^0}\left[\exp\left(\frac{A^2}{2}\int_0^T |p(\tau)|^2\,d\tau\right)\right] < \infty,$$

and show that

$$z \in B_{\mathbb{C}}(0,A^2) \longmapsto \mathbb{E}^{\mathbb{P}^0}\left[\exp\left(-\frac{z}{2}\int_0^T |p(\tau)|^2\,d\tau\right)\right] \in \mathbb{C}$$

must be a halomorphic function. After combining this with (iii), conclude that $A \leq \frac{\pi}{2T}$. That is,

$$A > \frac{\pi}{2T} \implies \mathbb{E}^{\mathbb{P}^0}\left[\exp\left(\frac{A^2}{2}\int_0^T |p(\tau)|^2\,d\tau\right)\right] = \infty. \qquad \cdot$$

(v) The preceding can be used to bring out the inherent weakness in criteria like (cf. Exercise 5.3.4) Novikov's. Namely, show that when $\theta(t) =$

$p(t)$ and $E_\theta(t)$ is defined as in (5.1.10) relative to the Brownian motion $(p(t), \overline{B}_t, \mathbb{P}^0)$, $(E_\theta(t), \overline{B}_t, \mathbb{P}^0)$ is a martingale but that

$$T > \frac{\pi}{2} \implies \mathbb{E}^{\mathbb{P}^0}\left[\exp\left(\frac{1}{2}\int_0^T |p(\tau)|^2 \, d\tau\right)\right] = \infty.$$

EXERCISE 6.2.12. The purpose of this exercise is to give an example of a θ for which the corresponding $E_\theta(t)$ is definitely not a martingale. Throughout, $n = 1$.

(i) Given $p \in C([0, \infty); \mathbb{R})$, suppose that $X(\cdot, p) \in C([0, T(p)]; \mathbb{R})$ satisfies

$$X(t, p) = p(t) + \int_0^t \left(2 + X(\tau, p)\right)^2 d\tau \quad \text{for } t \in [0, T(p)].$$

Assuming that $p(t) \geq -1$ for $t \in [0, T(p)]$, show that $X(t, p) \geq \frac{1}{1-t} - 2$, and conclude that $T(p) < 1$.

Hint: First note that $X(\cdot, p)$ must be obtainable from p via Picard's iteration method starting with $X_0(\cdot, p) = p$. Proceeding by induction, show that $X_N(\cdot, p) \geq Y_N(\cdot)$, where $\{Y_N : N \geq 0\}$ corresponds to $X(\cdot, p)$ when $p \equiv -1$.

(ii) Take $\theta(t) = \left(2 + p(t)\right)^2$, and define $E_\theta(t)$ relative to $(p(t), \overline{B}_t, \mathbb{P}^0)$. Show that

$$\mathbb{E}^{\mathbb{P}^0}\left[E_\theta(1)\right] \leq \mathbb{P}^0\left(\inf_{t \in [0,1]} p(t) \leq -1\right) < 1.$$

In particular, $\left(E_\theta(t), \overline{B}_t^{\mathbb{P}^0}, \mathbb{P}^0\right)$ is not a martingale, and so, by Novikov's criterion,

$$\mathbb{E}^{\mathbb{P}^0}\left[\exp\left(\frac{1}{2}\int_0^1 \left(2 + p(t)\right)^4 dt\right)\right] = \infty.$$

(iii) Show that if $\theta(t) = -\text{sgn}(p(t))\left(2 + p(t)\right)^2$, then $\left(E_\theta(t), \overline{B}_t, \mathbb{P}^0\right)$ is a martingale. Hence, in view of the last part of (ii), we see once again that Novikov's criterion is too weak.

EXERCISE 6.2.13. Set $\Omega = \mathbb{R}^n \times C([0, \infty); \mathbb{R}^n)$, and consider the Borel probability measure $\mathbb{Q} = \Gamma_T \times \mathbb{P}^0$ on Ω, where Γ_T is the centered Gauss measure on $\mathbb{R}^n$ with covariance TI.

(i) Referring to Theorem 6.2.7, define

$$(y, p) \in \Omega \longmapsto Y(\cdot, (y, p)) \equiv X_y(\cdot, p) \in C([0, T); \mathbb{R}^n),$$

and show that, $\mathbb{Q}$-almost surely, $\lim_{t \nearrow T} Y(t, (y, p)) = y$. Thus, $\mathbb{Q}$-almost surely, $Y(\cdot, \omega)$ extends as a continuous path on $[0, T]$.

(ii) Show that the $\mathbb{Q}$-distribution of $\omega \in \Omega \longmapsto Y(\,\cdot\,, \omega) \in C([0,T]; \mathbb{R}^n)$ is the same as the $\mathbb{P}^0$-distribution of $p \rightsquigarrow p \restriction [0,T]$. Equivalently, if

$$\beta(t, \omega) = Y(t \wedge T, \omega) + \big(p(t) - p(t \wedge T)\big) \quad \text{for } (t, \omega) = \big(t, (y, p)\big) \in [0, \infty) \times \Omega,$$

then

$$\Big(\beta(t), \overline{\sigma\big(\{\beta(\tau) : \tau \in [0, t]\}\big)}, \mathbb{Q}\Big)$$

is a Brownian motion.

(iii) Remember (cf. part **(ii)** of Exercise 2.4.16) that almost every Brownian path is Hölder continuous of every order strictly less than $\frac{1}{2}$. Thus, if $\big(\beta(t), \mathcal{F}_t, \mathbb{P}\big)$ is an $\mathbb{R}^n$-valued Brownian motion, then

$$t \in [0, \infty) \longmapsto \tilde{\beta}(t) \equiv \beta(t) - \int_0^{t \wedge T} \frac{\beta(T) - \beta(\tau)}{T - \tau}\, d\tau$$

is $\mathbb{P}$-almost surely well defined. Show that

$$\Big(\tilde{\beta}(t), \overline{\sigma\big(\{\tilde{\beta}(\tau) : \tau \in [0, t]\}\big)}, \mathbb{P}\Big)$$

is again a Brownian motion. Hence, if $C^{\frac{1}{4}}\big([0, \infty); \mathbb{R}^n\big)$ is the space of paths which are Hölder continuous of order $\frac{1}{4}$ and we use $\mathbb{P}^0$ again to denote $\mathbb{P}^0 \restriction C^{\frac{1}{4}}\big([0, \infty); \mathbb{R}^n\big)$, then the map on $C^{\frac{1}{4}}\big([0, \infty); \mathbb{R}^n\big)$ into itself given by

$$[\Phi(p)](t) = p(t) - \int_0^{t \wedge T} \frac{p(T) - p(\tau)}{T - \tau}\, d\tau$$

is $\mathbb{P}^0$-measure preserving.

6.3 Homogeneous Chaos

Anyone who has read the second part, *I Am a Mathematician*, of Norbert Wiener's remarkable autobiography knows that Wiener attributed much of his own mathematical achievement to his deep insight into spectral theory. Wiener's insight into the subject was not only deep, it was highly imaginative. Indeed, he understood that the basic principles underlying spectral decomposition apply to situations where the rest of us would probably not have seen that they might apply. One of his most imaginative applications of spectral decomposition was to *randomness*. More precisely, he had the marvelous idea that random noise should be decomposable into components of what he, with his flair for words, called spaces of *homogeneous chaos*.

Wiener's own treatment of this subject is fraught with difficulties,[5] all of which were resolved by Itô. Thus, we, once again, will be guided by Itô.

---

[5] One reason why Wiener is difficult to read is that he insisted on doing all integration theory with respect to Lebesgue measure on the interval $[0, 1]$. He seems to have thought that this decision would make engineers and other non-mathematicians happier.

§6.3.1. Multiple Stochastic Integrals: Throughout, we will be working with the canonical $\mathbb{R}^n$-valued Brownian motion $\left(p(t), \overline{\mathcal{B}}_t, \mathbb{P}^0\right)$.

Our first order of business is to define *multiple stochastic integrals*. That is, if for $m \geq 1$ and $t \in [0, \infty]$, $\square^{(m)}(t) \equiv [0, t)^m$ and $\square^{(m)} \equiv \square^{(m)}(\infty)$, we want to assign a meaning to expressions like

$$\tilde{I}_F^{(m)}(t) = \int_{\square^{(m)}(t)} \left(F(\vec{\tau}), d\vec{p}(\vec{\tau})\right)_{(\mathbb{R}^n)^m}$$

when $F \in L^2\left(\square^{(m)}; (\mathbb{R}^n)^m\right)$.

With this goal in mind, when $m = 1$ and $F = f \in L^2\left([0, \infty); \mathbb{R}^n\right)$, we take $I_F^{(1)}(t) = I_f(t)$, where $I_f(t)$ is the Paley–Wiener integral of f. When $m \geq 2$ and $F = f_1 \otimes \cdots \otimes f_m$ for some $f_1, \ldots, f_m \in L^2\left([0, \infty); \mathbb{R}^n\right),$[6] we use induction to define $I_F^{(m)}(t)$ so that

(6.3.1) $$I_{f_1 \otimes \cdots \otimes f_m}^{(m)}(t) = \int_0^t I_{f_1 \otimes \cdots \otimes f_{m-1}}^{(m-1)}(\tau)\left(f_m(\tau), dp(\tau)\right)_{\mathbb{R}^n},$$

where now we need Itô's integral. Of course, in order to do so, we are obliged to check that $\tau \rightsquigarrow f_m(\tau) I_{f_1 \otimes \cdots \otimes f_{m-1}}^{(m-1)}(t)$ is square integrable. But, assuming that $I_{f_1 \otimes \cdots \otimes f_{m-1}}^{(m-1)}$ is well defined, we have that

$$\mathbb{E}^{\mathbb{P}^0}\left[\int_0^T \left|f_m(\tau) I_{f_1 \otimes \cdots \otimes f_{m-1}}^{(m-1)}(\tau)\right|^2 d\tau\right]$$

$$= \int_0^T |f_m(\tau)|^2 \mathbb{E}^{\mathbb{P}^0}\left[\left|I_{f_1 \otimes \cdots \otimes f_{m-1}}^{(m-1)}(\tau)\right|^2\right] d\tau.$$

Hence, at each step in our induction procedure, we can check that

$$\mathbb{E}^{\mathbb{P}^0}\left[\left|I_{f_1 \otimes \cdots \otimes f_m}^{(m)}(T)\right|^2\right] = \int_{\Delta^{(m)}(T)} |f_1(\tau_1)|^2 \cdots |f_m(\tau_m)|^2 d\tau_1 \cdots d\tau_m,$$

where $\Delta^{(m)}(t) \equiv \left\{(t_1, \ldots, t_m) \in \square^{(m)} : \ \leq t_1 < \cdots < t_m < t\right\}$; and so, after polarization, we arrive at

$$\mathbb{E}^{\mathbb{P}^0}\left[I_{f_1 \otimes \cdots \otimes f_m}^{(m)}(T) I_{f_1' \otimes \cdots \otimes f_m'}^{(m)}(T)\right]$$

$$= \int_{\Delta^{(m)}(T)} \left(f_1(\tau_1), f'(\tau_1)\right)_{\mathbb{R}^n} \cdots \left(f_m(\tau_m), f_m'(\tau_m)\right)_{\mathbb{R}^n} d\tau_1 \cdots d\tau_m.$$

---

[6] Here we are identifying $f_1 \otimes \cdots \otimes f_m$ with the $(\mathbb{R}^n)^m$-valued function F on $[0, \infty)^m$ such that $(\Xi, F(t_1, \ldots, t_m))_{(\mathbb{R}^n)^m} = (\xi_1, f_1(t_1))_{\mathbb{R}^N} \cdots (\xi_m, f_m(t_m))_{\mathbb{R}^n}$ for $\Xi = (\xi_1, \ldots, \xi_m)$.

We next introduce

(6.3.2) $$\tilde{I}^{(m)}_{f_1 \otimes \cdots \otimes f_m}(t) \equiv \sum_{\pi \in \Pi_m} I^{(m)}_{f_{\pi(1)} \otimes \cdots \otimes f_{\pi(m)}}(t),$$

where Π_m is the symmetric group (i.e., the group of permutations) on $\{1, \ldots, m\}$. By the preceding, we see that

$$\mathbb{E}^{\mathbb{P}^0}\left[\tilde{I}^{(m)}_{f_1 \otimes \cdots \otimes f_m}(T) \tilde{I}^{(m)}_{f'_1 \otimes \cdots \otimes f'_m}(T)\right]$$

$$= \sum_{\pi, \pi' \in \Pi_m} \int_{\Delta^{(m)}(T)} \prod_{\ell=1}^{m} \left(f_{\pi(\ell)}(\tau_\ell), f'_{\pi'(\ell)}(\tau_\ell)\right)_{\mathbb{R}^n} d\tau_1 \cdots d\tau_m$$

$$= \sum_{\pi \in \Pi_m} \int_{\square^{(m)}(T)} \prod_{\ell=1}^{m} \left(f_\ell(\tau_\ell), f'_{\pi(\ell)}(\tau_\ell)\right)_{\mathbb{R}^n} d\tau_1 \cdots d\tau_m$$

$$= \sum_{\pi \in \Pi_m} \prod_{\ell=1}^{m} \left(f_\ell, f'_{\pi(\ell)}\right)_{L^2([0,T];\mathbb{R}^n)}.$$

In preparation for the next step, let $\{g_j : j \geq 1\}$ be an orthonormal basis in $L^2([0,\infty); \mathbb{R}^n)$, and note that $\{g_{j_1} \otimes \cdots \otimes g_{j_m} : (j_1, \ldots, j_m) \in (\mathbb{Z}^+)^m\}$ is an orthonormal basis in $L^2(\square^{(m)}(\infty); (\mathbb{R}^n)^m)$. Next, for $\mu \in \mathbb{N}^{\mathbb{Z}^+}$, set $\|\mu\|_1 = \sum_1^\infty \mu_j$, and take $\mathcal{A}$ to be the set of $\mathbb{N}^{\mathbb{Z}^+}$ with $\|\mu\|_1 < \infty$. Finally, given $\mu \in \mathcal{A}$ with $\|\mu\|_1 = m$, set $G_\mu = g^{\otimes \mu_1} \otimes \cdots \otimes g^{\otimes \mu_m}$, where the meaning here is determined by the convention that $g^{\otimes \mu_1} \otimes \cdots \otimes g^{\otimes \mu_\ell} = g^{\otimes \mu_1} \otimes \cdots \otimes g^{\otimes \mu_{\ell-1}}$ if $\mu_\ell = 0$. In particular, $G_\mu \in L^2(\square^{(m)}; (\mathbb{R}^n)^m)$, and, for $\mu, \nu \in \mathcal{A}$,

(6.3.3) $$\left(\tilde{I}^{(\|\mu\|_1)}_{G_\mu}, \tilde{I}^{(\|\nu\|_1)}_{G_\nu}\right)_{L^2(\mathbb{P}^0;\mathbb{R})} = \delta_{\mu,\nu}\mu!.$$

Now, given $F \in L^2(\square^{(m)}; (\mathbb{R}^n)^m)$, set

(6.3.4) $$\tilde{F}(t_1, \ldots, t_m) \equiv \sum_{\pi \in \Pi_m} F(t_{\pi(1)}, \ldots, t_{\pi(m)}),$$

and observe that

$$\|\tilde{F}\|^2_{L^2(\square^{(m)};(\mathbb{R}^n)^m)} = \sum_{J \in (\mathbb{Z}^+)^m} \left(\tilde{F}, g_{j_1} \otimes \cdots \otimes g_{j_m}\right)^2_{L^2(\square^{(m)};(\mathbb{R}^n)^m)}$$

$$= \sum_{\|\mu\|_1 = m} \binom{m}{\mu} \left(\tilde{F}, G_\mu\right)^2_{L^2(\square^{(m)};(\mathbb{R}^n)^m)},$$

where $\binom{m}{\mu}$ is the multinomial coefficient $\frac{m!}{\mu_1! \cdots \mu_m!}$. Hence, after combining this with calculation in (6.3.3), we have that

$$\mathbb{E}^{\mathbb{P}^0}\left[\left|\sum_{\|\mu\|_1 = m} \frac{\left(\tilde{F}, G_\mu\right)_{L^2(\square^{(m)};(\mathbb{R}^n)^m)}}{\mu!} \tilde{I}^{(m)}_{G_\mu}(\infty)\right|^2\right] = \frac{1}{m!} \|\tilde{F}\|^2_{L^2(\square^{(m)};(\mathbb{R}^n)^m)}.$$

With these considerations, we have proved the following.

THEOREM 6.3.5. *There is a unique linear map*

$$F \in L^2\big(\square^{(m)}; (\mathbb{R}^n)^m\big) \longmapsto \tilde{I}_F^{(m)} \in \mathcal{M}^2(\mathbb{P}^0; \mathbb{R})$$

such that $\tilde{I}_{f_1 \otimes \cdots \otimes f_m}^{(m)}$ *is given as in* (6.3.2) *and*

$$\mathbb{E}^{\mathbb{P}^0}\Big[\tilde{I}_F^{(m)}(\infty)\tilde{I}_{F'}^{(m)}(\infty)\Big] = \frac{1}{m!}\big(\tilde{F}, \tilde{F}'\big)_{L^2(\square^{(m)}; (\mathbb{R}^n)^m)}.$$

In fact, if $\{g_j : j \geq 1\}$ *is an orthonormal basis in* $L^2\big([0,\infty); \mathbb{R}^n\big)$ *and* $\{G_\mu : \mu \in \mathcal{A}\}$ *is defined as above, then*

$$\tilde{I}_F^{(m)} = \sum_{\|\mu\|_1 = m} \frac{\big(\tilde{F}, G_\mu\big)_{L^2(\square^{(m)}; (\mathbb{R}^n)^m)}}{\mu!}\tilde{I}_{G_\mu},$$

where $\mu! \equiv \prod_1^\infty {}^{\|\mu\|_1} \mu_j!$ *and the convergence is in* $L^2(\mathbb{P}^0; \mathbb{R})$.

Although it is somewhat questionable to do so, we will, as indicated at the beginning of this subsection, adopt the suggestive notation

$$\int_{\square^{(m)}(t)} \big(F(\vec{\tau}), d\vec{p}(\vec{\tau})\big)_{(\mathbb{R}^n)^m}$$

for $\tilde{I}_F^{(m)}(t)$. The reason why this notation is questionable is that, although it is suggestive, it may suggest the wrong thing. Specifically, in order to avoid stochastic integrals with non-progressively measurable integrands, our definition of $\tilde{I}_F^{(m)}$ carefully avoids integration across diagonals, whereas the preceding notation gives no hint of that fact.

§6.3.2. The Spaces of Homogeneous Chaos: Take $Z^{(0)}$ to be the subspace of $L^2(\mathbb{P}^0; \mathbb{R})$ consisting of the constant functions, and, for $m \geq 1$, take

$$Z^{(m)} = \big\{\tilde{I}_F^{(m)}(\infty) : F \in L^2\big(\square^{(m)}; (\mathbb{R}^n)^m\big)\big\}.$$

Clearly, each $Z^{(m)}$ is a linear subspace of $L^2(\mathbb{P}^0; \mathbb{R})$. Furthermore, if $\{F_k : k \geq 1\} \subseteq L^2\big(\square^{(m)}; (\mathbb{R}^n)^m\big)$ and $\big\{\tilde{I}_{F_k}^{(m)}(\infty) : k \geq 1\big\}$ converges in $L^2(\mathbb{P}^0; \mathbb{R})$, then $\{\tilde{F}_k : k \geq 1\}$ converges in $L^2\big(\square^{(m)}; (\mathbb{R}^n)^m\big)$ to some symmetric function G. Hence, since $\tilde{G} = m!G$, we see that $\tilde{I}_{F_k}^{(m)}(\infty) \longrightarrow \tilde{I}_F^{(m)}(\infty)$ in $L^2(\mathbb{P}^0; \mathbb{R})$ where $F = \frac{1}{m!}G$. That is, each $Z^{(m)}$ is a closed linear subspace of $L^2(\mathbb{P}^0; \mathbb{R})$. Finally, $Z^{(m)} \perp Z^{(m')}$ when $m' \neq m$. This is completely obvious if either m or m' is 0. Thus, suppose that $1 \leq m < m'$. Then

$$\mathbb{E}^{\mathbb{P}^0}\Big[I_{f_1 \otimes \cdots \otimes f_m}^{(m)}(\infty)I_{f_1' \otimes \cdots \otimes f_{m'}'}^{(m')}(\infty)\Big]$$

$$= \int_{\Delta^{(m'-m)}} \prod_{\ell=0}^{m'-m-1} \big(f_{m-\ell}(\tau_\ell), f_{m'-\ell}'(\tau_\ell)\big)_{\mathbb{R}^n}$$

$$\times \mathbb{E}^{\mathbb{P}^0}\big[I_{f_1' \otimes \cdots \otimes f_{m'-m}'}^{(m'-m)}(\tau_{m'-m})\big]\, d\tau_1 \cdots d\tau_{m'-m} = 0,$$

which completes the proof.

The space $Z^{(m)}$ is the space of *mth order homogeneous chaos*. The reason why elements of $Z^{(0)}$ are said to be of 0th order chaos is clear: constants are non-random. To understand why $\tilde{I}_F^{(m)}(\infty)$ is of mth order chaos when $m \geq 1$, it is helpful to replace $dp(\tau)$ by the much more ambiguous $\dot{p}(\tau)\, d\tau$ and write

$$\tilde{I}_F^{(m)}(\infty) = \int_{\square^{(m)}} \Big(F(\tau_1,\ldots,\tau_m), \big(\dot{p}(\tau_1),\cdots,\dot{p}(\tau_m)\big)\Big)_{(\mathbb{R}^n)^m} d\tau_1\cdots d\tau_m.$$

In the world of engineering and physics, $\tau \rightsquigarrow \dot{p}(\tau)$ is *white noise*.[7] Thus, $Z^{(m)}$ is the space built out of homogeneous mth order polynomials in white noises evaluated at different times.[8] In other words, the order of chaos is the order of the white noise polynomial.

The result of Wiener, alluded to at the beginning of this section, now becomes the assertion that

$$(6.3.6) \qquad\qquad L^2(\mathbb{P}^0;\mathbb{R}) = \bigoplus_{m=0}^{\infty} Z^{(m)}.$$

The key to Itô's proof of (6.3.6) is found in the following. See Exercise 6.3.18 for an alternative approach.

LEMMA 6.3.7. *If $f \in L^2\big([0,\infty);\mathbb{R}^n\big)$, then, for each $\lambda \in \mathbb{C}$,*

$$e^{\lambda I_f(\infty)-\frac{\lambda^2}{2}\|f\|^2_{L^2([0,\infty);\mathbb{R}^n)}} = \sum_{m=0}^{\infty}\frac{\lambda^m}{m!}\tilde{I}_{f^{\otimes m}}^{(m)}(\infty) \equiv \lim_{M\to\infty}\sum_{m=0}^{M}\frac{\lambda^m}{m!}\tilde{I}_{f^{\otimes m}}^{(m)}(\infty)$$

$\mathbb{P}^0$-*almost surely and in* $L^2(\mathbb{P}^0;\mathbb{R})$. *In fact, if*

$$R_f^M(\infty,\lambda) \equiv e^{\lambda I_f(\infty)-\frac{\lambda^2}{2}\|f\|^2_{L^2([0,\infty);\mathbb{R}^n)}} -\sum_{m=0}^{M-1}\frac{\lambda^m}{m!}\tilde{I}_{f^{\otimes m}}^{(m)}(\infty),$$

then

$$\mathbb{E}^{\mathbb{P}^0}\Big[\big|R_f^M(\infty,\lambda)\big|^2\Big] = e^{|\lambda|^2\|f\|^2_{L^2([0,\infty);\mathbb{R}^n)}} - \sum_{m=0}^{M-1}\frac{\big(|\lambda|\|f\|_{L^2([0,\infty);\mathbb{R}^n)}\big)^{2m}}{m!}.$$

---

[7] The terminology comes from the observation that, no matter how one interprets $t \rightsquigarrow \dot{p}(t)$, it is a stationary, centered Gaussian process whose covariance is the Dirac delta function times the identity. In particular, $\dot{p}(t_1)$ is independent of $\dot{p}(t_2)$ when $t_1 \neq t_2$.
[8] Remember that our integrals stay away from the diagonal.

Proof: Set $E(t, \lambda) = e^{\lambda I_f(t) - \frac{\lambda^2}{2} \|\mathbf{1}_{[0,t)} f\|^2_{L^2([0,\infty);\mathbb{R}^n)}}$. Using Itô's formula, as in part **(i)** of Exercise 5.3.4, one sees that, for each $\lambda \in \mathbb{C}$,

$$E(t, \lambda) = 1 + \lambda \int_0^t E(\tau, \lambda)\big(f(\tau), dp(\tau)\big)_{\mathbb{R}^n}.$$

Thus, if (cf. (6.3.1)) $I^{(m)}(t) \equiv I^{(m)}_{f^{\otimes m}}(t) = \frac{1}{m!} \tilde{I}^{(m)}_{f^{\otimes m}}$ and

$$R^0(t, \lambda) \equiv E(t, \lambda) \quad \text{and} \quad R^{M+1}(t, \lambda) \equiv \lambda \int_0^t R^M(\tau, \lambda)\big(f(\tau), dp(\tau)\big)_{\mathbb{R}^n}$$

for $M \geq 0$, then, by induction, one sees that

$$E(t, \lambda) = 1 + \sum_{m=1}^{M} \lambda^m I^{(m)}(t) + R^{M+1}(t, \lambda)$$

for all $M \geq 0$. Finally, if $\alpha = \mathfrak{Re}(\lambda)$ and $F(t) = \int_0^t |f(\tau)|^2 \, d\tau$, then

$$\mathbb{E}^{\mathbb{P}^0}\big[\big|R^0(t, \lambda)\big|^2\big] = e^{|\lambda|^2 F(t)} \mathbb{E}^{\mathbb{P}^0}\big[E(t, 2\alpha)\big] = e^{|\lambda|^2 F(t)},$$

and

$$\mathbb{E}^{\mathbb{P}^0}\big[\big|R^{M+1}(t, \lambda)\big|^2\big] = |\lambda|^2 \int_0^t \mathbb{E}^{\mathbb{P}^0}\big[\big|R^M(\tau, \lambda)\big|^2\big] \dot{F}(\tau) \, d\tau.$$

Hence, the asserted estimate follows by induction on $M \geq 0$. □

Theorem 6.3.8. *The span of*

$$\mathbb{R} \oplus \big\{\tilde{I}^{(m)}_{f^{\otimes m}}(\infty) : m \geq 1 \ \& \ f \in L^2\big([0,\infty); \mathbb{R}^n\big)\big\}$$

is dense in $L^2(\mathbb{P}^0; \mathbb{R})$. *In particular, (6.3.6) holds.*

Proof: Let $\mathbb{H}$ denote smallest closed subspace of $L^2(\mathbb{P}^0; \mathbb{R})$ containing all constants and all the functions $\tilde{I}^{(m)}_{f^{\otimes m}}(\infty)$. By the preceding, we know that $\cos \circ I_f(\infty)$ and $\sin \circ I_f(\infty)$ are in $\mathbb{H}$ for all $f \in L^2([0,\infty); \mathbb{R}^n)$.

Next, observe that the space of functions $\Phi : C\big([0,\infty); \mathbb{R}^n\big) \longrightarrow \mathbb{R}$ which have the form

$$\Phi = F\big(I_{f_1}(\infty), \ldots, I_{f_L}(\infty)\big)$$

for some $L \geq 1$, Schwartz class test function $F : \mathbb{R}^L \longrightarrow \mathbb{R}$, and $f_1, \ldots, f_L \in L^2\big([0,\infty); \mathbb{R}^n\big)$ is dense in $L^2(\mathbb{P}; \mathbb{R}^n)$. Indeed, this follows immediately from the density in $L^2(\mathbb{P}^0; \mathbb{R})$ of the space of functions of the form

$$p \rightsquigarrow F\big(p(t_1), \ldots, p(t_L)\big),$$

where $L \geq 1$, $F : (\mathbb{R}^n)^L \longrightarrow \mathbb{R}$ is in the Schwartz test function class, and $0 \leq t_0 < \cdots < t_L$.

Now suppose that $F : \mathbb{R}^L \longrightarrow \mathbb{R}$ is given, and let $\hat{F}$ denote its Fourier transform. Then, by elementary Fourier analysis,

$$F_N(x) \equiv \left(2(4^N + 1)\pi\right)^{-L} \sum_{\|\mathbf{m}\|_\infty \leq 4^N} e^{-\sqrt{-1}\,(2^{-N}\mathbf{m},x)_{\mathbb{R}^L}} \hat{F}(\mathbf{m}2^{-N}) \longrightarrow F(x),$$

both uniformly and boundedly, where $\mathbf{m} = (m_1, \ldots, m_L) \in \mathbb{Z}^L$ and $\|\mathbf{m}\|_\infty = \max_{1 \leq \ell \leq L} |m_\ell|$. Finally, since $\hat{F}(\Xi) = \hat{F}(-\Xi)$ for all $\Xi \in \mathbb{R}^L$, we can write

$$\left(2(4^N + 1)\pi\right)^L F_N\left(I_{f_1}(\infty), \ldots, I_{f_L}(\infty)\right)$$
$$= \sum_{\{\mathbf{m} \in \mathbb{N}^L : \|\mathbf{m}\|_\infty \leq 4^N\}} 2\left(\mathfrak{Re}\left(\hat{F}(\mathbf{m}2^{-N})\right) \cos \circ I_{\mathbf{m},N}(\infty)\right.$$
$$\left. + \mathfrak{Im}\left(\hat{F}(\mathbf{m}2^{-N})\right) \sin \circ I_{\mathbf{m},N}(\infty)\right) \in \mathbb{H},$$

where $I_{\mathbf{m},N} \equiv 2^{-N} I_{f_{\mathbf{m}}}$ and $f_{\mathbf{m}} = \sum_{\ell=1}^{L} m_\ell f_\ell$. $\quad\square$

The following corollary is an observation made by Itô after he cleaned up Wiener's treatment of (6.3.6). It is often called *Itô's representation theorem* and turns out to play an important role in applications of stochastic analysis to, of all things, models of financial markets.[9]

COROLLARY 6.3.9. *The map*

$$(\alpha, \theta) \in \mathbb{R} \times \Theta^2(\mathbb{P}^0; \mathbb{R}^n) \longmapsto \alpha + I_\theta(\infty) \in L^2(\mathbb{P}^0; \mathbb{R})$$

is a linear, isometric surjection. Hence, for each $\Phi \in L^2(\mathbb{P}^0; \mathbb{R})$ *there is a* $\mathbb{P}^0$*-unique* $\theta \in \Theta^2(\mathbb{P}^0; \mathbb{R}^n)$ *such that*

$$\Phi = \mathbb{E}^{\mathbb{P}^0}[\Phi] + \int_0^\infty \left(\theta(\tau), dp(\tau)\right)_{\mathbb{R}^n} \quad \mathbb{P}^0\text{-almost surely.}$$

In particular,

$$\mathbb{E}^{\mathbb{P}^0}\left[\Phi \mid \overline{\mathcal{B}}_t\right] = \mathbb{E}^{\mathbb{P}^0}[\Phi] + \int_0^t \left(\theta(\tau), dp(\tau)\right)_{\mathbb{R}^n} \quad \mathbb{P}^0\text{-almost surely for each } t \geq 0,$$

and therefore $(t, p) \rightsquigarrow \mathbb{E}^{\mathbb{P}^0}\left[\Phi \mid \overline{\mathcal{B}}_t\right](p)$ *can be chosen so that* $t \rightsquigarrow \mathbb{E}^{\mathbb{P}^0}\left[\Phi \mid \overline{\mathcal{B}}_t\right](p)$ *is* $\mathbb{P}^0$*-almost surely continuous.*

---

[9] In fact, it shares with Itô's formula responsibility for the widespread misconception in the financial community that Itô is an economist.

Proof: Since it is clear that the map is linear and isometric, the first assertion will be proved once we check that the map is onto. But, because it is a linear isometry, we know that its image is a closed subspace, and so we need only show that its image contains a set whose span is dense. However, for each $f \in L^2([0,\infty); \mathbb{R}^n)$ and $m \geq 1$,

$$\tilde{I}_{f^{\otimes m}}^{(m)}(\infty) = m \int_0^\infty \left(\tilde{I}_{f^{\otimes(m-1)}}^{(m-1)}(\tau) f(\tau), dp(\tau) \right)_{\mathbb{R}^n},$$

and so, by the first part of Theorem 6.3.8, we are done.

Given the first assertion, the other assertions become essentially trivial. $\square$

§6.3.3. Exercises

EXERCISE 6.3.10. Although our treatment of Wiener's homogeneous chaos decomposition makes no use of it, Wiener's own treatment rested on the connection between the decomposition in (6.3.6) and the Hermite polynomials. In order to understand this connection, recall the Hermite polynomials $\{H_m : m \geq 0\}$ described in part (i) of Exercise 5.1.27. Next, given $L \geq 1$ and $\mu \in \mathbb{N}^L$, define $H_\mu : \mathbb{R}^L \longrightarrow \mathbb{R}$ so that

$$H_\mu(x_1, \ldots, x_L) = \prod_{\ell=1}^L H_{\mu_\ell}(x_\ell).$$

(i) Let Γ^L denote the centered Gauss measure on $\mathbb{R}^L$ with covariance $I_{\mathbb{R}^L}$. Show that

$$\|H_\mu\|_{L^2(\Gamma^L; \mathbb{R})}^2 = \mu!$$

and that $\{H_\mu : \mu \in \mathbb{N}^L\}$ is an orthogonal basis in $L^2(\Gamma^L; \mathbb{R})$.

(ii) Given $\xi \in \mathbb{R}^L$, show that

$$e^{(\xi, x)_{\mathbb{R}^L} - \frac{1}{2}|\xi|^2} = \sum_{\mu \in \mathbb{N}^L} \frac{\xi^\mu}{\mu!} H_\mu(x),$$

where $\xi^\mu \equiv \prod_{\ell=1}^L \xi_\ell^{\mu_\ell}$.

(iii) Suppose that $\{f_1, \ldots, f_L\}$ is an orthonormal subset of $L^2([0,\infty); \mathbb{R}^n)$, and, given $\mu \in \mathbb{N}^L$ with $m = \|\mu\|_1 \equiv \sum_{\ell=1}^L \mu_\ell$, show that

$$\tilde{I}_{f_1^{\otimes \mu_1} \otimes \cdots \otimes f_L^{\otimes \mu_L}}^{(m)}(\infty) = H_\mu\left(I_{f_1}(\infty), \ldots, I_{f_L}(\infty)\right),$$

where $f_1^{\otimes \mu_1} \otimes \cdots \otimes f_L^{\otimes \mu_L}$ is the element of $L^2(\square^{(m)}; (\mathbb{R}^n)^m)$ which is determined by the convention introduced in §6.3.1 just before (6.3.3).

EXERCISE 6.3.11. In this exercise, we will use the preceding to show that (6.3.6) can be interpreted as the spectral resolution for a nice self-adjoint operator. For this purpose, let $\{g_j : j \geq 1\}$ be an orthonormal basis for $L^2\big([0,\infty);\mathbb{R}^n\big)$, and assume each $g_j \in C_c^\infty\big((0,\infty);\mathbb{R}^n\big)$. Then $(t,p) \rightsquigarrow I_{g_j}(t,p)$ is well defined as a continuous function on the whole of $[0,\infty] \times C\big([0,\infty);\mathbb{R}^n\big)$. Next, for

$$\mu \in \mathcal{A} \equiv \left\{ \mu \in \mathbb{N}^{\mathbb{Z}^+} : \|\mu\|_1 \equiv \sum_1^\infty \mu_j < \infty \right\},$$

set

$$H_\mu(x) = \prod_{j=1}^{\|\mu\|_1} H_{\mu_j}(x_j) \quad \text{when } x = (x_1,\dots,x_{\|\mu\|_1}) \in \mathbb{R}^{\|\mu\|_1}$$

and

$$G_\mu \equiv g_1^{\otimes\mu_1} \otimes \cdots \otimes g_{\|\mu\|_1}^{\otimes\mu_{\|\mu\|_1}}.$$

Finally, define

$$\mathcal{H}_\mu(p) = H_\mu\big(I_{g_1}(\infty,p),\dots,I_{g_{\|\mu\|_1}}(\infty,p)\big) \quad \text{and} \quad \tilde{I}_\mu(t) = \tilde{I}_{G_\mu}^{\|\mu\|_1}(t).$$

(i) Use parts (ii) and (iii) of the preceding exercise to see that, for each $\mu \in \mathcal{A}$,

$$\tilde{I}_\mu(\infty) = \mathcal{H}_\mu \quad \mathbb{P}^0\text{-almost surely},$$

and conclude that, for each $m \in \mathbb{Z}^+$, $\{\mathcal{H}_\mu : \|\mu\|_1 = m\}$ is an orthogonal basis for $Z^{(m)}$.

(ii) Let $\mathcal{D}$ denote the space of continuous functions $\Phi : C\big([0,\infty);\mathbb{R}^n\big) \longrightarrow \mathbb{R}$ with the property that

(*) $$\Phi(p) = F\big(I_{g_1}(\infty,p),\dots,I_{g_L}(\infty,p)\big)$$

for some $L \geq 1$ and smooth function $F : \mathbb{R}^L \longrightarrow \mathbb{R}$ which, together with all its derivatives, has tempered (i.e., at most polynomial) growth. Given $f \in L^2\big([0,\infty);\mathbb{R}^n\big)$, take $h_f \in C\big([0,\infty);\mathbb{R}^n\big)$ so that $h_f(t) = \int_0^t f(\tau)\,d\tau$ and define $D_f\Phi \in C\big(C\big([0,\infty);\mathbb{R}^n\big);\mathbb{R}\big)$ for $\Phi \in \mathcal{D}$ so that

$$D_f\Phi(p) = \frac{d}{ds}\Phi(p + sh_f)\Big|_{s=0} \quad \text{for } p \in C\big([0,\infty);\mathbb{R}^n\big).$$

Observe that when Φ is given by (*),

$$D_f\Phi(p) = \sum_{j=1}^L \big(f,g_j\big)_{L^2([0,\infty);\mathbb{R}^n)}\,\partial_j F\big(I_{g_1}(\infty,p),\dots,I_{g_L}(\infty,p)\big),$$

which shows that D_f maps $\mathcal{D}$ into itself. Use these observations to check that *number operator*[10]

$$\mathcal{N}\Phi \equiv \sum_{j=1}^{\infty}\left(I_{g_j}(\infty)D_{g_j}\Phi - D_{g_j}^2\Phi\right)$$

is a well defined operator which takes $\mathcal{D}$ into itself.

(iii) After verifying that $\partial_j H_\mu = \mu_j H_{\mu(-j)}$, where

$$\mu(-j)_i \equiv \begin{cases} \mu_i & \text{if } i \neq j \text{ or } \mu_j = 0 \\ \mu_j - 1 & \text{if } i = j \text{ and } \mu_j \geq 1, \end{cases}$$

and that $(x_j - \partial_j)H_\mu = H_{\mu(+j)}$, where

$$\mu(+j)_i \equiv \begin{cases} \mu_i & \text{if } i \neq j \\ \mu_j + 1 & \text{if } i = j, \end{cases}$$

show that

$$\mathcal{N}\mathcal{H}_\mu = \|\mu\|_1 \mathcal{H}_\mu, \quad \mu \in \mathcal{A},$$

and conclude that $\mathcal{N}$ is an essentially self-adjoint operator and for which $Z^{(m)}$ is the eigenspace with eigenvalue m of the unique self-adjoint extension $\bar{\mathcal{N}}$ of $\mathcal{N}$. See part (iv) in Exercise 6.3.17 below for more information.

EXERCISE 6.3.12. Continue with the notation in the preceding exercise, take $D_j = D_{g_j}$, and define $D_j^\top$ on $\mathcal{D}$ so that $D_j^\top \Phi = I_{g_j}(\infty)\Phi - D_j\Phi$.

(i) Show that $D_j^\top$ is the restriction to $\mathcal{D}$ of the adjoint of D_j. That is, show that

(6.3.13) $$\left(D_j^\top\Psi, \Phi\right)_{L^2(\mathbb{P}^0;\mathbb{R})} = \left(\Psi, D_j\Phi\right)_{L^2(\mathbb{P}^0;\mathbb{R})} \quad \text{for } \Phi, \Psi \in \mathcal{D}.$$

Hint: Using the Cameron–Martin formula, show that

$$\left(\Phi D_j, \Psi\right)_{L^2(\mathbb{P}^0;\mathbb{R})} = \frac{d}{ds}\int \Phi(p)e^{I_{sg_j}(\infty,p) - \frac{s^2}{2}}\Psi\left(p - sh_j\right)\mathbb{P}^0(dp)\bigg|_{s=0},$$

where $h_j(t) = \int_0^t g_j(\tau)\, d\tau$.

(ii) Define $D^\mu : \mathcal{D} \longrightarrow \mathcal{D}$ inductively so that $D^{\mu(+j)} = D_j \circ D^\mu$. When $\Phi \in \mathcal{D}$, use part (i) in Exercise 6.3.11 and (iv) above to show that

$$\mathbb{E}^{\mathbb{P}^0}\left[\Phi \tilde{I}_\mu(\infty)\right] = \mathbb{E}^{\mathbb{P}^0}\left[D^\mu\Phi\right],$$

---

[10] The terminology here comes from Euclidean quantum field theory.

and conclude that

$$\Phi = \mathbb{E}^{\mathbb{P}^0}[\Phi] + \sum_{m=1}^{\infty} \tfrac{1}{m!} \tilde{I}_{F_m}^{(m)}(\infty),$$

where (cf. (*) in (ii))

$$F_m \equiv \sum_{\|\mu\|_1 = m} \mathbb{E}^{\mathbb{P}^0}[D^\mu \Phi] G_\mu.$$

Thus, the resolution of Φ into its components of homogeneous chaos gives a sort of Taylor's expansion of Φ.

(iii) Set $\Lambda = \mathrm{Leb}_{[0,\infty)} \times \mathbb{P}^0$. Given $\theta \in \Theta^2(\mathbb{P}^0; \mathbb{R}^n)$, show that the Itô integral $I_\theta(\infty) = \int_0^\infty (\theta(\tau), dp(\tau))_{\mathbb{R}^n}$ is the unique element of $L^2(\mathbb{P}^0; \mathbb{R})$ with the property that[11]

$$\big(\Psi, I_\theta(\infty)\big)_{L^2(\mathbb{P}^0; \mathbb{R})} = \sum_{j=1}^{\infty} \big(g_j \otimes D_j \Psi, \theta\big)_{L^2(\Lambda; \mathbb{R}^n)} \quad \text{for all } \Psi \in \mathcal{D}.$$

Hint: As an application of Corollary 6.2.2, show that the right hand side of the preceding is equal to the derivative at $s = 0$ of (cf. (5.1.10)) $s \rightsquigarrow \int \Phi(p) E_{s\theta}(\infty, p) \, \mathbb{P}^0(dp)$.

(iv) Define the *gradient operator* D from $\mathcal{D}$ into $L^2(\Lambda; \mathbb{R}^n)$ so that $D\Phi = \sum_{j=1}^{\infty} g_j \otimes D_j \Phi$. Since $\mathcal{D}$ is dense in $L^2(\mathbb{P}^0; \mathbb{R})$, D admits a well defined adjoint $D^\top$ on $L^2(\Lambda; \mathbb{R}^n)$. To the extent that D deserves to be called a gradient, $D^\top$ has to be called a *divergence operator*. Show that the result in **(iii)** can be summarized by the statement of $\Theta^2(\mathbb{P}^0; \mathbb{R}^n) \subseteq \mathrm{Dom}(D^\top)$ and that $I_\theta = D^\top \theta$ for $\theta \in \Theta^2(\mathbb{P}^0; \mathbb{R}^n)$.

EXERCISE 6.3.14. The considerations in **(iv)** of the preceding give a natural way of extending to non-progressively measurable integrands Itô's theory of stochastic integration with respect to Brownian motion. Namely, given $\theta \in \mathrm{Dom}(D^\top)$, the extension we have in mind is provided by $D^\top \theta$. The purpose of the present exercise is to examine the domain of $D^\top$ and show that $D^\top$ determines the same extension of Itô's theory as the one suggested by A.V. Skorohod in [32].

(i) Given $j \in \mathbb{Z}^+$, $m \in \mathbb{N}$, and $F \in L^2\big(\square^{(m)}; (\mathbb{R}^n)^m\big)$, show that $g_j \otimes \tilde{I}_F^{(m)} \in \mathrm{Dom}(D^\top)$, and that $D^\top(g_j \otimes \tilde{I}_F^{(m)}) = \tilde{I}_{g_j \otimes F}^{(m+1)}$.

---

[11] Here we are using $f \otimes \Phi$ to denote the function on $[0, \infty) \times C([0, \infty); \mathbb{R}^n)$ which is equal to $f(t)\Phi(p)$ at (t, p).

Hint: Show that

$$
\left(\tilde{I}^{(m+1)}_{g_j\otimes F},\mathcal{H}_\nu\right)_{L^2(\mathbb{P}^0;\mathbb{R})} = \left(\widetilde{g_j\otimes F},G_\nu\right)_{L^2(\square^{(m+1)};(\mathbb{R}^n)^{m+1})}
$$

$$
= \nu_j\left(\tilde{F},G_{\nu(-j)}\right)_{L^2(\square^{(m)};(\mathbb{R}^n)^m)} = \left(g_j\otimes\tilde{I}_F,DF\right)_{L^2(\Lambda;\mathbb{R}^n)}
$$

when $\|\nu\|_1 = m+1$.

(ii) Given $\theta \in \Theta^2(\mathbb{P}^0;\mathbb{R}^n)$ and $m \in \mathbb{N}$, show that, for almost every $t \in [0,\infty)$,

$$
\theta^{(m)}(t) \equiv \sum_{j=1}^\infty \sum_{\|\mu\|_1=m} \frac{\left(\theta,g_j\otimes\mathcal{H}_\mu\right)_{L^2(\Lambda;\mathbb{R}^n)}}{\mu!} g_j\otimes\mathcal{H}_\mu
$$

exists (i.e., the series converges) in $L^2(\mathbb{P}^0;\mathbb{R})$. In fact, if $\Pi_{Z^{(m)}}$ denotes orthogonal projection onto $Z^{(m)}$, show that $\Pi_{(Z^{(m)})^n}\theta(t) = \theta^{(m)}$ for almost every t in the sense that $\Pi_{Z^{(m)}}\left(\xi,\theta(t)\right)_{\mathbb{R}^n} = \left(\xi,\theta^{(m)}(t)\right)_{\mathbb{R}^n}$ for all $\xi \in \mathbb{R}^n$ and almost every $t \geq 0$.

(iii) Again let $\theta \in L^2(\Lambda;\mathbb{R}^n)$ be given. Set

$$
F^\theta_{m+1} = \sum_{j=1}^\infty \sum_{\|\mu\|_1=1} \frac{\left(\theta,g_j\otimes\mathcal{H}_\mu\right)_{L^2(\Lambda;\mathbb{R}^n)}}{\mu!} g_j\otimes G_\mu,
$$

show that $F^\theta_{m+1} \in L^2\left(\square^{(m+1)};(\mathbb{R}^n)^{m+1}\right)$, and both that $\theta^{(m)} \in \mathrm{Dom}(D^\top)$ and $D^\top\theta^{(m)} = \tilde{I}^{(m+1)}_{F^\theta_{m+1}}$.

(iv) Given $\theta \in L^2(\mathbb{P}^0;\mathbb{R}^n)$, show that $\theta \in \mathrm{Dom}(D^T)$ if and only if (cf. part **(iii)**)

$$
(6.3.15) \qquad \sum_{m=0}^\infty \frac{1}{(m+1)!}\left\|\tilde{F}^\theta_{m+1}\right\|^2_{L^2(\square^{(m+1)};(\mathbb{R}^n)^{m+1})} < \infty,
$$

in which case $D^\top\theta = \sum_{m=0}^\infty \tilde{I}^{(m+1)}_{F^\theta_{m+1}}$. This expression is the one which Skorohod gave for what is often called the *Skorohod integral* of θ.

(v) Notice that

$$
\left\|\tilde{F}^\theta_{m+1}\right\|_{L^2(\square^{(m+1)};(\mathbb{R}^n)^{m+1})}
$$

$$
\leq (m+1)\left\|\sum_{j=1}^\infty \sum_{\|\mu\|_1=m} \frac{\left(\theta,g_j\otimes\mathcal{H}_\mu\right)_{L^2(\Lambda;\mathbb{R}^n)}}{\mu!} g_j\otimes\widetilde{G}_\mu\right\|_{L^2(\square^{(m+1)};(\mathbb{R}^n)^{m+1})}
$$

$$
= (m+1)(m!)^{\frac{1}{2}}\left\|\theta^{(m)}\right\|_{L^2(\Lambda;\mathbb{R}^n)},
$$

and conclude that $\theta \in \text{Dom}(D^\top)$ if

(6.3.16) $$\sum_{m=0}^{\infty}(m+1)\int_0^\infty \left\|\Pi_{(Z^{(m)})^n}\theta(t)\right\|_{L^2(\mathbb{P}^0;\mathbb{R}^n)}^2 \, dt < \infty.$$

EXERCISE 6.3.17. The condition in (6.3.15) is nearly impossible to check directly in general, even though there are special cases in which the one do so. In addition, (6.3.16) is reasonably tractable. The present exercise examines some aspects of these issues. Because we know that its adjoint $D^\top$ is densely defined, we know that D itself is closable. Let $\bar{D}$ denote the closure of D in $L^2(\mathbb{P}^0)$.

(i) In the case when $\theta \in \Theta^2(\mathbb{P}^0; \mathbb{R}^n)$, we know that $D^\top\theta = I_\theta$ and therefore

$$\sum_{m=0}^{\infty}\frac{\left\|\tilde{F}_{m+1}^\theta\right\|_{L^2(\square^{(m+1)};(\mathbb{R}^n)^{m+1})}^2}{(m+1)!} = \|D^\top\theta\|_{L^2(\mathbb{P}^0;\mathbb{R})}^2 = \|\theta\|_{L^2(\Lambda;\backslash Rn)}^2.$$

Give a proof of this based on the observation that $F_{m+1}^\theta(t,\tau_1,\dots,\tau_m) = 0$ for almost every $(t,\tau_1,\dots,\tau-m)$ with $t < \tau_1 \vee \cdots \vee \tau_m$. Thus, even without knowing how $D^\top$ acts on $\Theta^2(\mathbb{P}^0;\mathbb{R}^n)$, one can see that $\Theta^2(\mathbb{P}^0;\mathbb{R}^n) \subseteq \text{Dom}(D^\top)$.

(ii) In connection with idea in (i), set $\mathcal{B}^t = \sigma(\{p(\tau) - p(t) : \tau \in [t,\infty)\})$, and introduce the space $\underleftarrow{\Theta}^2(\mathbb{P}^0;\mathbb{R}^n)$ of $\theta \in L^2(\Lambda;\mathbb{R}^n)$ which are *reverse progressively measurable* in the sense that $\theta \restriction [t,\infty) \times C([0,\infty);\mathbb{R}^n)$ is $\mathcal{B}_{[t,\infty)} \times \mathcal{B}^t$-measurable for each $t \geq 0$. Given $\theta \in \underleftarrow{\Theta}^2(\mathbb{P}^0;\mathbb{R}^n)$, show that $F_{m+1}^\theta(t,\tau_1,\dots,\tau_m) = 0$ for almost everywhere $(t,\tau_1,\dots,\tau_m)$ with $t > \tau_1 \wedge \cdots \wedge \tau_m$, and conclude that $\underleftarrow{\Theta}^2(\mathbb{P}^0;\mathbb{R}^n) \subseteq \text{Dom}(D^\top)$ and $\|D^\top\theta\|_{L^2(\mathbb{P}^0;\mathbb{R})} = \|\theta\|_{L^2(\Lambda;\mathbb{R}^n)}$ for $\theta \in \underleftarrow{\Theta}^2(\mathbb{P}^0;\mathbb{R}^n)$. Given these observations, it is not hard to introduce the appropriate notion of *backward Itô stochastic integration* which gives $D^\top\theta$ for $\theta \in \underleftarrow{\Theta}^2(\mathbb{P}^0;\mathbb{R}^n)$.

(iii) In the rest of this exercise we will be developing another way of thinking about the condition in (6.3.16). To this end, note that, because its adjoint $D^\top$ is densely defined, D admits a closure $\bar{D}$. Show that $Z^{(m)} \subseteq \text{Dom}(\bar{D})$ for every $m \in \mathbb{N}$ and that

$$\bar{D}\Phi = \sum_{j=1}^{\infty} \sum_{\{\mu:\|\mu\|_1=m \ \& \ \mu_j \geq 1\}} \frac{(\Phi,\mathcal{H}_\mu)_{L^2(\mathbb{P}^0;\mathbb{R})}}{\mu(-j)!} g_j \otimes \mathcal{H}_{\mu(-j)}$$

for $m \in \mathbb{Z}^+$ and $\Phi \in Z^{(m)}$. In particular, conclude that, for all $m \in \mathbb{N}$,

$$\Phi \in Z^{(m)} \implies \|\bar{D}\Phi\|_{L^2(\mathbb{P}^0;\mathbb{R})}^2 = m\|\Phi\|_{L^2(\mathbb{P}^0;\mathbb{R})}^2.$$

(iv) Given $\Phi \in L^2(\mathbb{P}^0; \mathbb{R})$, show that $\Phi \in \mathrm{Dom}(\bar{D})$ if and only if

$$\sum_{m=1}^{\infty} m \|\Pi_{Z^{(m)}} \Phi\|^2_{L^2(\mathbb{P}^0;\mathbb{R})} < \infty,$$

in which case the preceding sum is equal to the $\|\bar{D}\Phi\|^2_{L^2(\Lambda;\mathbb{R}^n)}$. Also, show that (cf. part (iii) of Exercise 6.3.11) $\Phi \in \mathrm{Dom}(\bar{\mathcal{N}})$ if and only if $\Phi \in \mathrm{Dom}(\bar{D})$ and $\bar{D}\Phi \in S(\mathbb{P}^0; \mathbb{R}^n)$, in which case $\bar{\mathcal{N}}\Phi = D^\top \circ D\Phi$.

(v) If $\theta \in L^2(\Lambda; \mathbb{R}^n)$ and $\theta(t) \in \mathrm{Dom}(\bar{D})^n$ for almost every $t \in [0, \infty)$, show that $t \rightsquigarrow \bar{D}\theta(t)$ is Lebesgue measurable and that (6.3.16) holds if

$$\int_{[0,\infty)} \|\bar{D}\theta(t)\|^2_{L^2(\mathbb{P}^0;\mathbb{R}^n)} \, dt < \infty.$$

For other applications of the same condition to Skorohod integration, see the article [24].

EXERCISE 6.3.18. The route which we have taken to Itô's representation theorem, Corollary 6.3.9, is not very efficient and does not give much insight into how one might go about constructing the θ corresponding to a given Φ. In this exercise we will develop another, more straightforward, approach.

(i) Let $\{g_\ell : 1 \le \ell \le L\} \subseteq C^1([0,\infty); \mathbb{R}^n)$ be an orthonormal subset of $L^2([0,\infty); \mathbb{R}^n)$, and define $t \in [0,\infty) \longmapsto A(t) \in \mathrm{Hom}(\mathbb{R}^L; \mathbb{R}^L)$ so that

$$A_{\ell,\ell'}(t) = \int_t^\infty \left(g_\ell(\tau), g_{\ell'}(\tau)\right)_{\mathbb{R}^n} d\tau.$$

Next, given $F \in C_b(\mathbb{R}^L; \mathbb{R})$, set

$$U_F(t, x) = \int_{\mathbb{R}^L} F(x + y) \, \Gamma_{A(t)}(dy) \quad \text{for } (t, x) \in [0, \infty) \times \mathbb{R}^L,$$

where Γ_A denotes the centered Gauss measure on $\mathbb{R}^L$ with covariance A, and show that

$$\partial_t U_F(t, x) + \frac{1}{2} \sum_{\ell,\ell'=1}^{L} \left(g_\ell(t), g_{\ell'}(t)\right)_{\mathbb{R}^n} \partial_\ell \partial_{\ell'} U_F(t, x) = 0$$

with $\lim_{t \nearrow \infty} U_F(t, x) = F(x)$ and $U_F(0, x) = \int_{\mathbb{R}^L} F(y) \, \Gamma_I(dy)$.

(ii) Referring to part (i), define $t \in [0,\infty) \longmapsto \sigma(t) \in \mathrm{Hom}(\mathbb{R}^n; \mathbb{R}^L)$ so that $\sigma_{k\ell}(t)$ is the kth coordinate of $g_\ell(t)$, note that $\sigma(t)\sigma(t)^\top = -\dot{A}(t)$, and show that

$$\mathbb{E}^{\Gamma_I}[F^2] - \mathbb{E}^{\Gamma_I}[F]^2 = \int_0^\infty \left(\int_{\mathbb{R}^L} |\sigma(t)^\top \mathrm{grad}_y U_F(t, \cdot)|^2 \, \Gamma_{I-A(t)}(dy)\right) dt.$$

Hint: A proof of the preceding can be based on the observation that

$$\frac{d}{dt}\mathbb{E}^{\Gamma_I - A(t)}\left[U_F(t,\,\cdot\,)^2\right] = \mathbb{E}^{\Gamma_I - A(t)}\left[\left|\sigma(t)^\top \text{grad}U_F(t,\,\cdot\,)\right|^2\right].$$

(iii) Referring to parts (i) and (ii), suppose that $\Phi = F(I_{g_1}(\infty),\ldots,$ $I_{g_L}(\infty))$, and show that

$$\Phi = \mathbb{E}^{\mathbb{P}^0}[\Phi] + \int_0^\infty \left(\theta_\Phi(t), dp(t)\right)_{\mathbb{R}^n},$$

where

$$\theta_\Phi(t,p) \equiv \sum_{j=1}^L \partial_j F\big(I_{g_1}(t,p),\ldots,I_{g_L}(t)\big)g_j(t).$$

In particular, notice that this result can substitute for Theorem 6.3.8 in the argument used to prove Corollary 6.3.9.

EXERCISE 6.3.19. Let $X \in L^1(\mathbb{P}^0;\mathbb{R})$ be given, and set $M(t) = \mathbb{E}^{\mathbb{P}^0}\left[X \mid \mathcal{B}_t\right]$ for each $t \geq 0$.

(i) Show that $(t,p) \in [0,\infty) \times C\big([0,\infty);\mathbb{R}^n\big) \longmapsto M(t,p) \in \mathbb{R}$ can be chosen so that $\big(M(t),\overline{\mathcal{B}}_t,\mathbb{P}^0\big)$ is a continuous martingale. (See part (iii) of Exercise 7.3.9 below for more information.)

Hint: Choose $\{X_k\}_1^\infty \subseteq L^2(\mathbb{P}^0;\mathbb{R})$ so that $X_k \longrightarrow X$ in $L^1(\mathbb{P}^0;\mathbb{R})$, let $t \rightsquigarrow M_k(t)$ be a $\mathbb{P}^0$-almost surely continuous version of $t \rightsquigarrow \mathbb{E}^{\mathbb{P}^0}\left[X_k \mid \mathcal{B}_t\right]$, and use Doob's inequality to see that

$$\lim_{k\to\infty}\sup_{\ell \geq k}\mathbb{P}^0\big(\|M_\ell - M_k\|_{[0,\infty)} \geq \epsilon\big) = 0 \quad \text{for all } \epsilon > 0.$$

(ii) As an application of (i) above and the martingale convergence theorem, show that if $X : [0,\infty) \times C\big([0,\infty);\mathbb{R}^n\big) \longrightarrow \mathbb{R}$ is a $\{\overline{\mathcal{B}}_t : t \geq 0\}$-progressively measurable function with the properties that $\{X(t) : t \geq 0\}$ is uniformly $\mathbb{P}$-integrable and $X(s) = \mathbb{E}^{\mathbb{P}^0}\left[X(t) \mid \mathcal{B}_s\right]$ for all $0 \leq s \leq t$, then there exists a $\{\overline{\mathcal{B}}_t : t \geq 0\}$-progressively measurable $\tilde{X} : [0,\infty) \times C\big([0,\infty);\mathbb{R}^n\big) \longrightarrow \mathbb{R}$ such that $\tilde{X}(\,\cdot\,,p)$ is continuous for each $p \in C\big([0,\infty);\mathbb{R}^n\big)$ and $\tilde{X}(t) = X(t)$ $\mathbb{P}$-almost surely for each $t \in [0,\infty)$.

CHAPTER 7

The Kunita–Watanabe Extension

A careful examination of the results in §§5.1 and 5.3 reveals that they depend very little on detailed properties of Brownian motion and, in fact, that analogous results can be derived about any square-integrable martingale $(M(t), \mathcal{F}_t, \mathbb{P})$ with the properties that

(1) the $t \rightsquigarrow M(t)$ is $\mathbb{P}$-almost surely continuous;
(2) there is an $\{\mathcal{F}_t : t \geq 0\}$-progressively measurable $A : [0, \infty) \times \Omega \longrightarrow [0, \infty)$ such that $t \rightsquigarrow A(t)$ is $\mathbb{P}$-almost surely continuous and nondecreasing, $A(0) = 0$, and $(M(t)^2 - A(t), \mathcal{F}_t, \mathbb{P})$ is a martingale.

In the case of an $\mathbb{R}$-valued Brownian motion $(\beta(t), \mathcal{F}_t, \mathbb{P})$, $A(t) = t$. In the case when $t \rightsquigarrow \beta(t)$ is $\mathbb{R}^n$-valued and $X(t) = (\xi, \beta(t))_{\mathbb{R}^n}$ for some $\xi \in \mathbb{R}^n$, $A(t) = t|\xi|^2$. More generally, if $\theta \in \Theta^2(\mathbb{P}; \mathbb{R}^n)$ and $M = I_\theta$, then $A(t) = \int_0^t |\theta(\tau)|^2 \, d\tau$.

Although J.L. Doob (cf. Chapter 6 of [6]) was the first to recognize that these are the only ingredients which are essential for Itô's theory, it was Kunita and Watanabe [21] who first accomplished the elegant extension of Itô's theory which we will present here. However, before we can do so, we need to have a special, and particularly simple, case of the renowned *Doob–Meyer decomposition theorem* for submartingales.[1]

Throughout, $(\Omega, \mathcal{F}, \mathbb{P})$ is a complete probability space and $\{\mathcal{F}_t : t \geq 0\}$ is a nondecreasing family of $\mathbb{P}$-complete sub σ-algebras of $\mathcal{F}$. Also, when I say that a stochastic process X on $[0, \infty) \times \Omega$ with values in a topological space is $\mathbb{P}$-almost surely *right continuous* or *continuous*, I will mean that $t \rightsquigarrow X(t, \omega)$ is right continuous or continuous for $\mathbb{P}$-almost every ω.

7.1 Doob–Meyer for Continuous Martingales

Recall (cf. Lemma 5.2.18 in [36]) Doob's decomposition lemma for discrete parameter, integrable submartingales $(X(m), \mathcal{F}_m, \mathbb{P})$: if $A_0 \equiv 0$ and

---

[1] It should be recognized that A.V. Skorohod demonstrated in [30] and [31] that he already understood most of the ideas discussed below. What makes his treatment less palatable than Kunita and Watanabe's is his ignorance of the Doob–Meyer theorem.

$A(m) - A(m - 1) \equiv \mathbb{E}^{\mathbb{P}}[X(m) - X(m - 1) \,|\, \mathcal{F}_{m-1}] \vee 0$ for $m \geq 1$, then $\{A(m) : m \geq 0\}$ is the $\mathbb{P}$-almost surely uniquely determined by the facts that $A(0) \equiv 0$, $A(m-1)$ is $\mathcal{F}_{m-1}$-measurable for each $m \geq 1$, and $\big(M(m) - A(m), \mathcal{F}_m, \mathbb{P}\big)$ is a martingale. Although, aside from recognizing its potential importance, this lemma requires no effort in the discrete setting, even formulating its generalization to the continuous parameter setting was a major achievement of P.A. Meyer and can be seen as the cornerstone of what became the *Strasbourg School of Probability*.

Fortunately for us, most of the difficulties Meyer had to overcome disappear when the submartingale is the square of a continuous martingale. Indeed, in this case the program is really an application of the Itô's ideas. In fact, it was Itô who gave me the outline for the existence proof given below.

§7.1.1. Uniqueness: Even without delving into the details, it is easy to appreciate the major difficulty confronting Meyer. Namely, when the time parameter is continuous, what plays the role of $\mathcal{F}_{m-1}$? That is, what is the measurability property which one has to impose on the process A? Loosely speaking, Meyer's answer was that $t \rightsquigarrow A(t)$ must be amenable to the reasoning contained in the corollary to the following theorem. What this corollary shows is that continuity is sufficient. One of the key observations made by Meyer is that continuity is not necessary. However, its replacement is subtle. (Cf. Exercise 7.1.4 below.)

THEOREM 7.1.1. *Suppose that $V : [0, \infty) \times \Omega \longrightarrow \mathbb{R}$ is a progressively function with the properties that, $\mathbb{P}$-almost every ω, $t \rightsquigarrow V(t, \omega)$ is a right continuous function of locally bounded variation; and use $|V|(t, \omega)$ to denote the variation of $V(\,\cdot\,, \omega) \upharpoonright [0, t]$. Then $|V|$ is again progressively measurable. Next, suppose that $\big(M(t), \mathcal{F}_t, \mathbb{P}\big)$ is a right continuous martingale with the property that, for every $t \geq 0$,*

$$\mathbb{E}^{\mathbb{P}}\Big[\big\| M(\,\cdot\,) \big\|_{[0,t]} \big(|V(0)| + |V|(t) \big) \Big] < \infty.$$

Then $\big(M(t)V(t) - B(t), \mathcal{F}_t, \mathbb{P} \big)$ is a martingale when

$$B(t, \omega) \equiv \begin{cases} \int_{(0,t]} M(\tau, \omega)\, V(d\tau, \omega) & \text{if } \big\| M(\,\cdot\,, \omega) \big\|_{[0,t]} |V|(t, \omega) < \infty \\ 0 & \text{otherwise,} \end{cases}$$

where the $V(d\tau, \omega)$ is meant to be Lebesgue integration with respect to the (signed) measure on $[0, \infty)$ determined by $V(\,\cdot\,, \omega)$.

PROOF: To see that $|V|$ is progressively measurable, simply observe that, because of right continuity,

$$|V|(t, \omega) = \sup_{N \in \mathbb{N}} \sum_{m=0}^{\infty} \Big| V\big(t \wedge (m + 1)2^{-N} \big) - V\big(t \wedge m 2^{-N} \big) \Big|.$$

Knowing this, it is easy to see that B is progressively measurable and (cf. Exercise 1.2.29 in [34]) $\mathbb{P}$-almost surely right continuous. Finally, using the assumed integrability properties, one can easily justify the computation:

$$\mathbb{E}^{\mathbb{P}}\big[B(t_2) - B(t_1), A\big]$$

$$= \lim_{N\to\infty} \sum_{m=[2^N t_1]}^{[2^N t_2]} \mathbb{E}^{\mathbb{P}}\Big[M\big(t_2 \wedge (m+1)2^{-N}\big)$$

$$\times \Big(V\big(t_2 \wedge (m+1)2^{-N}\big) - V\big(t_1 \vee m2^{-N}\big)\Big), A\Big]$$

$$= \lim_{N\to\infty} \sum_{m=[2^N t_1]}^{[2^N t_2]} \mathbb{E}^{\mathbb{P}}\Big[M(t_2+1)\Big(V\big(t_2 \wedge (m+1)2^{-N}\big) - V\big(t_1 \vee m2^{-N}\big)\Big), A\Big]$$

$$= \mathbb{E}^{\mathbb{P}}\big[M(t_2+1)\big(V(t_2) - V(t_1)\big), A\big] = \mathbb{E}^{\mathbb{P}}\big[M(t_2)V(t_2) - M(t_1)V(t_1), A\big]$$

for all $0 \le t_1 < t_2$ and $A \in \mathcal{F}_{t_1}$. $\square$

COROLLARY 7.1.2. *Suppose* $(M(t), \mathcal{F}_t, \mathbb{P})$ *is a continuous local martingale, and let* $|M|(t,\omega)$ *denote the variation of* $M(\cdot\,,\omega) \upharpoonright [0,t]$. *Then*

$$\mathbb{P}\big(\exists \in [0,\infty)\; 0 < |M|(t) < \infty\big) = 0.$$

In particular, if $X : [0,\infty) \times \Omega \longrightarrow \mathbb{R}$ *is progressively measurable, then there is, up to a* $\mathbb{P}$-*null set, at most one progressively measurable* $A : [0,\infty) \times \Omega \longrightarrow \mathbb{R}$ *with the properties that* $t \rightsquigarrow A(t)$ *is* $\mathbb{P}$-*almost surely continuous and of locally bounded variation,* $A(0) \equiv 0$, *and* $(X(t) - A(t), \mathcal{F}_t, \mathbb{P})$ *is a local martingale.*

PROOF: Without loss in generality, assume that $M(0) = 0$. Next, given $R > 0$, set $\zeta_R(\omega) = \sup\{t \ge 0 : |M|(t,\omega) \le R\}$, observe[2] that ζ_R is a stopping time, and set $M_R(t) = M(t \wedge \zeta_R)$. By Doob's stopping time theorem, $(M_R(t), \mathcal{F}_t, \mathbb{P})$ is a continuous martingale. At the same time, $|M_R|(t,\omega) \le R$. Hence, by the preceding theorem,

$$\left(M_R(t)^2 - \int_0^t M_R(\tau)\, M_R(d\tau), \mathcal{F}_t, \mathbb{P}\right)$$

is a continuous martingale. In particular, this means that

$$\mathbb{E}^{\mathbb{P}}\big[M(t \wedge \zeta_R)^2\big] = \mathbb{E}^{\mathbb{P}}\left[\int_0^t M_R(\tau)\, M_R(d\tau)\right].$$

---

[2] Remember that we have adopted $\{\zeta < t\} \in \mathcal{F}_t$ as the condition which determines whether ζ is a stopping time.

On the other hand, because $M_R(\,\cdot\,,\omega)$ is continuous, as well as of bounded variation, integration by parts leads to the pathwise identity

$$M_R(t,\omega)^2 = 2\int_0^t M_R(\tau,\omega)\,M_R(d\tau,\omega) \quad \text{for } \mathbb{P}\text{-almost every } \omega.$$

Hence, after combining this with the above, we conclude that $\mathbb{E}^{\mathbb{P}}\big[M_R(t)^2\big] = 0$. Finally, suppose that $\mathbb{P}\big(\exists\, t \geq 0\ 0 < |M|(t) < \infty\big) > 0$. Then there would exist an $R > 0$ and $t \in (0,\infty)$ such that $\mathbb{P}(\zeta_R \leq t) > 0$, which would lead to the contradiction $\mathbb{E}^{\mathbb{P}}\big[M_R(t)^2\big] \geq \tfrac{1}{4}\mathbb{E}^{\mathbb{P}}\big[\|M\|_{[0,T]}^2\big] > 0$.

To complete the proof, suppose that A and A' are two functions with the described properties. Then $\big(A(t) - A'(t), \mathcal{F}_t, \mathbb{P}\big)$ is a continuous local martingales whose paths are of locally bounded variation. Hence, by what we have just proved, this means that $A = A'$ $\mathbb{P}$-almost surely. $\quad\square$

§7.1.2. **Existence:** In this subsection, we will show that if $\big(M(t), \mathcal{F}_t, \mathbb{P}\big)$ is a continuous, $\mathbb{R}$-valued local martingale, then there exists a $\mathbb{P}$-almost surely unique progressively measurable $A : [0,\infty) \times \Omega \longrightarrow [0,\infty)$ such that $A(0) = 0$, $t \rightsquigarrow A(t)$ is $\mathbb{P}$-almost surely continuous and nondecreasing, and $\big(M(t)^2 - A(t), \mathcal{F}_t, \mathbb{P}\big)$ is a local martingale.

To begin, notice that Corollary 7.1.2 provides us with the required uniqueness. Next, observe that it suffices to prove existence in the case when $\big(M(t), \mathcal{F}_t, \mathbb{P}\big)$ is a bounded martingale with $M(0) \equiv 0$. Indeed, if this is not already the case, we can take $\zeta_m(\omega) = \inf\{t \geq 0 : |M(t,\omega)| \geq m\}$ and set $M_m(t) = M(t \wedge \zeta_m)$. Assuming that A_m exists for each $m \in \mathbb{Z}^+$, we would know, by Doob's stopping time theorem and uniqueness, that $A_{m+1} \restriction [0,\zeta_m) = A_m \restriction [0,\zeta_m)$ $\mathbb{P}$-almost surely for all $m \geq 1$. Hence, we could construct A by taking $A(t) = \sup\{A_m(t) : m \text{ with } \zeta_m \geq t\}$.

Now assume that $\big(M(t), \mathcal{F}_t, \mathbb{P}\big)$ is a bounded, continuous martingale with $M(0) = 0$. For convenience, we will assume that $M(\,\cdot\,,\omega)$ is continuous for every $\omega \in \Omega$. The idea behind Itô's construction of A is to realization that, if A exists, then Itô's formula would hold when t is systematically replaced by $A(t)$. In particular, one would have $M(t)^2 = 2\int_0^t M(\tau)\,dM(\tau) + A(t)$. Thus, it is reasonable to see what happens when we take $A(t) \equiv M(t)^2 - 2\int_0^t M(\tau)\,M(d\tau)$. Of course, this line of reasoning might seem circular since we want A in order to construct stochastic integrals with respect to M, but entry into the circle turns out to be easy.

Set $\zeta_{m,0}(\omega) = m$ for $m \in \mathbb{N}$. Next, proceeding by induction, define $\{\zeta_{m,N}\}_{m=0}^\infty$ for $N \in \mathbb{Z}^+$ so that $\zeta_{0,N} \equiv 0$ and, for $m \in \mathbb{Z}^+$, $\zeta_{m,N}(\omega)$ is equal to

$$\zeta_{\ell,N-1}(\omega) \wedge \inf\left\{t \geq \zeta_{m-1,N}(\omega) : \big|M(t,\omega) - M\big(\zeta_{m-1,N}(\omega),\omega\big)\big| \geq \tfrac{1}{N}\right\}$$

for the $\ell \in \mathbb{Z}^+$ with $\zeta_{\ell-1,N-1}(\omega) \leq \zeta_{m-1,N}(\omega) < \zeta_{\ell,N-1}(\omega)$.

For each $N \in \mathbb{N}$, $\{\zeta_{m,N} : m \geq 0\}$ is a nondecreasing sequence of bounded stopping times which tend to ∞ as $m \to \infty$. Further, these sequences are nested in the sense that $\{\zeta_{m,N-1} : m \geq 0\} \subseteq \{\zeta_{m,N} : m \geq 0\}$ for every $N \in \mathbb{Z}^+$.

Now set

$$M_{m,N}(\omega) = M\big(\zeta_{m,N}(\omega), \omega\big) \quad \text{and}$$
$$\Delta_{m,N}(t,\omega) = M\big(t \wedge \zeta_{m,N}(\omega), \omega\big) - M\big(t \wedge \zeta_{m-1,N}(\omega), \omega\big),$$

and observe that

$$M(t,\omega)^2 - M(0,\omega)^2 = 2Y_N(t,\omega) + A_N(t,\omega),$$

where

$$Y_N(t,\omega) \equiv \sum_{m=1}^{\infty} M_{m-1,N}(\omega)\Delta_{m,N}(t,\omega) \text{ and } A_N(t,\omega) \equiv \sum_{m=1}^{\infty} \Delta_{m,N}(t,\omega)^2.$$

Furthermore, $\big(Y_N(t), \mathcal{F}_t, P\big)$ is a continuous martingale, and $A_N : [0,\infty) \times \Omega \longrightarrow [0,\infty)$ is a progressively measurable function with the properties that, for each $\omega \in \Omega$: $A_N(0,\omega) = 0$, $A_N(\cdot, \omega)$ is continuous, and $A_N(t,\omega) + \frac{1}{N^2} \geq A_N(s,\omega)$ whenever $0 \leq s < t$. Thus, we will be done if we can prove that, for each $T \in [0,\infty)$, $\{A_N : N \geq 0\}$ converges in $L^2\big(\mathbb{P}; C([0,T];\mathbb{R})\big)$, which is equivalent to showing that $\{Y_N : N \geq 0\}$ converges there.

With this in mind, for each $0 \leq N < N'$ and $m \in \mathbb{Z}^+$, define

$$M_{m,N'}^{(N)}(\omega) = M_{\ell,N}(\omega) \quad \text{when } \zeta_{\ell-1;N}(\omega) \leq \zeta_{m,N'}(\omega) < \zeta_{\ell,N}(\omega),$$

and note that

$$Y_{N'}(t,\omega) - Y_N(t,\omega) = \sum_{m=1}^{\infty} \Big(M_{m,N'}(\omega) - M_{m,N'}^{(N)}(\omega)\Big)\Delta_{m,N'}(t,\omega).$$

Because $\big|M_{m,N'}(\omega) - M_{m,N'}^{(N)}(\omega)\big| \leq \frac{1}{N}$ and the terms in the series are orthogonal,

$$\mathbb{E}^P\Big[\big(Y_{N'}(t) - Y_N(t)\big)^2\Big] \leq N^{-2}\mathbb{E}^P\big[M(t)^2\big].$$

In particular, as an application of Doob's inequality, we see first that, for each $T \geq 0$,

$$\lim_{N \to \infty} \sup_{N' > N} \mathbb{E}^{\mathbb{P}}\Big[\|Y_{N'} - Y_N\|_{[0,T]}^2\Big] = 0,$$

and then that there exists a continuous martingale $(Y(t), \mathcal{F}_t, \mathbb{P})$ with the property that

$$\lim_{N \to \infty} \mathbb{E}^P\left[\|Y_N - Y\|^2_{[0,T]}\right] = 0 \quad \text{for each } T \in [0, \infty).$$

To complete the proof at this point, define the function $A : [0, \infty) \times \Omega \longmapsto [0, \infty)$ so that

$$A(t, \omega) = 0 \vee \sup\left\{M(s, \omega)^2 - 2Y(s, \omega) : s \in [0, t]\right\},$$

and check that A has the required properties. Hence, we have now proved the following version of the Doob–Meyer decomposition theorem.

THEOREM 7.1.3. *If* $(M(t), \mathcal{F}_t, \mathbb{P})$ *is a continuous,* $\mathbb{R}$-*valued local martingale, then there exists a* $\mathbb{P}$-*almost surely unique progressively measurable function* $A : [0, \infty) \times \Omega \longrightarrow [0, \infty)$ *with the properties that* $A(0) \equiv 0$, $t \rightsquigarrow A(t)$ *is* $\mathbb{P}$-*almost surely continuous and nondecreasing, and* $(M(t)^2 - A(t), \mathcal{F}_t, \mathbb{P})$ *is a local martingale.*

From now on, we will use the notation $\langle M \rangle$ to denote the process A described in Theorem 7.1.3.

§7.1.3. Exercises:

EXERCISE 7.1.4. Because we have not considered martingales with discontinuities, the most subtle aspects of Meyer's theorem are not apparent in our treatment. To get a feeling for what these subtleties are, consider a simple Poisson process (cf. §1.4.2) $N(t)$ on some probability space $(\Omega, \mathcal{F}, \mathbb{P})$, let $\mathcal{F}_t$ be the $\mathbb{P}$-completion of $\sigma(N(\tau) : \tau \in [0, t])$, set $M(t) = N(t) - t$, and check that $(M(t), \mathcal{F}_t, \mathbb{P})$ is a non-constant martingale. At the same time, $t \rightsquigarrow M(t)$ $\mathbb{P}$-almost surely has locally bounded variation. Hence, the first part of Corollary 7.1.2 is, in general, *false* unless one imposes some condition on the paths $t \rightsquigarrow M(t)$. The condition which we imposed was continuity. However, a look at the proof reveals that the only place where we used continuity was when we integrated by parts to get $M_R(t) = 2 \int_0^t M_R(\tau) \, M_R(d\tau)$. This is the point alluded to in the rather cryptic remark preceding Theorem 7.1.1.

EXERCISE 7.1.5. Let $(M(t), \mathcal{F}_t, \mathbb{P})$ be a continuous local martingale, ζ a stopping time, and set $M^\zeta(t) = M(t \wedge \zeta)$.

(i) Show that $\langle M^\zeta \rangle(t) = \langle M \rangle(t \wedge \zeta)$.

(ii) If $\langle M \rangle(\zeta) \in L^1(\mathbb{P}; \mathbb{R})$, show that

$$\left(M^\zeta(t) - M(0), \mathcal{F}_t, \mathbb{P}\right) \quad \text{and} \quad \left(\left(M^\zeta(t) - M(0)\right)^2 - \langle M^\zeta \rangle(t), \mathcal{F}_t, \mathbb{P}\right)$$

are martingales.

(iii) Suppose $\alpha : \Omega \longrightarrow \mathbb{R}$ is an $\mathcal{F}_\zeta$-measurable function, and set $M'(t) = \alpha\big(M(t) - M(t \wedge \zeta)\big)$. After checking that $\big(M'(t), \mathcal{F}_t, \mathbb{P}\big)$ is a continuous local martingale, show that

$$\langle M' \rangle(t) = \alpha^2 \big(\langle M \rangle(t) - \langle M \rangle(t \wedge \zeta)\big).$$

EXERCISE 7.1.6. Let $\big(M(t), \mathcal{F}_t, \mathbb{P}\big)$ be a continuous, $\mathbb{R}$-valued local martingale.

(i) Show that $M(\infty) \equiv \lim_{t \to \infty} M(t)$ exists $\mathbb{P}$-almost surely on the set $\{\langle M \rangle(\infty) < \infty\}$.

Hint: Let $\zeta_R = \inf\{t \geq 0 : \langle M \rangle \geq R\}$, show that $\big(M(t \wedge \zeta_R) - M(0), \mathcal{F}_t, \mathbb{P}\big)$ is a continuous martingale whose second moment is bounded by R, and apply the Martingale Convergence Theorem (cf. Theorem 7.1.16 in [36]) to conclude that $\lim_{t \to \infty} M(t \wedge \zeta_R)$ exists $\mathbb{P}$-almost surely.

(ii) Let ζ be a stopping time with the property that $\langle M \rangle(\zeta) < \infty$ $\mathbb{P}$-almost surely, and, using part **(i)**, define $M(\zeta)$ on $\{\zeta = \infty\}$ equal to be $\mathbb{P}$-almost surely equal to $\lim_{t \nearrow \infty} M(t)$. Show that $\langle M \rangle(\zeta) \in L^1(\mathbb{P}; \mathbb{R})$ if and only if $M(\zeta) \in L^2(\mathbb{P}; \mathbb{R})$, in which case $\mathbb{E}^{\mathbb{P}}\big[\big(M(\zeta) - M90)\big)^2\big] = \mathbb{E}^{\mathbb{P}}\big[\langle M \rangle(\zeta)\big]$.

EXERCISE 7.1.7. Suppose that $\{M_k\}_1^\infty \subseteq \mathcal{M}_{\mathrm{loc}}(\mathbb{P}; \mathbb{R})$ and that ζ is a stopping time. If $\langle M_k \rangle(\zeta) \longrightarrow 0$ in $\mathbb{P}$-probability, show that $\|M_k - M_k(0)\|_{[0,\zeta)} \longrightarrow 0$ in $\mathbb{P}$-probability.

EXERCISE 7.1.8. Let a continuous, $\mathbb{R}$-valued, local martingale $\big(M(t), \mathcal{F}_t, \mathbb{P}\big)$ be given, and, for each ω, use $G(\omega)$ to denote the set of all $t \in (0, \infty)$ for which there exist $0 \leq a < t < b < \infty$ with $\langle M \rangle(b, \omega) = \langle M \rangle(a, \omega)$. Clearly, $G(\omega)$ is an open subset of $\mathbb{R}$, and as such its connected components are open intervals. The goal of this exercise is to show that, for $\mathbb{P}$-almost every ω, $M(\cdot, \omega)$ is constant on each connected component of $G(\omega)$.

(i) For each $t \in [0, \infty)$, define $\sigma(t, \omega) = \sup\{\tau \geq t : \langle M \rangle(\tau, \omega) = \langle M \rangle(t, \omega)\}$. Show that $M(\cdot, \omega)$ is constant on each connected component of $G(\omega)$ if and only if $M(t, \omega) = M\big(\sigma(t, \omega), \omega\big)$ for each rational number $t \in [0, \infty)$.

(ii) In view of **(i)** and the $\mathbb{P}$-almost sure continuity of $M(\cdot, \omega)$, we will have reached our goal once we show that, for each $0 \leq t < \tau < \infty$, $M\big(\tau \wedge \sigma(t, \omega), \omega\big) = M(t, \omega)$ $\mathbb{P}$-almost surely. Prove this first in the case when M is a square-integrable martingale, and then reduce to this case by a stopping time argument.

7.2 Kunita–Watanabe Stochastic Integration

Recall (cf. §5.1.3) the notation $\mathcal{M}_{\mathrm{loc}}(\mathbb{P}; \mathbb{R})$ for the space of all $\mathbb{R}$-valued, continuous local martingales on the complete probability space $(\Omega, \mathcal{F}, \mathbb{P})$

relative to the nondecreasing family $\{\mathcal{F}_t : t \geq 0\}$ of $\mathbb{P}$-complete sub σ-algebras. Given an $M \in \mathcal{M}_{\text{loc}}(\mathbb{P}; \mathbb{R})$, let $\Theta^2_{\text{loc}}(\langle M \rangle, \mathbb{P}; \mathbb{R})$ denote the space of progressively measurable $\theta : [0, \infty) \longrightarrow \mathbb{R}$ with the property that

$$\int_0^T |\theta(\tau)|^2 \langle M \rangle(d\tau) < \infty \quad \mathbb{P}\text{-almost surely for all } T \in [0, \infty).$$

Following Kunita and Watanabe, we will define in this section the stochastic integral $I_\theta^M \in \mathcal{M}_{\text{loc}}(\mathbb{P}; \mathbb{R})$ of $\theta \in \Theta^2_{\text{loc}}(\langle M \rangle, \mathbb{P}; \mathbb{R})$ with respect to $M \in \mathcal{M}_{\text{loc}}(\mathbb{P}; \mathbb{R})$.

§**7.2.1. The Hilbert Structure of** $\mathcal{M}_{\text{loc}}(\mathbb{P}; \mathbb{R})$**:** Clearly $\mathcal{M}_{\text{loc}}(\mathbb{P}; \mathbb{R})$ is a vector space and $M \rightsquigarrow \langle M \rangle$ is some sort of non-negative, quadratic functional on this vector space. In particular, these trivial observations, in conjuction with Corollary 7.1.2, lead immediately to the conclusion that for each pair $(M_1, M_2) \in \mathcal{M}_{\text{loc}}(\mathbb{P}; \mathbb{R})^2$ there is a $\mathbb{P}$-almost surely unique progressively measurable $\langle M_1, M_2 \rangle : [0, \infty) \times \Omega \longrightarrow \mathbb{R}$ with the properties that $\langle M_1, M_2 \rangle(0) = 0$, $t \rightsquigarrow \langle M_1, M_2 \rangle(t)$ is $\mathbb{P}$-almost surely continuous and of locally bounded variation, and

$$\left(M_1(t)M_2(t) - \langle M_1, M_2 \rangle(t), \mathcal{F}_t, \mathbb{P} \right) \quad \text{is a local martingale.}$$

Indeed, the uniqueness is immediate from Corollary 7.1.2 and the existence is an application of polarization[3]:

$$\langle M_1, M_2 \rangle = \frac{\langle M_1 + M_2 \rangle - \langle M_1 - M_2 \rangle}{4}.$$

THEOREM 7.2.1. *The map* $(M_1, M_2) \rightsquigarrow \langle M_1, M_2 \rangle$ *is symmetric, bilinear, and non-negative in the sense that,* $\mathbb{P}$*-almost surely,* $\langle M_1, M_2 \rangle = \langle M_2, M_1 \rangle$*,* $\langle \alpha_1 M_1 + \alpha_2 M_2, M_3 \rangle = \alpha_1 \langle M_1, M_3 \rangle + \alpha_2 \langle M_2, M_3 \rangle$*, and* $\langle M, M \rangle \geq 0$*. Moreover,*

(7.2.2)
$$\begin{aligned} \big| \langle M_1, M_2 \rangle(t_2) &- \langle M_1, M_2 \rangle(t_1) \big| \\ &\leq \sqrt{\langle M_1 \rangle(t_2) - \langle M_1 \rangle(t_1)} \sqrt{\langle M_2 \rangle(t_2) - \langle M_2 \rangle(t_1)} \end{aligned}$$
for all $0 \leq t_1 < t_2$ $\mathbb{P}$*-almost surely.*

Equivalently, $\langle M_1, M_2 \rangle$ *is* $\mathbb{P}$*-almost surely absolutely continuous with respect to* $\mu_\omega \equiv \langle M_1 \rangle(\cdot, \omega) + \langle M_2 \rangle(\cdot, \omega)$*, and if* $f_{i,j}(\cdot, \omega)$ *denotes the Radon–Nikodym derivative of* $\langle M_i, M_j \rangle(\cdot, \omega)$ *with respect to* μ_ω*, then, for* $\mathbb{P}$*-almost every* $\omega \in \Omega$*,*

$$|f_{1,2}(\cdot, \omega)| \leq \sqrt{f_{1,1}(\cdot, \omega) f_{2,2}(\cdot, \omega)} \quad \mu_\omega\text{-almost everywhere.}$$

---

[3] It must be admitted that the notation here is a little confusing. Namely, we now have two closely related notations for one object: $\langle M \rangle = \langle M, M \rangle$.

In particular, $\mathbb{P}$-almost surely,

$$\left\| \langle M_2 \rangle^{\frac{1}{2}} - \langle M_1 \rangle^{\frac{1}{2}} \right\|_{[0,T]} \leq \langle M_2 - M_1 \rangle(T) \quad \text{for all } T \geq 0.$$

PROOF: The first assertion requiring comment is the inequality in (7.2.2). To prove it, first note that it suffices to show that for each $0 \leq t_1 < t_2$ and $\alpha > 0$,

(*)
$$2\left| \langle M_1, M_2 \rangle(t_2) - \langle M_1, M_2 \rangle(t_1) \right|$$
$$\leq \alpha\left(\langle M_1 \rangle(t_2) - \langle M_1 \rangle(t_1) \right) + \frac{1}{\alpha}\left(\langle M_2 \rangle(t_2) - \langle M_2 \rangle(t_1) \right)$$

$\mathbb{P}$-almost surely. Indeed, given (*), one can easily argue that, $\mathbb{P}$-almost surely, the same inequality holds simultaneously for all $\alpha > 0$ and $0 \leq t_1 < t_2$; and once this is known, (7.2.2) follows by the usual minimization procedure with which one proves Schwartz's inequality. But (*) is a trivial consequence of non-negative bilinearity. Namely, for any $\alpha > 0$,

$$0 \leq \langle \alpha^{\frac{1}{2}} M_1 \pm \alpha^{-\frac{1}{2}} M_2, \alpha^{\frac{1}{2}} M_1 \pm \alpha^{-\frac{1}{2}} M_2 \rangle(t_2)$$
$$- \langle \alpha^{\frac{1}{2}} M_1 \pm \alpha^{-\frac{1}{2}} M_2, \alpha^{\frac{1}{2}} M_1 \pm \alpha^{-\frac{1}{2}} M_2 \rangle(t_1)$$
$$= \alpha\left(\langle M_1 \rangle(t_2) - \langle M_1 \rangle(t_1) \right) \pm 2\left(\langle M_1, M_2 \rangle(t_2) - \langle M_1, M_2 \rangle(t_2) \right)$$
$$+ \alpha^{-1}\left(\langle M_2 \rangle(t_2) - \langle M_2 \rangle(t_1) \right)$$

$\mathbb{P}$-almost surely.

Knowing the Schwarz inequality for $\langle M_1, M_2 \rangle$, the triangle inequality

$$\left| \sqrt{\langle M_2 \rangle(t)} - \sqrt{\langle M_1 \rangle(t)} \right| \leq \langle M_2 - M_1 \rangle(t) \leq \langle M_2 - M_1 \rangle(T) \quad , 0 \leq t \leq T,$$

$\mathbb{P}$-almost surely follows immediately. Hence, completing the proof from here comes down to showing that if μ_1 and μ_2 are finite, non-negative, non-atomic Borel measures on $[0, T]$ and ν is a signed Borel measure on $[0, T]$ satisfying $|\nu(I)| \leq \sqrt{\mu_1(I)\mu_2(I)}$ for all half-open intervals $I = [a, b) \subseteq [0, T]$, then $\nu \ll \mu \equiv \mu_1 + \mu_2$ and $|g| \leq \sqrt{f_1 f_2}$ μ-almost everywhere, where $g = \frac{d\nu}{d\mu}$ and $f_i = \frac{d\mu_i}{d\mu}$. Because, for all $\alpha > 0$, $2\sqrt{\mu_1(I)\mu_2(I)} \leq \alpha\mu_1(I) + \alpha^{-1}\mu_2(I)$, the absolute continuity statement is clear. In addition, we have

$$2\left| \int_0^T \varphi g \, d\mu \right| \leq \alpha \int_0^T |\varphi| f_1 \, d\mu + \alpha^{-1} \int_0^T |\varphi| f_2 \, d\mu$$

first for φ's which are indicator functions of intervals $[a, b)$, next for linear combinations of such functions, then for continuous φ's, and finally for all Borel bounded measurable φ's. But this means that, μ-almost everywhere, $2|g| \leq \alpha f_1 + \alpha^{-1} f_2$ for all $\alpha > 0$, and therefore that $|g| \leq \sqrt{f_1 f_2}$. $\square$

In the following, and throughout, we will say that a sequence $\{M_k\}_1^\infty$ in $\mathcal{M}_{\mathrm{loc}}(\mathbb{P}; \mathbb{R})$ *converges in* $\mathcal{M}_{\mathrm{loc}}(\mathbb{P}; \mathbb{R})$ to $M \in \mathcal{M}_{\mathrm{loc}}(\mathbb{P}; \mathbb{R})$ if, for each $T \geq 0$, $\langle M_k - M \rangle(T) \longrightarrow 0$ in $\mathbb{P}$-probability.

COROLLARY 7.2.3. *If* $M_k \longrightarrow M$ *in* $\mathcal{M}_{\mathrm{loc}}(\mathbb{P}; \mathbb{R})$, *then*

$$\big\| \big(M_k - M_k(0) \big) - \big(M - M(0) \big) \big\|_{[0,T]} \vee \big\| \langle M_k \rangle - \langle M \rangle \big\|_{[0.T]} \longrightarrow 0$$
in $\mathbb{P}$-*probability for each* $T \geq 0$.

Moreover, if $\{M_k\}_1^\infty \subseteq \mathcal{M}_{\mathrm{loc}}(\mathbb{P}; \mathbb{R})$ *and*

$$\lim_{k \to \infty} \sup_{\ell \geq k} \langle M_\ell - M_k \rangle (T) = 0 \quad \text{in } \mathbb{P}\text{-}\textit{probability for each } T \geq 0,$$

then there exists a $M \in \mathcal{M}_{\mathrm{loc}}(\mathbb{P}; \mathbb{R})$ *to which* $\{M_k - M(0)\}_1^\infty$ *converges to* M *in* $\mathcal{M}_{\mathrm{loc}}(\mathbb{P}; \mathbb{R})$.

PROOF: Without loss in generality, we will assume that $M_k(0) = 0 = M(0)$ for $k \geq 1$.

In view of the triangle inequality proved in Theorem 7.2.1, the only part of the first assertion which requires comment is the proof that $\big\| M_k - M \big\|_{[0,T]} \longrightarrow 0$ in $\mathbb{P}$-probability for all $T \geq 0$. However, if

$$\zeta_R \equiv \inf \left\{ t \geq 0 : \sup_{k \geq 1} \langle M_k \rangle (t) \geq R \right\},$$

then $\zeta_R \nearrow \infty$ $\mathbb{P}$-almost surely as $R \to \infty$ and, by Exercise 7.1.5 and Doob's inequality,

$$\mathbb{E}^\mathbb{P} \left[\big\| M_k - M \big\|_{[0,T \wedge \zeta_R]}^2 \right] \leq 4 \mathbb{E}^\mathbb{P} \left[\langle M_k - M \rangle (T \wedge \zeta_R) \right] \longrightarrow 0$$

as $k \to \infty$ for each $R > 0$.

Turning to the Cauchy criterion in the second assertion, define ζ_R as in the preceding paragraph, and observe that the argument given there also shows that
$$\lim_{k \to \infty} \sup_{\ell \geq k} \mathbb{E}^\mathbb{P} \left[\big\| M_\ell - M_k \big\|_{[0,T \wedge \zeta_R]}^2 \right] = 0$$

for each $R > 0$. Hence, there exists an $M \in \mathcal{M}_{\mathrm{loc}}(\mathbb{P}; \mathbb{R})$ such that $\| M_k - M \|_{[0,T]} \longrightarrow 0$ in $\mathbb{P}$-probability for all $T > 0$. At the same time, we know that, for each $R > 0$ and $T > 0$,

$$\mathbb{E}^\mathbb{P} \left[\langle M_k - M \rangle (T \wedge \zeta_R) \right] = \mathbb{E}^\mathbb{P} \left[\big| (M_k - M)(T \wedge \zeta_R) \big|^2 \right] \longrightarrow 0$$

as $k \to \infty$. Hence, for each $T > 0$, $\langle M_k - M \rangle (T) \longrightarrow 0$ in $\mathbb{P}$-probability. $\square$

§7.2.2. **The Kunita–Watanabe Stochastic Integral:** The idea of Kunita and Watanabe is to base the definition of stochastic integration on the Hilbert structure described in the preceding subsection. Namely, given $\theta \in \Theta^2_{\mathrm{loc}}(\langle M \rangle, \mathbb{P}; \mathbb{R})$, they say that I^M_θ should be the element of $\mathcal{M}_{\mathrm{loc}}(\mathbb{P}; \mathbb{R})$ with the properties that

(7.2.4)
$$I^M_\theta(0) = 0 \text{ and } \langle I^M_\theta, M' \rangle(t) = \int_0^t \theta(\tau) \langle M, M' \rangle(d\tau)$$
$$\text{for all } M' \in \mathcal{M}_{\mathrm{loc}}(\mathbb{P}; \mathbb{R}).$$

Before adopting this definition, one must check that (7.2.4) makes sense and that, up to a $\mathbb{P}$-null set, it determines a unique element of $\mathcal{M}_{\mathrm{loc}}(\mathbb{P}; \mathbb{R})$. To handle the first of these, observe that, by Theorem 7.2.1,

$$\int_0^T |\theta(\tau)| \, |\langle M, M' \rangle|(d\tau) \le \sqrt{\int_0^T |\theta(\tau)|^2 \, \langle M \rangle(d\tau) \, \langle M' \rangle(T)}.$$

Hence, $\theta \in \Theta^2_{\mathrm{loc}}(\langle M \rangle, \mathbb{P}; \mathbb{R})$ implies that, $\mathbb{P}$-almost surely, θ is locally integrable with respect to the signed measure $\langle M, M' \rangle$. As for the uniqueness question, suppose that I and J both satisfy (7.2.4), and set $\Delta = I - J$. Then $\langle \Delta \rangle \equiv 0$, and so there exists a nondecreasing sequence $\{\zeta_m\}_1^\infty$ of stopping times such that $\zeta_m \nearrow \infty$ and $\left(\Delta(\cdot \wedge \zeta_m)^2, \mathcal{F}_t, \mathbb{P} \right)$ is a bounded martingale for each m, which, since $\Delta(0) \equiv 0$, means that $\mathbb{E}^{\mathbb{P}}\left[\Delta(t \wedge \zeta_m)^2 \right] = 0$ for all $m \ge 1$ and $t \ge 0$.

Having verified that (7.2.4) makes sense and uniquely determines I^M_θ, what remains is for us to prove that I^M_θ always exists, and, as should come as no surprise, this requires us to return (cf. §5.1.2) to Itô's technique for constructing his integral. Namely, given $M \in \mathcal{M}_{\mathrm{loc}}(\mathbb{P}; \mathbb{R})$ and a bounded, progressively measurable $\theta : \Omega \longrightarrow \mathbb{R}$ with the property that $\theta(t) = \theta([t]_N)$ for some $N \in \mathbb{N}$, set

$$I^M_\theta(t) = \sum_{m=0}^\infty \theta(m 2^{-N}) \left(M(t \wedge (m+1) 2^{-N}) - M(t \wedge m 2^{-N}) \right).$$

Clearly (cf. part (ii) of Exercise 7.1.5), if ζ is a stopping time for which $\langle M \rangle(\zeta) \in L^1(\mathbb{P}; \mathbb{R})$, then $I^M_\theta(t \wedge \zeta)$ is $\mathbb{P}$-square integrable for all $t \ge 0$ and, for all $m \in \mathbb{N}$ and $m 2^{-N} \le t_1 < t_2 \le (m+1) 2^{-N}$,

$$\mathbb{E}^{\mathbb{P}}\left[I^M_\theta(t_2 \wedge \zeta) - I^M_\theta(t_1 \wedge \zeta) \,\middle|\, \mathcal{F}_{t_1} \right]$$
$$= \theta(m 2^{-N}) \mathbb{E}^{\mathbb{P}}\left[M(t_2 \wedge \zeta) - M(t_1 \wedge \zeta) \,\middle|\, \mathcal{F}_{t_1} \right] = 0.$$

Thus $I^M_\theta \in \mathcal{M}_{\mathrm{loc}}(\mathbb{P}; \mathbb{R})$. In addition, if $M' \in \mathcal{M}_{\mathrm{loc}}(\mathbb{P}; \mathbb{R})$ and $\langle M' \rangle(\zeta)$ is also

$\mathbb{P}$-integrable, then

$$\mathbb{E}^{\mathbb{P}}\left[I_\theta^M(t_2 \wedge \zeta)M'(t_2 \wedge \zeta) - I_\theta^M(t_1 \wedge \zeta)M'(t_1 \wedge \zeta) \,\middle|\, \mathcal{F}_{t_1}\right]$$
$$= \theta(m2^{-N})\mathbb{E}^{\mathbb{P}}\left[M(t_2 \wedge \zeta)M'(t_2 \wedge \zeta) - M(t_1 \wedge \zeta)M'(t_1 \wedge \zeta) \,\middle|\, \mathcal{F}_{t_1}\right]$$
$$= \theta(m2^{-N})\mathbb{E}^{\mathbb{P}}\left[\langle M, M'\rangle(t_2 \wedge \zeta) - \langle M, M'\rangle(t_1 \wedge \zeta) \,\middle|\, \mathcal{F}_{t_1}\right]$$
$$= \mathbb{E}^{\mathbb{P}}\left[\int_{t_1 \wedge \zeta}^{t_2 \wedge \zeta} \theta(\tau)\langle M, M'\rangle(d\tau) \,\middle|\, \mathcal{F}_{t_1}\right],$$

which proves that $\langle I_\theta^M, M'\rangle(dt) = \theta(t)\langle M, M'\rangle(dt)$. Hence, we now know that I_θ^M exists for all bounded, progressively measurable $\theta : [0, \infty) \times \Omega \longrightarrow \mathbb{R}$ with the property that $\theta(t) = \theta([t]_N)$ for some $N \in \mathbb{N}$. Furthermore, by Corollary 7.2.3, we know that if $\{\theta_N\}_1^\infty \cup \{\theta\} \subseteq \Theta_{\mathrm{loc}}^2(\langle M\rangle, \mathbb{P}; \mathbb{R})$ and

$$(7.2.5) \qquad \int_0^T \left|\theta_N(\tau) - \theta(\tau)\right|^2 \langle M\rangle(d\tau) \longrightarrow 0 \quad \text{in } \mathbb{P}\text{-probability for all } T > 0,$$

then I_θ^M exists. Hence, we will be done once we prove the following lemma.

LEMMA 7.2.6. *For each $\theta \in \Theta_{\mathrm{loc}}^2(\langle M\rangle, \mathbb{P}; \mathbb{R})$ there exists a sequence $\{\theta_N\}_1^\infty$ of bounded, $\mathbb{R}$-valued, progressively measurable functions such that $\theta_N(t) = \theta_N([t]_N)$ and (7.2.5) holds.*

PROOF: Clearly, it suffices to handle θ's which are bounded and vanish off $[0, T]$ for some $T > 0$. In addition, we may assume that $t \rightsquigarrow \langle M\rangle(t, \omega)$ is bounded, continuous, and nondecreasing for each ω. Thus, we will make these assumptions.

There is no problem if $t \rightsquigarrow \theta(t, \omega)$ is continuous for all $\omega \in \Omega$, since we can then take $\theta_N(t) = \theta([t]_N)$. Hence, what must be shown is that for each bounded, progressively measurable θ there exists a sequence $\{\theta_N\}_1^\infty$ of bounded, progressively, $\mathbb{R}$-valued functions with the properties that $t \rightsquigarrow \theta(t, \omega)$ is continuous for each ω and (7.2.5) holds. To this end, set $A(t, \omega) = t + \langle M\rangle(t, \omega)$, and, for each $s \in [0, \infty)$, determine $\omega \in \Omega \longmapsto \zeta(s, \omega) \in [0, \infty)$ so that $A(\zeta(s, \omega), \omega) = s$. Clearly, for each ω, $t \rightsquigarrow A(t, \omega)$ is a homeomorphism from $[0, \infty)$ onto itself, and so, an elementary change of variables yields

$$(*) \qquad \int_{[0,\infty)} f(t)\, A(dt, \omega) = \int_{[0,\infty)} f \circ \zeta(s, \omega)\, ds$$

for any non-negative, Borel measurable f on $[0, \infty)$..

To take the next step, notice that, for each s, $\omega \rightsquigarrow \zeta(s, \omega)$ is a stopping time, and set $\mathcal{F}_s'$ equal to the $\mathbb{P}$-completion of $\mathcal{F}_{\zeta(s)}$. Thus, if $\theta'(s, \omega) \equiv$

$\theta(\zeta(s,\omega),\omega)$, then $\theta' : [0,\infty) \times \Omega \longrightarrow \mathbb{R}$ is a bounded function which vanishes off of $[0, A(T)] \times \Omega$ and is progressively measurable with respect to the filtration $\{\mathcal{F}'_s : s \geq 0\}$. Hence, by the argument given to prove the density statement in Lemma 5.1.8, we can find a sequence $\{\theta'_N\}_1^\infty$ of $\{\mathcal{F}'_s : s \geq 0\}$-progressively measurable such that $t \rightsquigarrow \theta'_N(t,\omega)$ is bounded, continuous, and supported on $[0, A(1+T,\omega)]$ for each ω, and

$$\lim_{N\to\infty} \mathbb{E}^{\mathbb{P}}\left[\int_0^{A(1+T)} |\theta'_N(s) - \theta'(s)|^2\, ds\right] = 0.$$

Finally, set $\theta_N(t,\omega) = \theta'_N(A(t,\omega),\omega)$, and note that each θ_N is a bounded, $\{\mathcal{F}_t : t \geq 0\}$-progressively measurable function with the properties that $t \rightsquigarrow \theta_N(t,\omega)$ is continuous and vanishes off $[0, 1+T] \times \Omega$ for each ω. Further, by (*)

$$\lim_{N\to\infty} \mathbb{E}^{\mathbb{P}}\left[\int_0^{1+T} |\theta_N(t) - \theta(t)|^2\, A(dt)\right] = 0.$$

Hence, (7.2.5) holds. $\square$

Summarizing the results proved in this subsection, we state the following theorem.

THEOREM 7.2.7. *For each $M \in \mathcal{M}_{\mathrm{loc}}(\mathbb{P};\mathbb{R})$ there is a linear map $\theta \in \Theta^2_{\mathrm{loc}}(\langle M\rangle, \mathbb{P};\mathbb{R}) \longmapsto I_\theta^M \in \mathcal{M}_{\mathrm{loc}}(\mathbb{P};\mathbb{R})$ such that (7.2.4) holds.*

Just as we did in the case treated in Chapter 5, we will use the notation $\int_0^t \theta(\tau)\, dM(\tau)$ interchangeably with $I_\theta^M(t)$. More generally, given stopping times $\zeta_1 \leq \zeta_2$, we define

$$\int_{t\wedge\zeta_1}^{t\wedge\zeta_2} \theta(\tau)\, dM(\tau) = I_\theta^M(t \wedge \zeta_2) - I_\theta^M(t \wedge \zeta_1).$$

Starting from Exercise 7.1.5, it is an easy matter to check that

$$(7.2.8) \qquad \int_{t\wedge\zeta_1}^{t\wedge\zeta_2} \theta(\tau)\, dM(\tau) = \int_0^t \mathbf{1}_{[\zeta_1,\zeta_2)}(\tau)\theta(\tau)\, dM(\tau).$$

§7.2.3. **General Itô's Formula:** Because our proof of Itô 's formula in §5.3 was modeled on the argument given by Kunita and Watanabe, its adaptation to their stochastic integral defined in §7.2.3 requires no substantive changes. Indeed, because, by part (i) of Exercise 7.1.5, we already know that, for bounded stopping times $\zeta_1 \leq \zeta_2$

$$\int_{\zeta_1}^{\zeta_2} \theta(\tau)\, dM(\tau) = \int_0^\infty \mathbf{1}_{[0,\zeta)}(\tau)\theta(\tau)\, dM(\tau),$$

the same argument as we used to prove Theorem 5.3.1 allows us to prove the following extension.

THEOREM 7.2.9. *Let* $X = (X_1, \ldots, X_k) : [0, \infty) \times \Omega \longrightarrow \mathbb{R}^k$ *and* $Y = (Y_1, \ldots, Y_\ell) : [0, \infty) \times \Omega \longrightarrow \mathbb{R}^\ell$ *be progressively measurable maps with the properties that, for each* $1 \leq i \leq k$ *and* ω, $t \leadsto X_i(t, \omega)$ *is continuous and of locally bounded variation and, for each* $1 \leq j \leq \ell$, $Y_j \in \mathcal{M}_{\text{loc}}(\mathbb{P}; \mathbb{R})$; *and set* $Z = (X, Y)$. *Then, for each* $F \in C^{1,2}(\mathbb{R}^k \times \mathbb{R}^\ell; \mathbb{R})$,

$$
F(Z(t)) - F(Z(0))
$$

$$
= \sum_{i=1}^{k} \int_0^t \partial_{x_i} F(Z(\tau)) \, dX_i(\tau) + \sum_{j=1}^{\ell} \int_0^t \partial_{y_j} F(Z(\tau)) \, dY_j(\tau)
$$

$$
+ \frac{1}{2} \sum_{j,j'=1}^{\ell} \int_0^t \partial_{y_j} \partial_{y_{j'}} F(Z(\tau)) \, \langle Y_j, Y_{j'} \rangle(d\tau), \quad t \geq 0,
$$

$\mathbb{P}$-*almost surely. Here, the* dX_i-*integrals are taken in the sense of Riemann–Stieltjes and the* dY_j-*integrals are taken in the sense of Itô , as described in Theorem 7.2.7.*

We will again refer to this extension as *Itô's formula*, and, not surprisingly, there are myriad applications of it. For example, as Kunita and Watanabe pointed out, it gives an elegant proof of the following famous theorem of Paul Lévy.

COROLLARY 7.2.10. *Suppose that* $\{M_j : 1 \leq j \leq n\} \subseteq \mathcal{M}_{\text{loc}}(\mathbb{P}; \mathbb{R})$ *and that* $M_j(0) = 0$ *for each* $1 \leq j \leq n$. *If* $\beta = (M_1, \ldots, M_n)$, *then* $(\beta(t), \mathcal{F}_t, \mathbb{P})$ *is an* $\mathbb{R}^n$-*valued Brownian motion if and only if* $\langle M_j, M_{j'} \rangle(t) = t\delta_{j,j'}$.

PROOF: We need only discuss the sufficiency. Given $\xi \in \mathbb{R}^n$, set

$$
F_\xi(t, y) = \exp\left(\sqrt{-1}\,(\xi, y)_{\mathbb{R}^n} + \tfrac{t}{2}|\xi|^2\right),
$$

apply Itô 's formula to see that

$$
F_\xi(t, M(t)) = 1 + \sqrt{-1} \sum_{j=1}^{n} \int_0^t \xi_j F_\xi(\tau, M(\tau)) \, dM_j(\tau),
$$

and conclude that if $E_\xi(t) \equiv F_\xi(t, M(t))$ then $(E_\xi, \mathcal{F}_t, \mathbb{P})$ is a continuous, $\mathbb{C}$-valued, local martingale. Thus, because $E_\xi \upharpoonright [0, T] \times \Omega$ is bounded for each $T > 0$, $(E_\xi, \mathcal{F}_t, \mathbb{P})$ is a continuous martingale. In particular,

$$
\mathbb{E}^{\mathbb{P}}\left[\exp\left(\sqrt{-1}\,(\xi, M(s+t) - M(s))_{\mathbb{R}^n}\right) \middle| \mathcal{F}_s\right] = e^{-\frac{t}{2}|\xi|^2},
$$

which, together with $M(0) = 0$, is enough to see that $(M(t), \mathcal{F}_t, \mathbb{P})$ is a Brownian motion. $\square$

§**7.2.4. Exercises**

EXERCISE 7.2.11. Suppose that M_1, $M_2 \in \mathcal{M}_{\mathrm{loc}}(\mathbb{P}; \mathbb{R})$, and assume that $\sigma(\{M_1(\tau) : \tau \geq 0\})$ is $\mathbb{P}$-independent of $\sigma(\{M_2(\tau) : \tau \geq 0\})$. Show that the product $M_1 M_2$ is again an element of $\mathcal{M}_{\mathrm{loc}}(\mathbb{P}; \mathbb{R})$, and conclude that $\langle M_1, M_2 \rangle \equiv 0$ $\mathbb{P}$-almost everywhere.

EXERCISE 7.2.12. Given $M \in \mathcal{M}_{\mathrm{loc}}(\mathbb{P}; \mathbb{R})$, $\theta \in \Theta^2_{\mathrm{loc}}(\langle M \rangle, \mathbb{P}; \mathbb{R})$, and $\eta \in \Theta^2_{\mathrm{loc}}(\langle I^M_\theta \rangle, \mathbb{P}; \mathbb{R})$, check that $\eta\theta \in \Theta^2_{\mathrm{loc}}(\langle M \rangle, \mathbb{P}; \mathbb{R})$ and that

$$I^M_{\eta\theta}(t) = \int_0^t \eta(\tau) \, dI^M_\theta(\tau) \quad \mathbb{P}\text{-almost surely.}$$

EXERCISE 7.2.13. When $M \in \mathcal{M}_{\mathrm{loc}}(\mathbb{P}; \mathbb{R})$ is a Brownian motion, and therefore $\langle M \rangle(t) = t$, it is an elementary exercise to check that $\langle M \rangle(t)$ is $\mathbb{P}$-almost everywhere equal to the square variation

$$\lim_{N \to \infty} \sum_{m=0}^{\infty} \left(M\big(t \wedge (m+1)2^{-N}, \omega\big) - M(t \wedge m2^{-N}, \omega) \right)^2$$

of $M(\cdot, \omega) \upharpoonright [0, t]$. The purpose to this exercise is to show that if $\mathbb{P}$-almost everywhere convergence is replaced by convergence in $\mathbb{P}$-probability, then the analogous result is easy to derive in general. In fact, show that, for each pair $M_1, M_2 \in \mathcal{M}_{\mathrm{loc}}(\mathbb{P}; \mathbb{R})$ and all $T \in [0, \infty)$,

$$\lim_{N \to \infty} \sup_{t \in [0,T]} \left| \sum_{m=0}^{\infty} \left(\big(M_1(t \wedge (m+1)2^{-N}) - M_1(t \wedge m2^{-N})\big) \right. \right.$$
$$\left. \times \big(M_2(t \wedge (m+1)2^{-N}) - M_2(t \wedge m2^{-N})\big) \right)$$
$$\left. - \langle M_1, M_2 \rangle(t) \right| = 0$$

in $\mathbb{P}$-probability.

Hint: First, use polarization to reduce to the case when $M_1 = M = M_2$. Next, do a little algebraic manipulation, and apply Itô's formula to see that

$$\sum_{m=0}^{\infty} \left(M\big(t \wedge (m+1)2^{-N}, \omega\big) - M(t \wedge m2^{-N}, \omega) \right)^2 - \langle M \rangle(t)$$
$$= 2 \int_0^t \big(M(\tau) - M([\tau]_N)\big) \, dM(\tau).$$

Finally, check that

$$\int_0^T \big(M(\tau) - M([\tau]_N)\big)^2 \langle M \rangle(d\tau) \longrightarrow 0$$

$\mathbb{P}$-almost surely, and use this, together with Exercise 7.1.7, to get the desired conclusion.

EXERCISE 7.2.14. The following should be comforting to those who worry about such niceties. Namely, given $M \in \mathcal{M}_{\mathrm{loc}}(\mathbb{P}; \mathbb{R})$, set $\mathcal{A}_t$ equal to the $\mathbb{P}$-completion of $\sigma(\{M(\tau) : \tau \in [0, t]\})$, and use the preceding exercise to see that $\langle M \rangle$ is progressively measurable with respect to $\{\mathcal{A}_t : t \geq 0\}$. Conclude, in particular, that no matter which filtration $\{\mathcal{F}_t : t \geq 0\}$ is the one with respect to which M was introduced, the $\langle M \rangle$ relative $\{\mathcal{F}_t : t \geq 0\}$ is the same as it is relative to $\{\mathcal{A}_t : t \geq 0\}$.

EXERCISE 7.2.15. In Exercise 5.1.27, we gave a rather clumsy, and incomplete, derivation of Burkholder's inequality. The full statement, including the extensions due to Burkholder and Gundy, is that, for each $q \in (0, \infty)$, there exist $0 < c_q < C_q < \infty$ such that, for any $M \in \mathcal{M}_{\mathrm{loc}}(\mathbb{P}; \mathbb{R})$ with $M(0) = 0$ and any stopping time ζ,

$$(7.2.16) \qquad c_q \mathbb{E}^{\mathbb{P}}\left[\langle M \rangle(\zeta)^{\frac{q}{2}}\right]^{\frac{1}{q}} \leq \mathbb{E}^{\mathbb{P}}\left[\|M\|_{[0,\zeta)}^q\right]^{\frac{1}{q}} \leq C_q \mathbb{E}^{\mathbb{P}}\left[\langle M \rangle(\zeta)^{\frac{q}{2}}\right]^{\frac{1}{q}}.$$

Here, following A. Garsia (as recorded by Getoor and Sharpe), we will outline steps which lead to a proof (7.2.16) for $q \in [2, \infty)$. As explained in Theorem 3.1 of [16], Garsia's line of reasoning can be applied to handle the general case, but trickier arguments are required.

(i) The first step is to show that it suffices to treat the case in which both M and $\langle M \rangle$ are uniformly bounded and ζ is equal to some constant T.

(ii) Let $q \in [2, \infty)$ be given, and set $C_q = \sqrt{\frac{q^{q+1}}{(q-1)^{q-1}}}$. Prove that the right hand side of (7.2.16) holds with this choice of C_q.

Hint: Begin by making the reductions in (i). Given $\epsilon > 0$, set $F_\epsilon(x) = \left(x^2 + \epsilon^2\right)^{\frac{q}{2}}$, and apply Doob's inequality plus Itô's formula to see that

$$\mathbb{E}^{\mathbb{P}}\left[\|M\|_{[0,T]}^q\right] \leq (q')^q \mathbb{E}^{\mathbb{P}}\left[F_\epsilon(M(T))\right]$$

$$= \frac{(q')^q q(q-1)}{2} \mathbb{E}^{\mathbb{P}}\left[\int_0^T F_\epsilon''(M(\tau)) \langle M \rangle(d\tau)\right]$$

$$\leq \frac{(q')^q q(q-1)}{2} \mathbb{E}^{\mathbb{P}}\left[\left(\|M\|_{[0,T]}^2 + \epsilon^2\right)^{\frac{q}{2}-1} \langle M \rangle(T)\right].$$

Now let $\epsilon \searrow 0$, and apply Hölder's inequality.

(iii) Assume that M and $\langle M \rangle$ are bounded, set $\theta(t) = \langle M \rangle(t)^{\frac{q}{4}-\frac{1}{2}}$, and take $M' = I_\theta^M$. After noting that

$$\langle M' \rangle(T) = \int_0^T \langle M \rangle^{\frac{q}{2}-1}(\tau) \langle M \rangle(d\tau) = \frac{2}{q} \langle M \rangle(T)^{\frac{q}{2}},$$

conclude that $\mathbb{E}^{\mathbb{P}}\left[\langle M \rangle(T)^{\frac{q}{2}}\right] = \frac{q}{2} \mathbb{E}^{\mathbb{P}}\left[M'(T)^2\right]$.

(iv) Continuing part **(iii)**, apply Itô's formula to see that

$$M(T)\langle M\rangle(T)^{\frac{q}{4}-\frac{1}{2}} = M'(T) + \int_0^T M(\tau)\, d\langle M\rangle(\tau)^{\frac{q}{4}-\frac{1}{2}},$$

and conclude that $\|M'\|_{[0,T]} \leq 2\|M\|_{[0,T]}\langle M\rangle(T)^{\frac{q}{4}-\frac{1}{2}}$. Now combine this with the result in **(iii)** to get the left hand side of (7.2.16) with $c_q = (2q)^{-\frac{1}{2}}$.

EXERCISE 7.2.17. Suppose that $M = (M_1,\dots,M_n) \in \mathcal{M}_{\mathrm{loc}}(\mathbb{P};\mathbb{R})^n$ and set

$$\langle\langle M\rangle\rangle(t) = \big((\langle M_i, M_j\rangle(t))\big)_{1\leq i,j\leq n}.$$

(i) Show that, $\mathbb{P}$-almost surely, $\langle\langle M\rangle\rangle(t) - \langle\langle M\rangle\rangle(s)$ is symmetric and non-negative definite for all $0 \leq s < t$. Next, set $A(t)$ equal to the trace of $\langle\langle M\rangle\rangle(t)$, and show that there exists a progressively measurable, symmetric, non-negative definite-valued function $a : [0,\infty) \times \Omega \longmapsto \mathrm{Hom}(\mathbb{R}^n;\mathbb{R}^n)$ such that

$$\langle\langle M\rangle\rangle(t) = \int_0^t a(\tau)\, A(d\tau) \quad \mathbb{P}\text{-almost surely for all } t \in [0,\infty).$$

(ii) Referring to part **(i)**, let $\Theta^2_{\mathrm{loc}}(\langle\langle M\rangle\rangle, \mathbb{P};\mathbb{R}^n)$ be the set of progressively measurable $\theta : [0,\infty) \times \Omega \longrightarrow \mathbb{R}^n$ such that

$$\int_0^T \big(\theta(t), a(t)\theta(t)\big)_{\mathbb{R}^n} A(dt) < \infty \quad \mathbb{P}\text{-almost surely for all } T \in [0,\infty).$$

Show that there is a unique linear map

$$\theta \in \Theta^2_{\mathrm{loc}}(\langle\langle M\rangle\rangle, \mathbb{P};\mathbb{R}^n) \longmapsto I_\theta^M = \int_0^\cdot \big(\theta(\tau), dM(\tau)\big)_{\mathbb{R}^n} \in \mathcal{M}_{\mathrm{loc}}(\mathbb{P};\mathbb{R})$$

such that

$$\langle I_\theta^M, M'\rangle(dt) = \sum_{j=1}^n \theta_j(t)\langle M_j, M'\rangle(dt)$$

for all $M' \in \mathcal{M}_{\mathrm{loc}}(\mathbb{P};\mathbb{R})$.

7.3 Representations of Continuous Martingales

The considerations in this chapter lead to various representations of continuous local martingales in terms of Brownian motion, and this section contains some samples of these. In one way or another, all these results lend credence to the notion that there really is only one continuous local martingale: Brownian motion.

§7.3.1. Representation via Random Time Change: This section is devoted to showing that *if $M \in \mathcal{M}_{\mathrm{loc}}(\mathbb{P}; \mathbb{R})$, then $M - M(0)$ is a Brownian motion run with clock $\langle M \rangle$.*

In the case when, $\mathbb{P}$-almost surely, $t \rightsquigarrow \langle M \rangle(t)$ is strictly increasing and $\langle M \rangle(\infty) = \infty$, this assertion is easy to verify. Namely, one takes

$$(7.3.1) \qquad \zeta(s, \omega) = \sup\{t \geq 0 : \langle M \rangle(t, \omega) \leq s\},$$

notes that, for each s, $\zeta(s)$ is a stopping time (in the sense described at the beginning of §4.1.2), and sets $\mathcal{F}'_s$ equal to the $\mathbb{P}$-completion of $\mathcal{F}_{\zeta(s)}$. Further, observe that, for $\mathbb{P}$-almost every ω, $s \rightsquigarrow \zeta(s, \omega)$ is a strictly increasing, continuous $[0, \infty)$-valued function which satisfies

$$\zeta\big(\langle M \rangle(t, \omega), \omega\big) = t \quad \text{and} \quad \langle M \rangle\big(\zeta(s, \omega), \omega\big) = s.$$

In particular, if

$$\beta(s, \omega) = \begin{cases} M\big(\zeta(s, \omega)\big) - M(0) & \text{when } \langle M \rangle(\infty, \omega) > s \\ 0 & \text{otherwise,} \end{cases}$$

then $\beta : [0, \infty) \times \Omega \longrightarrow \mathbb{R}$ is a $\mathbb{P}$-almost surely continuous, $\{\mathcal{F}'_s : s \geq 0\}$-progressively measurable function, and $M(\,\cdot\,, \omega) - M(0) = \beta\big(\langle M \rangle(\,\cdot\,, \omega), \omega\big)$ $\mathbb{P}$-almost surely. Thus, all that remains is to check that $\big(\beta(s), \mathcal{F}'_s, \mathbb{P}\big)$ is a Brownian motion. But (cf. part **(iii)** of Exercise 7.1.5) for each $s \in [0, \infty)$, $\big(M(t \wedge \zeta(s)) - M(0), \mathcal{F}_t, \mathbb{P}\big)$ is a martingale whose second moment is uniformly bounded. Hence, by Hunt's theorem (cf. Theorem 7.1.14 in [36]), for $0 \leq s_1 < s_2$ and $A \in \mathcal{F}'_{s_1}$,

$$\mathbb{E}^{\mathbb{P}}\Big[M\big(t \wedge \zeta(s_2)\big) - M\big(\zeta(s_1)\big), A \cap \{\zeta(s_1) \leq t\}\Big] = 0$$

and

$$\mathbb{E}^{\mathbb{P}}\Big[M\big(t \wedge \zeta(s_2)\big)^2 - M\big(\zeta(s_1)\big)^2, A \cap \{\zeta(s_1) \leq t\}\Big]$$
$$= \mathbb{E}^{\mathbb{P}}\Big[\big(\langle M \rangle\big(t \wedge \zeta(s_2)\big) - s_1\big), A \cap \{\zeta(s_1) \leq t\}\Big]$$

Furthermore, $\big\|M\big(t \wedge \zeta(s_2)\big) - M\big(\zeta(s_2)\big)\big\|_{L^2(\mathbb{P};\mathbb{R})} \longrightarrow 0$ and $\langle M \rangle\big(t \wedge \zeta(s_2)\big) \nearrow s_2$ $\mathbb{P}$-almost surely as $t \nearrow \infty$, which, together with the preceding, shows that $\big(\beta(s), \mathcal{F}'_s, \mathbb{P}\big)$ is a continuous local martingale and that $\langle \beta \rangle(s) = s$. Now apply Lévy's theorem (Theorem 7.2.10).

The preceding already contains the essential idea. However, there are technical difficulties which arise when $\langle M \rangle$ fails to be either strictly increasing or $\langle M \rangle(\infty) < \infty$ with positive probability. Actually, the first of these, which brings into question the continuity of $\beta(\,\cdot\,,\omega)$, causes no real problem because, by Exercise 7.1.8, for $\mathbb{P}$-almost every ω, $M(\,\cdot\,,\omega)$ is constant on each interval $[\zeta_-(s,\omega),\zeta(s,\omega))$, where $\zeta_-(s,\omega) \equiv \inf\{t \geq 0 : \langle M\rangle(t,\omega) \geq s\}$. A more serious issue is the one which arises when $\langle M\rangle(\infty) < \infty$. In this case the probability space may be just too anæmic to support an entire Brownian motion. For example, if $\Omega = \{\omega\}$ and $\mathcal{F} = \mathcal{F}_t = \{\emptyset, \Omega\}$, then there is precisely one probability measure on $(\Omega, \mathcal{F})$ and the only martingales there are constant. Thus, in general, $\langle M\rangle(\infty) < \infty$ will necessitate our supplementing the original probability space in order to obtain the desired representation in terms of a Brownian motion.

THEOREM 7.3.2. *Given a continuous, local martingale $\big(M(t),\mathcal{F}_t,\mathbb{P}\big)$ on the probability space $(\Omega,\mathcal{F},\mathbb{P})$, there exists a Brownian motion $\big(\beta(t),\hat{\mathcal{F}}_t,\hat{\mathbb{P}}\big)$ on a probability space $(\hat{\Omega},\hat{\mathcal{F}},\hat{\mathbb{P}})$ and a progressively measurable function $A : [0,\infty) \times \hat{\Omega} \longrightarrow [0,\infty)$ such that*

(1) *$A(0,\hat{\omega}) = 0$ and $t \rightsquigarrow A(t,\hat{\omega})$ is continuous and nondecreasing for each $\hat{\omega} \in \hat{\Omega}$,*
(2) *$\hat{\omega} \rightsquigarrow A(t,\hat{\omega})$ is a stopping time relative to $\{\hat{\mathcal{F}}_s : s \geq 0\}$ for each $t \geq 0$,*
(3) *the $\hat{\mathbb{P}}$-distribution of*

$$\hat{\omega} \in \hat{\Omega} \longmapsto \Big(\beta\big(A(\,\cdot\,,\hat{\omega}),\hat{\omega}\big), A(\,\cdot\,,\hat{\omega})\Big) \in C\big([0,\infty);\mathbb{R}^2\big)$$

is the same as the $\mathbb{P}$-distribution of

$$\omega \in \Omega \longmapsto \Big(M(\,\cdot\,,\omega) - M(0,\omega), \langle M\rangle(\,\cdot\,,\omega)\Big) \in C\big([0,\infty);\mathbb{R}^2\big).$$

PROOF: Without loss in generality, we will assume that $M(0) \equiv 0$.

Choose (cf. Exercise 7.1.6) a $\sigma\big(\{M(t) : t \geq 0\}\big)$-random variable $M(\infty)$ so that $M(\infty,\omega) = \lim_{t \nearrow \infty} M(t,\omega)$ for $\mathbb{P}$-almost every $\omega \in \{\langle M\rangle(\infty) < \infty\}$. Next, define $M' : [0,\infty) \times \Omega \longrightarrow \mathbb{R}$ so that

$$M'(s,\omega) = \begin{cases} M\big(\zeta(s,\omega),\omega\big) & \text{if } 0 \leq s < \langle M\rangle(\infty,\omega) \\ M(\infty,\omega) & \text{if } \langle M\rangle(\infty,\omega) \leq s < \infty. \end{cases}$$

Then, by Hunt's theorem, $\big(M'(s),\mathcal{F}'_s,\mathbb{P}\big)$ and $\big(M'(s)^2 - s \wedge \langle M\rangle(\infty),\mathcal{F}'_s,\mathbb{P}\big)$ are martingales. In addition, by the reasoning given in the discussion above, $s \rightsquigarrow M'(s)$ is $\mathbb{P}$-almost surely continuous. Similarly, because, for each (t,ω), the interval $\big[t,\zeta\big(\langle M\rangle(t,\omega),\omega\big)\big)$ is contained in the closure of a connected component of (cf. Exercise 7.1.8) $G(\omega)$, we know that $M(\,\cdot\,,\omega) = M'\big(\langle M\rangle(\,\cdot\,,\omega),\omega\big)$ for $\mathbb{P}$-almost all ω. Finally, observe that, for each $(s,t) \in$

$[0,\infty)^2$, $\{\langle M\rangle(t) \le s\} = \{\zeta(s) \ge t\} \in \mathcal{F}'_s$, and conclude that $\langle M\rangle(t)$ is a $\{\mathcal{F}'_s : s \ge 0\}$-stopping time for each $t \ge 0$. Hence, when $\langle M\rangle(\infty) = \infty$ $\mathbb{P}$-almost surely, we can take $\hat{\Omega} = \Omega$, $\hat{\mathcal{F}} = \mathcal{F}$, $\hat{\mathbb{P}} = \mathbb{P}$, $\hat{\mathcal{F}}_s = \mathcal{F}'_s$, $\beta(\,\cdot\,,\omega) = M'(\zeta(s,\omega),\omega)$, and $A(t,\omega) = \langle M\rangle(t,\omega)$.

To handle the case when $\langle M\rangle(\infty) < \infty$ with positive probability, we take $\hat{\Omega} = \Omega \times C([0,\infty);\mathbb{R})$, $\hat{\mathcal{F}}$ equal the $\mathbb{P}\times\mathbb{P}^0$-completion of $\mathcal{F}\times\mathcal{B}$, $\hat{\mathbb{P}} = \mathbb{P}\times\mathbb{P}^0$, and $\hat{\mathcal{F}}_s$ equal to the $\hat{\mathbb{P}}$-completion of $\mathcal{F}'_s\times\mathcal{B}_s$. Then, if $B(s,\hat{\omega}) = p(s)$ and $\hat{M}(s,\hat{\omega}) = M'(s,\omega)$ for $s \ge 0$ and $\hat{\omega} = (\omega,p)$, $(B(s),\hat{\mathcal{F}}_s,\hat{\mathbb{P}})$ is a Brownian motion, $(\hat{M}(s),\hat{\mathcal{F}}_s,\hat{\mathbb{P}})$ is a continuous, local martingale, $\langle\hat{M}\rangle(s,\hat{\omega}) = s \wedge \langle M\rangle(\infty,\omega)$, and, by Exercise 7.2.11, $\langle B,\hat{M}\rangle \equiv 0$. Finally, for $\hat{\omega} = (\omega,p)$, we take $A(t,\hat{\omega}) = \langle M\rangle(t,\omega)$ for $t \ge 0$ and

$$\beta(s,\hat{\omega}) = \begin{cases} \hat{M}(s,\hat{\omega}) & \text{when } 0 \le s < \langle M\rangle(\infty,\omega) \\ B(s,\hat{\omega}) - B(\langle M\rangle(\infty,\omega),\hat{\omega}) & \text{when } \langle M\rangle(\infty,\omega) \le s < \infty. \end{cases}$$

By the reasoning in the first paragraph, we know that $A(t)$ is an $\{\hat{\mathcal{F}}_s : s \ge 0\}$-stopping time for each t. Furthermore, because an equivalent description of $\beta(s,\hat{\omega})$ is to say

$$\beta(s,\hat{\omega}) = \hat{M}(s,\hat{\omega}) + \int_0^s \mathbf{1}_{[A(\infty),\infty)}(\sigma)\,dB(\sigma),$$

we see that $(\beta(s),\hat{\mathcal{F}}_s,\hat{\mathbb{P}})$ is a continuous, local martingale with

$$\langle\beta\rangle(s) = \langle\hat{M}\rangle(s) + 2\int_0^s \mathbf{1}_{[A(\infty),\infty)}(\sigma)\,\langle\hat{M},B\rangle(d\sigma)$$

$$+ \int_0^s \mathbf{1}_{[A(\infty),\infty)}(\sigma)\,\langle B\rangle(d\sigma) = s \wedge A(\infty) + \big(s - s \wedge A(\infty)\big).$$

Hence, $(\beta(s),\hat{\mathcal{F}}_s,\hat{\mathbb{P}})$ is a Brownian motion. Finally, by the result in the first paragraph,

$$\beta\big(A(\,\cdot\,,\hat{\omega}),\hat{\omega}\big) = M'\big(\langle M\rangle(\,\cdot\,,\omega),\omega\big) = M(\,\cdot\,,\omega)$$

for $\hat{\mathbb{P}}$-almost every $\hat{\omega} = (\omega,p)$, and so we have completed the proof. $\square$

One of the most important implications of Theorem 7.3.2 is the content of the following corollary.

COROLLARY 7.3.3. If $M \in \mathcal{M}_{\mathrm{loc}}(\mathbb{P};\mathbb{R})$, then[4]

$$\varlimsup_{t\to\infty} \frac{M(t)}{\sqrt{2\langle M\rangle(t)\log_{(2)}\langle M\rangle(t)}} = 1 = -\varliminf_{t\to\infty} \frac{M(t)}{\sqrt{2\langle M\rangle(t)\log_{(2)}\langle M\rangle(t)}}$$

$\mathbb{P}$-almost surely on the set $\{\langle M\rangle(\infty) = \infty\}$. In particular, the set $\{\langle M\rangle(\infty) < \infty\}$ is $\mathbb{P}$-almost surely equal to the set of ω such that $\lim_{t\to\infty} M(t,\omega)$ exists in $\mathbb{R}$.

---

[4] Here we use $\log_{(2)}\tau$ to denote $\log(\log\tau)$ for $\tau > e$.

Proof: Using the notation in Theorem 7.3.2, what we have to do to prove the first assertion is to show that

$$\varlimsup_{t\to\infty} \frac{\beta\big(A(t)\big)}{\sqrt{2A(t)\log_{(2)} A(t)}} = 1 = -\varliminf_{t\to\infty} \frac{\beta\big(A(t)\big)}{\sqrt{2A(t)\log_{(2)} A(t)}}$$

$\hat{\mathbb{P}}$-almost surely on the set $\{A(\infty) = \infty\}$. But, by the law of the iterated logarithm for Brownian motion (cf. Theorem 4.1.6 in [36]), this is obvious.

Given the first assertion, the second assertion follows immediately from either Exercise 7.1.8 or by another application of the representation given by Theorem 7.3.2. □

§7.3.2. Representation via Stochastic Integration: Except for special cases (cf. F. Knight's theorem in Chapter V of [27]), representation via random time change does not work when dealing with more than one $M \in \mathcal{M}_{\mathrm{loc}}(\mathbb{P};\mathbb{R})$ at a time. By contrast, Brownian stochastic integral representations have no dimension restriction, although they do require that the $\langle M\rangle$'s be absolutely conditions. To make all of this precise, we will prove the following statement.

THEOREM 7.3.4. *Suppose that* $M = \big(M_1,\ldots,M_n\big) \in \big(\mathcal{M}_{\mathrm{loc}}(\mathbb{P};\mathbb{R})\big)^n$ *and that, for each* $1 \le i \le n$, $t \rightsquigarrow \langle M_i\rangle(t)$ *is* $\mathbb{P}$*-almost surely absolutely continuous. Then there exists progressively measurable map* $\alpha : [0,\infty) \times \Omega \longrightarrow \mathrm{Hom}(\mathbb{R}^n;\mathbb{R}^n)$ *with the properties that:* $\alpha(t,\omega)$ *is symmetric and non-negative definite for each* (t,ω), *and*

$$\langle M_i, M_j\rangle(t) = \int_0^t a_{ij}(\tau)\,d\tau, \quad t \ge 0, \quad \text{where } a(\tau) \equiv \alpha(\tau)^2.$$

Furthermore, there exist an $\mathbb{R}^n$*-valued Brownian motion* $\big(\beta(t),\hat{\mathcal{F}}_t,\hat{\mathbb{P}}\big)$ *on some probability space* $(\hat{\Omega},\hat{\mathcal{F}},\hat{\mathbb{P}})$ *and an* $\hat{\alpha} \in \Theta^2_{\mathrm{loc}}\big(\hat{\mathbb{P}}; \mathrm{Hom}(\mathbb{R}^n;\mathbb{R}^n)\big)$ *such that the* $\hat{\mathbb{P}}$*-distribution of*

$$\hat{\omega} \rightsquigarrow \big(\hat{\alpha}(\,\cdot\,,\hat{\omega}), I_{\hat{\alpha}}(\,\cdot\,,\hat{\omega})\big)$$

is the same as the $\mathbb{P}$*-distribution of*

$$\omega \rightsquigarrow \big(\alpha(\,\cdot\,,\omega), M(\,\cdot\,,\omega) - M(0,\omega)\big)$$

when

$$I_{\hat{\alpha}}(t) \equiv \int_0^t \hat{\alpha}(\tau)\,d\beta(\tau).$$

PROOF: The first step is to notice that, by Theorem 7.2.1, $t \rightsquigarrow \langle M_i, M_j \rangle(t)$ is $\mathbb{P}$-almost surely absolutely continuous for all $1 \leq i, j \leq n$. Hence, we can find (cf. Theorem 5.2.26 in [36]) a progressively measurable $a : [0, \infty) \times \Omega \longrightarrow \mathrm{Hom}(\mathbb{R}^n; \mathbb{R}^n)$ such that $\langle M_i, M_j \rangle(dt) = a_{ij}(t)\, dt$. Obviously, there is no reason to not take $a(t, \omega)$ to be symmetric. In addition, because

$$\sum_{i,j=1}^{n} \xi_i \xi_j \langle M_i, M_j \rangle(dt) = \left\langle \sum_{1}^{n} \xi_i M_i \right\rangle(dt) \geq 0,$$

$a(t, \omega)$ is non-negative definite for $\lambda_{[0,\infty)} \times \mathbb{P}$-almost every $(t, \omega) \in [0, \infty)$. Hence, without loss in generality, we will take $a(t, \omega)$ to be symmetric and non-negative definite for all (t, ω).

The next step is to take $\alpha(t, \omega)$ to be the non-negative definite, symmetric square root of $a(t, \omega)$. To see that α is progressively measurable, we need only apply Lemma 3.2.1 to see that the non-negative, symmetric square root $\alpha^\epsilon(t, \omega)$ of $a(t, \omega) + \epsilon I$ is progressively measurable for all $\epsilon > 0$ and then use $\alpha(t, \omega) = \lim_{\epsilon \searrow 0} \alpha^\epsilon(t, \omega)$.

Obviously, $\alpha(t, \omega)$ will not, in general, be invertible. Thus, we take $\pi(t, \omega)$ to denote orthogonal projection onto the null space $N(t, \omega)$ of $a(t, \omega)$ and $\alpha^{-1}(t, \omega) : \mathbb{R}^n \longrightarrow N(t, \omega)^\perp$ to be the symmetric, linear map for which $N(t, \omega)$ is the null space and $\alpha^{-1}(t, \omega) \restriction N(t, \omega)^\perp$ is the inverse of $\alpha(t, \omega) \restriction N(t, \omega)^\perp$. Again, both these maps are progressively measurable:

$$\pi(t, \omega) = \lim_{\epsilon \searrow 0} a(t, \omega)\big(a(t, \omega) + \epsilon I\big)^{-1} \ \& \ \alpha^{-1}(t, \omega) = \lim_{\epsilon \searrow 0} \alpha(t, \omega)\big(a(t, \omega) + \epsilon I\big)^{-1}.$$

Because (cf. Exercise 7.2.17) $\alpha^{-1}(t, \omega)\alpha(t, \omega) = \pi(t, \omega)^\perp$, and therefore

$$\max_{1 \leq i \leq n} \sum_{j,j'=1}^{n} \int_0^T (\alpha^{-1})_{ij}(\tau)(\alpha^{-1})_{ij'}(\tau) \langle M_j, M_{j'} \rangle(d\tau) \leq T,$$

we can take $B(t) = \int_0^t \alpha^{-1}(\tau)\, dM(\tau)$, in which case $\big(B(t), \mathcal{F}_t, \mathbb{P}\big)$ is an $\mathbb{R}^n$-valued, continuous local martingale and, if $B_\xi(t) \equiv \big(\xi, B(t)\big)_{\mathbb{R}^n}$, then

$$\langle B_\xi, B_\eta \rangle(t) = \int_0^t \big(\eta, \pi(\tau)^\perp \xi\big)_{\mathbb{R}^n} d\tau.$$

Furthermore, if $X(t) = \int_0^t \alpha(\tau)\, dB(\tau)$, then, by Exercise 7.2.12,

$$X_\xi(t) \equiv \big(\xi, X(t)\big)_{\mathbb{R}^n} = \int_0^t \big(\alpha(\tau, \omega)\xi, dB(\tau)\big)_{\mathbb{R}^n} = \int_0^t \big(\pi(\tau)^\perp \xi, dM(\tau)\big)_{\mathbb{R}^n},$$

and so, if $M_\xi(t) \equiv \big(\xi, M(t)\big)_{\mathbb{R}^n}$, then

$$\langle X_\xi - M_\xi \rangle(t) = \int_0^t \big(\xi, \pi(\tau)a(\tau)\pi(\tau)\xi\big)_{\mathbb{R}^n} \, d\tau = 0.$$

Clearly, in the case when $a(t,\omega) > 0$ for $\lambda_{[0,\infty)} \times \mathbb{P}$-almost every (t,ω), we are done. Indeed, in this case $(B(t), \mathcal{F}_t, \mathbb{P})$ is an $\mathbb{R}^n$-valued Brownian motion, and so we can take $(\hat{\Omega}, \hat{\mathcal{F}}, \hat{\mathbb{P}}) = (\Omega, \mathcal{F}, \mathbb{P})$, $\hat{\mathcal{F}}_t = \mathcal{F}_t$, $\hat{\alpha} = \alpha$, and $\beta = B$. To handle the general case, take $\hat{\Omega} = \Omega \times C\big([0,\infty); \mathbb{R}^n\big)$, $\hat{\mathcal{F}}$ and $\hat{\mathcal{F}}_t$ to be the $\mathbb{P} \times \mathbb{P}^0$-completions of $\mathcal{F} \times \mathcal{B}$ of $\mathcal{F}_t \times \mathcal{B}_t$, $\hat{\mathbb{P}} = \mathbb{P} \times \mathbb{P}^0$, $\hat{\alpha}(t,\hat{\omega}) = \alpha(t,\omega)$, and

$$\beta(t,\hat{\omega}) = B(t,\omega) + \int_0^t \pi(\tau,\omega) \, dp(\tau)$$

for $t \geq 0$ and $\hat{\omega} = (\omega, p)$. Because $\langle B_\xi \rangle(dt) = \pi\big|(t)^\perp \xi\big|^2 \, dt$ and (cf. Exercise 7.2.11) $\langle B_\xi, p_\xi \rangle(dt) = 0 \, dt$ when $p_\xi \equiv \big(\xi, p(\,\cdot\,)\big)_{\mathbb{R}^n}$, it is easy to check that these choices work. $\square$

By combining the ideas in this section with those in the preceding, we arrive at the following structure theorem, which, in a somewhat different form, was anticipated by A.V. Skorohod in [31].

COROLLARY 7.3.5. *Let* $M = (M_1, \ldots, M_n) \in \big(\mathcal{M}_{\mathrm{loc}}(\mathbb{P}; \mathbb{R})\big)^n$ *be given, and set* $A(t) = \sum_1^n \langle M_i \rangle(t)$. *Then there is a probability space* $(\hat{\Omega}, \hat{\mathcal{F}}, \hat{\mathbb{P}})$ *on which there exists a Brownian motion* $(\beta(t), \hat{\mathcal{F}}_t, \hat{\mathbb{P}})$ *and* $\{\hat{\mathcal{F}}_t : t \geq 0\}$-*progressively measurable maps* $\hat{\alpha} : [0,\infty) \times \hat{\Omega} \longrightarrow \mathrm{Hom}(\mathbb{R}^n; \mathbb{R}^n)$ *and* $\hat{A} : [0,\infty) \times \hat{\Omega} \longrightarrow [0,\infty)$ *such that:* $\hat{\alpha}(t,\hat{\omega})$ *is symmetric,* $0I_{\mathbb{R}^n} \leq \hat{\alpha}(t,\hat{\omega}) \leq I_{R^n}$ *for all* $(t,\hat{\omega})$, $\hat{A}(t)$ *is stopping time for each* $t \geq 0$, *and, the* $\hat{\mathbb{P}}$-*distribution of*

$$\hat{\omega} \rightsquigarrow \left(\hat{A}(t), \int_0^{\hat{A}(t)} \hat{\alpha}(\tau,\hat{\omega}) \, d\beta(\tau,\hat{\omega}) \right),$$

is the same as the $\mathbb{P}$-*distribution of* $\omega \rightsquigarrow \big(A(\,\cdot\,,\omega), M(\,\cdot\,,\omega) - M(0,\omega)\big)$.

There are essentially no new ideas here. Namely, define $\zeta(s,\omega) = \inf\{t \geq 0 : A(t,\omega) \geq s\}$. By the techniques used in the preceding section, we can define $M' : [0,\infty) \times \Omega \longrightarrow \mathbb{R}^n$ so that $M'(\,\cdot\,,\omega) = M\big(\zeta(\,\cdot\,,\omega),\omega\big)$ $\mathbb{P}$-almost surely and can show that $\big(M'(s), \mathcal{F}'_s, \mathbb{P}\big)$ is an $\mathbb{R}^n$-valued continuous martingale when $\mathcal{F}'_s$ is the $\mathbb{P}$-completion of $\mathcal{F}_{\zeta(s)}$ and that $\sum_1^n \langle M'_i \rangle(dt,\omega) \leq dt$ $\mathbb{P}$-almost surely. In addition, those same techniques show that, for each $t \geq 0$, $A(t)$ is an $\{\mathcal{F}'_s : s \geq 0\}$-stopping time and $M(t,\omega) = M'\big(A(t,\omega),\omega\big)$ $\mathbb{P}$-almost surely. Hence, all that remains is to apply Theorem 7.3.4 to $\big(M'(s), \mathcal{F}'_s, \mathbb{P}\big)$.

A more practical reason for wanting Theorem 7.3.4 is that it enables us to prove the following sort of uniqueness theorem.

COROLLARY 7.3.6. *Let $\sigma : [0, \infty) \times \mathbb{R}^n \longrightarrow \mathrm{Hom}(\mathbb{R}^n; \mathbb{R}^n)$ and $b : [0, \infty) \times \mathbb{R}^n \longrightarrow \mathbb{R}^n$ be measurable functions with the properties that $\sigma(t, x)$ is symmetric and non-negative definite for each (t, x), $t \rightsquigarrow \|\sigma(t, 0)\|_{\mathrm{H.S.}} \vee |b(t, 0)|$ is locally bounded, and*

$$\sup_{\substack{t \in [0,T] \\ x_2 \neq x_1}} \frac{\|\sigma(t, x_2) - \sigma(t, x_1)\|_{\mathrm{H.S.}} \vee |b(t, x_2) - b(t, x_1)|}{|x_2 - x_1|} < \infty$$

for each $T > 0$. Set $a(t, x) = \sigma^2(t, x)$, and define the time dependent operator $t \rightsquigarrow L_t$ on $C^2(\mathbb{R}^n; \mathbb{R})$ so that

$$L_t \varphi(x) = \frac{1}{2} \sum_{i,j=1}^n a_{ij}(t, x) \partial_i \partial_j \varphi(x) + \sum_{i=1}^n b_i(t, x) \partial_i \varphi(x).$$

Then, for each $(s, x) \in [0, \infty) \times \mathbb{R}^n$, there is precisely one solution $\mathbb{P}^L_{s,x}$ to the martingale problem for $t \rightsquigarrow L_t$ on $C^\infty_c(\mathbb{R}^n; \mathbb{R})$ starting from x at time s. In fact, $\mathbb{P}^L_{s,x}$ is the $\mathbb{P}^0$-distribution of $p \rightsquigarrow X(\,\cdot\,, (s, x), p)$, where (cf. Theorem 5.2.2) $X(\,\cdot\,, (s, x))$ is the $\mathbb{P}^0$-almost surely unique, progressively measurable solution to

$$X(t, (s, x), p) = x + \int_0^t \sigma(s + \tau, X(\tau, (s, x), p)) \, dp(\tau)$$
$$+ \int_0^t b(s + \tau, X(\tau, (s, x), p)) \, d\tau.$$

PROOF: Without loss in generality, we will assume that $s = 0$.

All that we have to do is show that if $\mathbb{P}$ solves the martingale problem for $t \rightsquigarrow L_t$ starting from x at time 0, then $\mathbb{P}$ is the $\mathbb{P}^0$-distribution of $p \rightsquigarrow X(\,\cdot\,, x, p) \equiv X(\,\cdot\,, (s, x), p)$. Thus, suppose that $\mathbb{P}$ is a solution.

Set

$$M(t, p) \equiv p(t) - x - \int_0^t b(\tau, p(\tau)) \, d\tau,$$

and observe that $(M(t), \overline{\mathcal{B}}_t, \mathbb{P})$ is a continuous, local $\mathbb{R}^n$-valued martingale for which $\langle M \rangle(t, p) = \int_0^t a(\tau, p(\tau)) \, d\tau$, in the sense that $\langle M_\xi \rangle(t, p) = \int_0^t (\xi, a(\tau, p(\tau))\xi) \, d\tau$ when $M_\xi(t) \equiv (\xi, M(t))_{\mathbb{R}^n}$.

Next, determine the map $\Psi : [0, \infty) \times C([0, \infty); \mathbb{R}^n) \longrightarrow \mathbb{R}^n$ so that

$$\Psi(t, p) = x + p(t) + \int_0^t b(\tau, \Psi(\tau, p)) \, d\tau, \quad t \geq 0.$$

Clearly Ψ is a progressively measurable. Furthermore, $p(t) = \Psi(t, M(\,\cdot\,,p))$ and $(t,p) \rightsquigarrow \alpha(t,p) \equiv \sigma(t, \Psi(t, M(\,\cdot\,,p)))$ is a progressively measurable, symmetric, non-negative definite $\text{Hom}(\mathbb{R}^n; \mathbb{R}^n)$-valued map for which $\langle M_i, M_j \rangle(dt) = (\alpha^2)_{ij}(t)\,dt$. Hence, by Theorem 7.3.4, we can find a probability space $(\hat{\Omega}, \hat{\mathcal{F}}, \hat{\mathbb{P}})$ on which there exists an $\mathbb{R}^n$-valued Brownian motion $(\beta(t), \hat{\mathcal{F}}_t, \hat{\mathbb{P}})$ and a symmetric, non-negative definite valued progressively measurable function $\hat{\alpha} : [0, \infty) \times \hat{\Omega} \longrightarrow \text{Hom}(\mathbb{R}^n; \mathbb{R}^n)$ such that $\hat{\omega} \rightsquigarrow (\hat{\alpha}(\,\cdot\,, \hat{\omega}), I_{\hat{\alpha}}(\,\cdot\,, \hat{\omega}))$ has the same distribution under $\hat{\mathbb{P}}$ as $p \rightsquigarrow (\alpha(\,\cdot\,,p), M(\,\cdot\,,p))$ has under $\mathbb{P}$. In particular, if $X(t, \hat{\omega}) \equiv \Psi(t, I_{\hat{\alpha}}(\,\cdot\,, \hat{\omega}))$, then $\mathbb{P}$ is the $\hat{\mathbb{P}}$-distribution of $\hat{\omega} \rightsquigarrow X(\,\cdot\,, \hat{\omega})$ and, for each $t \geq 0$, $\hat{\alpha}(t) = \sigma(t, X(t))$ $\hat{\mathbb{P}}$-almost surely. Finally, the second of these tells us that

$$X(t) = x + I_{\hat{\alpha}}(t) + \int_0^t b(\tau, X(\tau))\,d\tau$$

$$= x + \int_0^t \sigma(\tau, X(\tau))\,d\beta(\tau) + \int_0^t b(\tau, X(\tau))\,d\tau,$$

and, as we showed in Theorem 5.2.2, the solution to this equation can be written as the limit of the sequence $\{X_N\}_0^\infty$, where $X_0 \equiv x$ and

$$X_{N+1}(t) = x + \int_0^t \sigma(\tau, X_N(\tau))\,d\beta(\tau) + \int_0^t b(\tau, X_N(\tau))\,d\tau.$$

Since the $\hat{\mathbb{P}}$-distribution of each X_N is uniquely determined by the distribution of β, the proof is complete. $\square$

§7.3.3. Skorohod's Representation Theorem: Our final example of a representation is a particularly clever one due to A.V. Skorohod. Namely, Skorohod proved that, *for any centered, square-integrable, $\mathbb{R}$-valued random variable X with mean value 0, there exists an $\mathbb{R}$-valued Brownian motion $(\beta(t), \mathcal{F}_t, \mathbb{P})$ and a finite stopping time ζ such that the $\mathbb{P}$-distribution of $\omega \rightsquigarrow \beta(\zeta(\omega), \omega)$ is equal to the distribution of X.*

   Skorohod's own treatment (cf. Chapter 7 in [29] and Theorem 12.4.2 in [7])) is more beautiful and direct than the one presented here. On the other hand, given the contents of the preceding subsections, our approach is more elementary. Indeed, we will use Itô's formula, in much the same way as we did in Exercise 6.3.19, to prove that there is a continuous function $u'_X : [0, 1) \times \mathbb{R} \longrightarrow \mathbb{R}$ such that the $\mathbb{P}^0$-distribution of

$$p \in C([0, \infty); \mathbb{R}) \longmapsto \int_0^1 u'_X(\tau, p(\tau))\,dp(\tau)$$

equals the distribution of X. We will then apply Theorem 7.3.2 to find a Brownian motion $(\beta(s), \mathcal{F}_s, \mathbb{P})$ and a finite $\{\mathcal{F}_s : s \geq 0\}$-stopping time

214   7 THE KUNITA–WATANABE EXTENSION

ζ such that the $\mathbb{P}$-distribution of $\omega \rightsquigarrow (\zeta(\omega), \beta(\zeta(\omega), \omega))$ is equal to the $\mathbb{P}^0$-distribution of

$$p \rightsquigarrow \left(\int_0^1 u_X'(\tau, p(\tau))^2 \, d\tau, \int_0^1 u_X'(\tau, p(\tau)) \, dp(\tau) \right).$$

LEMMA 7.3.7. *If X is an $\mathbb{R}$-valued random variable with distribution function F_X and if*

$$\psi_X(x) \equiv \inf\{t \in \mathbb{R} : F_X(t) \geq \Gamma_1(-\infty, x])\},$$

where Γ_1 is the centered Gaussian measure with variance 1, then ψ_X is a nondecreasing, left-continuous map from $\mathbb{R}$ into itself, and the $\mathbb{P}^0$-distribution of $p \rightsquigarrow \psi_X(p(1))$ is that of X. Next, assume that X is square-integrable. Then $\|\psi_X\|_{L^2(\Gamma_1;\mathbb{R})} = \sqrt{\mathbb{E}[X^2]} < \infty$. Furthermore, if $\gamma_t(y) \equiv (2\pi t)^{-\frac{1}{2}} \exp(-\frac{y^2}{2t})$ is the density for the centered Gaussian measure Γ_t with variance t, then

$$\int_{|y-x|\geq \epsilon} |\psi_X(y)|\gamma_t(y-x) \, dy \leq \|\psi_X\|_{L^2(\Gamma_1;\mathbb{R})} \left(\tfrac{2}{t}\right)^{\frac{1}{3}} e^{x^2 - \frac{\epsilon^2}{2t}}$$

for all $\epsilon \geq 0$ and $(t, x) \in (0, 1] \times \mathbb{R}^n$. In particular,

$$u_X(t, x) \equiv \int_{\mathbb{R}} \psi_X(y)\gamma_{1-t}(y-x) \, dy \quad \text{for } (t, x) \in [0, 1) \times \mathbb{R}$$

is well defined. In fact, $u_X \in C([0, 1) \times \mathbb{R}; \mathbb{R}) \cap C^\infty((0, 1) \times \mathbb{R}; \mathbb{R})$, u_X solves the backward heat equation $(\partial_t + \frac{1}{2}\partial_x^2)u_X = 0$ in $(0, 1) \times \mathbb{R}$, and

$$u_X(t, p(t)) \longrightarrow \psi_X(p(1)) \quad \text{in } L^2(\mathbb{P}^0; \mathbb{R}) \text{ as } t \nearrow 1.$$

PROOF: The initial assertions about ψ_X are clear. Furthermore,

$$\mathbb{P}^0\left(\psi_X(p(1)) \leq x\right) = \mathbb{P}^0\left(\Gamma_1((-\infty, p(1)]) \leq F_X(x)\right) = F_X(x),$$

since Γ_1 is the $\mathbb{P}^0$-distribution of $p \rightsquigarrow p(1)$.

Next, assume that X is square integrable. Then, by the preceding, we know that $\|\psi_X\|_{L^2(\Gamma_1;\mathbb{R})}^2 = \mathbb{E}[X^2] < \infty$, and, for any $t \in (0, 1]$ and $k \geq 0$,

$$\int_{\mathbb{R}} |y - x|^k |\psi_X(y)|\gamma_t(y-x) \, dy \leq t^{-\frac{1}{2}} e^{-\frac{x^2}{2}} \int_{\mathbb{R}} |y - x|^k |\psi_X(y)| e^{xy} \gamma_1(y) \, dy$$

$$\leq t^{-\frac{1}{2}} e^{-\frac{x^2}{2}} \|\psi_X\|_{L^2(\Gamma_1;\mathbb{R})} \left(\int_{\mathbb{R}} |y - x|^{2k} e^{2xy} \gamma_1(y) \, dy \right)^{\frac{1}{2}}.$$

When $k = 0$, this proves that u_X is well defined as a continuous function in $[0,1) \times \mathbb{R}$. In addition, by using it for $0 \leq k \leq 2\ell$, it is easy to justify differentiation under the integral defining u_X and thereby prove that u_X on $(0,1) \times \mathbb{R}$ is 2ℓ times differentiable with respect to x, ℓ times differentiable with respect to t, and satisfies the backward heat equation there.

Turning to the asserted estimate, first observe that

$$\int_{|y-x| \geq \epsilon} |\psi_X(y)| \gamma_t(y-x)\, dy \leq \left(\int_{\mathbb{R}} |\psi_X(y)|^{\frac{3}{2}} \gamma_t(y-x)\, dy \right)^{\frac{2}{3}} \Gamma_t\big([-\epsilon, \epsilon] \complement \big)^{\frac{1}{3}}.$$

Proceeding as above, one sees that

$$\left(\int_{\mathbb{R}} |\psi_X(y)|^{\frac{3}{2}} \gamma_t(y-x)\, dy \right)^{\frac{2}{3}} \leq t^{-\frac{1}{3}} e^{x^2} \|\psi\|_{L^2(\Gamma_1;\mathbb{R})}.$$

At the same time, $\Gamma_t\big([-\epsilon, \epsilon]\complement \big) \leq 2\exp\big(-\frac{\epsilon^2}{2t}\big)$, which, when combined with the preceding, gives the asserted estimate. In particular, for $\epsilon > 0$ and $R \in [0, \infty)$,

$$\lim_{t \searrow 0} \sup_{|x| \leq R} \int_{|y-x| \geq \epsilon} |\psi_X(y)| \gamma_t(y-x)\, dy = 0,$$

from which it is an easy step to see that $\lim_{t \nearrow 1} \lim_{y \to x} u_X(t,y) = \psi_X(x)$ for each $x \in \mathbb{R}$ at which ψ_X is continuous. Thus, because ψ_X has at most countably many points of discontinuity, this means that $\lim_{t \nearrow 1} u\big(t, p(t)\big) = \psi_X\big(p(1)\big)$ $\mathbb{P}^0$-almost surely.

Since ψ_X is locally bounded, our estimates tell us that $\sup_{t \in [0,1)} |u_X(t, \cdot)|$ is also locally bounded. Hence, because we already know that the convergence takes place $\mathbb{P}^0$-almost everywhere, we will know that $u_X\big(t, p(t)\big) \longrightarrow \psi_X\big(p(1)\big)$ in $L^2(\mathbb{P}^0; \mathbb{R})$ as $t \nearrow 1$ once we show that

$$\lim_{R \to \infty} \sup_{t \in [0,1)} \mathbb{E}^{\mathbb{P}^0}\left[u_\varphi\big(t, p(t)\big)^2,\, |p(t)| \geq R \right] = 0.$$

But

$$\mathbb{E}^{\mathbb{P}^0}\left[u_\varphi\big(t, p(t)\big)^2,\, |p(t)| \geq R \right] = \int_{|x| \geq R} \left(\int_{\mathbb{R}} \psi_X(y) \gamma_{1-t}(y-x)\, dy \right)^2 \Gamma_t(dx)$$

$$\leq \int_{\mathbb{R}} \psi_X(y)^2 f_R(t,y)\, \Gamma_1(dy),$$

where

$$f_R(t,y) \equiv \frac{e^{\frac{y^2}{2}}}{\sqrt{2\pi t(1-t)}} \int_{|x| \geq R} e^{-\frac{(y-x)^2}{2(1-t)}} e^{-\frac{x^2}{2t}}\, dx = \int_{S(t,R)} \gamma_1(\xi)\, d\xi$$

$$\text{and } S(t,R) \equiv \left\{ \xi : \left| \xi + \sqrt{\tfrac{t}{1-t}}\, y \right| \geq \frac{R}{\sqrt{t(1-t)}} \right\}.$$

Since $0 \le f_R(t,y) \le 1$ and, for each y, $\sup_{t \in [0,1)} f_R(t,y) \longrightarrow 0$ as $R \nearrow \infty$, we are done. $\square$

Now suppose that X is a centered, square-integrable random variable. Then,

$$u_X(0,0) = \int_{\mathbb{R}} \psi_X(y)\, \Gamma_1(dy) = \mathbb{E}^{\mathbb{P}^0}\left[\psi_X(p(1))\right] = \mathbb{E}[X] = 0,$$

and so, because u_X satisfies the backward heat equation in $(0,1) \times \mathbb{R}$, Itô's formula says that

$$u_X\big(t,p(t)\big) = \int_0^t u'_X\big(\tau, p(\tau)\big)\, dp(\tau) \quad \mathbb{P}^0\text{-almost surely for } t \in [0,1).$$

Hence, since $u_X\big(t,p(t)\big) \longrightarrow \psi_X\big(p(1)\big)$ in $L^2\big(\mathbb{P}^0;\mathbb{R}\big)$ as $t \nearrow 1$, we conclude that

$$\psi_X\big(p(1)\big) = \int_0^1 u'_X\big(\tau, p(\tau)\big)\, dp(\tau) \quad \mathbb{P}^0\text{-almost surely.}$$

In particular, this means that if

$$M_X(t,p) \equiv \int_0^{t \wedge 1} u'_X\big(\tau, p(\tau)\big)\, dp(\tau),$$

then $\big(M_X(t), \overline{\mathcal{B}}_t, \mathbb{P}^0\big)$ is a square-integrable martingale for which

$$\mathbb{E}^{\mathbb{P}^0}\big[\langle M_X\rangle(1)\big] = \mathbb{E}^{\mathbb{P}^0}\big[M_X(1)^2\big] = \mathbb{E}[X^2].$$

In conjunction with Theorem 7.3.2, the preceding already leads to Skorohod's representation of X. However, for applications, it is better to carry this line of reasoning another step before formulating it as a theorem. Namely, set

$$\theta_X(t,p) = \begin{cases} u'_X\big(t - [t], p(t) - p([t])\big) & \text{for } t \in [0,\infty) \setminus \mathbb{N} \\ 0 & \text{for } t \in \mathbb{N}. \end{cases}$$

Then, because $p \rightsquigarrow p(\,\cdot\, + s) - p(s)$ is $\mathbb{P}^0$-independent of $\overline{\mathcal{B}}_s$ and again has the same $\mathbb{P}^0$-distribution as p itself, we see that the $\mathbb{P}^0$-distribution of

$$\left\{ \int_0^m \theta_X(\tau,p)\, dp(\tau) : m \in \mathbb{Z}^+ \right\}$$

is the same as the distribution of the partial sums of independent copies of X. Thus, we have now proved the following form of Skorohod's representation theorem.

THEOREM 7.3.8. *Let X be a centered, square-integrable, $\mathbb{R}$-valued random variable. Then there exists an $\mathbb{R}$-valued Brownian motion $(\beta(t), \mathcal{F}_t, \mathbb{P})$ and a nondecreasing sequence $\{\zeta_m\}_0^\infty$ of finite stopping times such that:*

(1) *$\zeta_0 \equiv 0$, the random variables $\{\zeta_m - \zeta_{m-1} : m \geq 1\}$ are mutually $\mathbb{P}$-independent and identically distributed, and $\mathbb{E}^{\mathbb{P}}[\zeta_1] = \mathbb{E}[X^2]$.*

(2) *The random variables $\{\beta(\zeta_m) - \beta(\zeta_{m-1}) : m \geq 1\}$ are mutually $\mathbb{P}$-independent and the $\mathbb{P}$-distribution of each equals the distribution of X.*

In fact, for each $q \in [2, \infty)$, (cf. (7.2.16))

$$c_q \mathbb{E}^{\mathbb{P}}\big[\zeta_1^{\frac{q}{2}}\big]^{\frac{1}{q}} \leq \mathbb{E}\big[|X|^q\big]^{\frac{1}{q}} \leq C_q \mathbb{E}^{\mathbb{P}}\big[\zeta_1^{\frac{q}{2}}\big]^{\frac{1}{q}}.$$

PROOF: There is essentially nothing left to do. Indeed, by Theorem 7.3.2, we know that there is a Brownian motion $(\beta(s), \mathcal{F}_s, \mathbb{P})$ on some probability space $(\Omega, \mathcal{F}, \mathbb{P})$ and a map $A : [0, \infty) \times \Omega \longrightarrow [0, \infty)$ such that $A(t)$ is an $\{\mathcal{F}_s : s \geq 0\}$-stopping time for each $t \geq 0$, the $\mathbb{P}$-distribution of

$$\omega \rightsquigarrow \Big(\beta\big(A(\cdot, \omega)\big), A(\cdot, \omega)\Big)$$

is the same as the $\mathbb{P}^0$-distribution of

$$p \rightsquigarrow \left(\int_0^t \theta_X(\tau, p)\, dp(\tau), \int_0^t \theta_X(\tau, p)^2\, d\tau\right).$$

Hence, all that remains is to set $\zeta_m = A(m)$. $\quad\square$

§7.3.4. Exercises

EXERCISE 7.3.9. Let $M \in \mathcal{M}_{\mathrm{loc}}(\mathbb{P}; \mathbb{R})$ be given.

(i) Show that the conclusion in Exercise 7.1.6 can be strengthened to say that, for any stopping time ζ,

$$\lim_{t \to \infty} M(t \wedge \zeta) \text{ exists in } \mathbb{R} \quad \mathbb{P}\text{-almost surely on } \{\langle M\rangle(\zeta) < \infty\}$$

and

$$\varlimsup_{t \to \infty} M(t \wedge \zeta) = \infty = -\varliminf_{t \to \infty} M(t \wedge \zeta) \quad \mathbb{P}\text{-almost surely on } \{\langle M\rangle(\zeta) = \infty\}.$$

(ii) If M is non-negative, show that $\langle M\rangle(\infty) < \infty$ $\mathbb{P}$-almost surely.

(iii) Refer to part (i) of Exercise 6.3.12, and show that there exists a $\theta \in \Theta_{\mathrm{loc}}^2(\mathbb{P}; \mathbb{R}^n)$ such that

$$M(t) = M(0) + \int_0^t \big(\theta(\tau), dp(\tau)\big)_{\mathbb{R}^n}, \quad \mathbb{P}\text{-almost surely for each } t \in [0, \infty).$$

Further, show that $\int_0^\infty |\theta(\tau)|^2\, d\tau < \infty$ $\mathbb{P}$-almost surely.

EXERCISE 7.3.10. The purpose of this exercise is to see how the considerations in this section can contribute to an understanding of the relationship
between dimension and explosion.

(i) Suppose that $\sigma : \mathbb{R} \longrightarrow \mathbb{R}$ is a locally Lipschitz continuous function,
and let $(\beta(t), \mathcal{F}_t, \mathbb{P})$ be a one-dimensional Brownian motion. Show that,
without any further conditions, the solution to the 1-dimensional stochastic
differential equation

$$dX(t) = \sigma\big(X(\tau)\big)\, d\beta(\tau)$$

exists for all time whenever $\mathbb{P}\big(|X(0)| < \infty\big) = 1$. That is, no matter what
the distribution of $X(0)$ is or how fast σ grows, $X(\,\cdot\,)$ will $\mathbb{P}$-almost surely
not explode.

Hint: The key observation is that, because its unparameterized trajectories
follow those of a 1-dimensional Brownian motion,

$$\langle X \rangle(\infty) = \int_0^\infty \sigma^2\big(X(\tau)\big)\, d\tau = \infty \implies X(\,\cdot\,) \text{ returns to } 0 \text{ infinitely often.}$$

Thus, if $\zeta_0 \equiv 0$ and if, for $m \geq 1$,

$$\zeta_m = \begin{cases} \infty \\ \inf\{t \geq \zeta_{m-1} : \exists \tau \in [\zeta_{m-1}, t]\ |X(\tau, x)| = 1\ \&\ X(t, x) = 0\}, \end{cases}$$

depending on whether $\zeta_{m-1} = \infty$ or $\zeta_{m-1} < \infty$, then, by the Markov property, $\{\zeta_{m+1} - \zeta_m : m \geq 1\}$ is a family of $\mathbb{P}$-mutually independent, identically
distributed, strictly positive random variables. Hence, $\mathbb{P}$-almost surely, either $\langle X \rangle(\infty) < \infty$ or for all $s \geq 0$ there exists a $t \geq s$ such that $X(t) = 0$.
In either case, $\langle X \rangle(T) < \infty$ $\mathbb{P}$-almost surely for all $T \geq 0$.

(ii) The analogous result holds for a diffusion in $\mathbb{R}^2$ associated with
$L = \frac{1}{2}a^2(x)\Delta$, where $\alpha : \mathbb{R}^2 \longrightarrow \mathbb{R}$ is a continuously differentiable function. Namely, such a diffusion never explodes. The reason is that such a
diffusion is obtained from a two-dimensional Brownian motion by a random-
time change and that two-dimensional Brownian motion is recurrent. Only
the fact that two-dimensional Brownian motion never actually returns to the
origin complicates the argument a little. To see that recurrence is the essential property here, show that, for any $\epsilon > 0$, the three-dimensional diffusion
corresponding to $L = \frac{1}{2}(1 + |x|^2)^{1+\epsilon}\Delta$ *does* explode.

Hint: Take

$$u(x) = \frac{1}{\pi} \int_{\mathbb{R}^3} \frac{1}{|x - y|} \frac{1}{(1 + |y|^2)^{1+\epsilon}}\, dy,$$

show that $Lu = -1$, and conclude that, for $R > |x|$, $u(x) \geq \mathbb{E}^{\mathbb{P}^L_x}[\zeta_R]$, where
ζ_R is the first exit time of $B_{\mathbb{R}^3}(0, R)$.

EXERCISE 7.3.11. V. Strassen made one of the most remarkable applications of Skorohod's representation theorem when he used it in [33] to prove a function space version of the law of the iterated logarithm. Here we will aim for much less. Namely, assume the law of the iterated logarithm for Brownian motion, as stated in the proof of Corollary 7.3.3,[5] and prove that for any sequence $\{X_m\}_0^\infty$ of mutually independent, identically distributed $\mathbb{R}$-valued random variables with variance 1,

$$\varlimsup_{n \to \infty} \frac{S_m}{\sqrt{2m \log_{(2)} m}} = 1 = -\varlimsup_{n \to \infty} \frac{S_m}{\sqrt{2m \log_{(2)} m}},$$

where $S_m \equiv \sum_1^m X_k$.

EXERCISE 7.3.12. Another direction in which Skorohod's representation theorem can be useful is in applications to central limit theory. Indeed, it can be seen as providing an ingenious coupling procedure for such results. The purpose of this exercise is to give examples of this sort of application. Throughout, $\{X_m\}_1^\infty$ will denote a sequence of mutually independent, centered $\mathbb{R}$-valued random variables with variance 1, $S_0 = 0$, and $S_m = \sum_{\ell=1}^m X_\ell$ for $m \geq 1$. In addition, $(\beta(t), \mathcal{F}_t, \mathbb{P})$ will be the Brownian motion and $\{\zeta_m : m \geq 0\}$ will be the stopping times described in Theorem 7.3.8.

(i) Set $\beta^m(t) = m^{-\frac{1}{2}} \beta(mt)$, and note that $(\beta^m(t), \mathcal{F}_{mt}, \mathbb{P})$ is again a Brownian motion and that the $\mathbb{P}$-distribution of $\beta^m \left(\frac{\zeta_m}{m} \right)$ is the same as the distribution of $m^{-\frac{1}{2}} S_m$. As an application of the weak laws of large numbers and the continuity of Brownian paths, conclude that

$$\lim_{m \to \infty} \mathbb{P} \left(\left| \beta^m \left(\frac{\zeta_m}{m} \right) - \beta^m(1) \right| \geq \epsilon \right) = 0 \quad \text{for all } \epsilon > 0,$$

and use this to derive the *central limit theorem*, which is that statement of the distribution of $m^{-\frac{1}{2}} S_m$ tends to the standard normal distribution.

(ii) The preceding line of reasoning can be improved to give *Donsker's invariance principal*. That is, define $t \rightsquigarrow S^m(t)$ so that $S^m(0) = 0$, $S^m \left(\frac{\ell}{m} \right) = m^{-\frac{1}{2}} S_m$ and $S^m \upharpoonright \left[\frac{\ell-1}{m}, \frac{\ell}{m} \right]$ is linear for each $\ell \in \mathbb{Z}^+$. Donsker's invariance principal is the statement that the distribution on $C\left([0, \infty); \mathbb{R}^n\right)$ of $S^m(\cdot)$ tends to Wiener measure. That is, $\mathbb{E} \left[\Phi \circ S^m(\cdot) \right] \longrightarrow \mathbb{E}^{\mathbb{P}^0} [\Phi]$ for all bounded, continuous $\Phi : C\left([0, \infty); \mathbb{R}\right) \longrightarrow \mathbb{R}$.

---

[5] It should be recognized that the law of the iterated logarithm for Brownian motion is essentially the same as the law of the iterated logarithm for centered Gaussian random variables with variance 1 and, as such, is *much* easier than the statement for general centered random variables with variance 1.

To prove this, define $Y^m(t)$ so that $Y^m\left(\frac{\ell}{m}\right) = \beta^m\left(\frac{\varsigma_\ell}{m}\right)$ and $Y^m \upharpoonright \left[\frac{\ell}{m}, \frac{\ell+1}{m}\right]$ is linear for $\ell \in \mathbb{Z}^+$. Note that the distribution of $Y^m(\,\cdot\,)$ is the same as the distribution of $S^m(\,\cdot\,)$, and show that, for any $L \in \mathbb{Z}^+$, $\delta > 0$, and $m \geq \frac{1}{\delta}$,

$$\left\|Y^m - \beta^m\right\|_{[0,L]} \leq 2 \sup_{\substack{0 \leq s < t \leq L \\ t - s \leq \delta}} \left|\beta^m(t) - \beta^m(s)\right|$$

on the set where $\max_{1 \leq \ell \leq mL}\left|\frac{\varsigma_\ell}{m} - \frac{\ell}{m}\right| \leq \delta$. Conclude that Donsker's result will follow once one shows that

$$\lim_{m \to \infty} \mathbb{P}\left(\max_{1 \leq \ell \leq mL}\left|\frac{\varsigma_\ell}{m} - \frac{\ell}{m}\right| \geq \delta\right) = 0$$

for each $\delta > 0$.

(iii) To complete the program begun in **(ii)**, note that it suffices to show that if $\{Z_m : m \geq 1\}$ is a sequence of independent, identically distributed, centered, integrable random variables, then

(*) $$\lim_{m \to \infty} \mathbb{P}\left(\sup_{0 \leq \ell \leq mL}\left|\frac{1}{m}\sum_{k=1}^{\ell} Z_k\right| \geq \epsilon\right) = 0$$

for each $\epsilon > 0$. When Z_1 is square integrable, (*) follows from Kolmogorov's inequality (cf. (1.4.4) in [36]):

$$\mathbb{P}\left(\sup_{0 \leq \ell \leq mL}\left|\frac{1}{m}\sum_{k=1}^{\ell} Z_k\right| \geq \epsilon\right) \leq \frac{L\,\mathrm{var}(Z_1)}{m\epsilon^2}.$$

To get the general case, first apply Doob's inequality (cf. Theorem 5.2.4 in [36]) to the martingale formed by the partial sums of the Z_k's to see that

$$\mathbb{P}\left(\sup_{0 \leq \ell \leq mL}\left|\frac{1}{m}\sum_{k=1}^{\ell} Z_k\right| \geq \epsilon\right) \leq \frac{L\,\mathbb{E}[|Z_1|]}{\epsilon},$$

and then use this together with an approximation procedure to reduce to the case when the Z_k's are square integrable.

CHAPTER 8

Stratonovich's Theory

From an abstract mathematical standpoint, Itô's theory of stochastic integration has as a serious flaw: it behaves dreadfully under changes of coordinates (cf. Remark 3.3.6). In fact, Itô's formula itself is the most dramatic manifestation of this problem.

The origin of the problems Itô's theory has with coordinate changes can be traced back to its connection with independent increment processes. Indeed, the very notion of an independent increment process is inextricably tied to the linear structure of Euclidean space, and anything but a linear change of coordinates will wreak havoc to that structure. Generalizations of Itô's theory like the one of Kunita and Watanabe do not really cure this problem, they only make it slightly less painful.

To make Itô's theory more amenable to coordinate changes, we will develop an idea which was introduced by R.L. Stratonovich. Stratonovich was motivated by applications to engineering, and his own treatment [34] had some mathematically awkward aspects. In fact, it is ironic, if not surprising, that Itô [12] was the one who figured out how to put Stratonovich's ideas on a firm mathematical foundation.

8.1 Semimartingales and Stratonovich Integrals

From a technical perspective, the coordinate change problem alluded to above is a consequence of the fact that nonlinear functions destroy the martingale property. Thus, our first step will be to replace martingales with a class of processes which is invariant under composition with nonlinear functions. Throughout, $(\Omega, \mathcal{F}, \mathbb{P})$ will be a complete probability space which comes equipped with a nondecreasing family $\{\mathcal{F}_t : t \geq 0\}$ of $\mathbb{P}$-complete σ-algebras.

§8.1.1. Semimartingales: We will say that $Z : [0, \infty) \times \Omega \longrightarrow \mathbb{R}$ is a continuous *semimartingale*[1] and will write $Z \in \mathcal{S}(\mathbb{P}; \mathbb{R})$ if Z can be written

---

[1] It is unfortunate that Doob originally adopted this term for the class of processes which are now called submartingales. However, the confusion caused by this terminological accident recedes along with the generation of probabilists for whom it was a problem.

in the form $Z(t, \omega) = M(t, \omega) + B(t, \omega)$, where $(M(t), \mathcal{F}_t, \mathbb{P})$ is a continuous local martingale and B is a progressively measurable function with the property that $B(\cdot, \omega)$ is a continuous function of locally bounded variation for $\mathbb{P}$-almost every ω.

Notice that when we insist that $M(0) \equiv 0$ the decomposition of a semi-martingale Z into a *martingale part* M and its *bounded variation part* B is almost surely unique. This is just an application of the first statement in Corollary 7.1.2. In addition, Itô's formula (cf. Theorem 7.2.9) shows that the class of continuous semimartingales is invariant under composition with twice continuously differentiable functions f. In fact, his formula says that if $Z(t) = (M_1(t) + B_1(t), \ldots, Z_n(t) + B_n(t))$ is an $\mathbb{R}^n$-valued, continuous semi-martingale and $F \in C^2(\mathbb{R}^n; \mathbb{R})$, then the martingale and bounded variation parts of $F \circ Z$ are, respectively,

$$\sum_{i=1}^{n} \int_0^t \partial_i F(Z(\tau)) \, dM_i(\tau) \quad \text{and}$$

$$F(Z(0)) + \frac{1}{2} \sum_{i,j=1}^{n} \int_0^t \partial_i \partial_j F(Z(\tau)) \langle M_i, M_j \rangle(d\tau) + \sum_{i=1}^{n} \int_0^t \partial_i F(Z(\tau)) \, B_i(d\tau),$$

where, of course, the integrals in the first line are taken in the sense of Itô and the ones in the second are Lebesgue (or Riemann) integrals.

In various circumstances it is useful to have defined the integral

$$\int_0^t \theta(\tau) \, dZ(t) = \int_0^t \theta(\tau) \, dM(\tau) + \int_0^t \theta(\tau) \, B(d\tau)$$
$$\text{for } Z = M + B \text{ and } \theta \in \Theta^2_{\text{loc}}(\langle M \rangle, \mathbb{P}; \mathbb{R}),$$

where the $dM(\tau)$-integral is taken in the sense of Itô and the $B(d\tau)$ integral is a taken *à la* Lebesgue. Also, we take

(8.1.1) $\langle Z_1, Z_2 \rangle(t) \equiv \langle M_1, M_2 \rangle(t)$ if $Z_i = M_i + B_i$ for $i \in \{1, 2\}$.

Indeed, with this notation, Itô's formula 7.2.9 becomes

$$F(Z(t)) - F(Z(0)) = \sum_{i=1}^{n} \int_0^t \partial_i F(Z(t)) \, dZ_i(t)$$

(8.1.2)

$$+ \frac{1}{2} \sum_{i,j=1}^{n} \int_0^t \partial_i \partial_j F(Z(\tau)) \langle Z_i, Z_j \rangle(d\tau).$$

The notational advantage of (8.1.2) should be obvious. On the other hand, the disadvantage is that it mixes the martingale and bounded variation parts of the right hand side, and this mixing has to be disentangled before expectation values are taken.

Finally, it should be noticed that the extension of the brackets given in (8.1.1) is more than a notational device and has intrinsic meaning. Namely, by Exercise 7.2.13 and the easily checked fact that, as $N \to \infty$,

$$\sum_{m=0}^{2^N t} \left| \left(Z_1\big((m+1)2^{-N}\big) - Z_1\big(m2^{-N}\big) \right) \left(Z_2\big((m+1)2^{-N}\big) - Z_2\big(m2^{-N}\big) \right) \right.$$

$$\left. - \left(M_1\big((m+1)2^{-N}\big) - M_1\big(m2^{-N}\big) \right) \left(M_2\big((m+1)2^{-N}\big) - M_2\big(m2^{-N}\big) \right) \right|,$$

tends to 0 $\mathbb{P}$-almost surely uniformly for t in compacts, we know that

$$(8.1.3) \quad \sum_{m=0}^{\infty} \left(Z_1\big(t \wedge (m+1)2^{-N}\big) - Z_1\big(t \wedge m2^{-N}\big) \right)$$

$$\times \left(Z_2\big(t \wedge (m+1)2^{-N}\big) - Z_2\big(t \wedge m2^{-N}\big) \right) \longrightarrow \langle Z_1, Z_2 \rangle(t)$$

in $\mathbb{P}$-probability uniformly for t in compacts.

§8.1.2. **Stratonovich's Integral:** Keeping in mind that Stratonovich's purpose was to make stochastic integrals better adapted to changes in variable, it may be best to introduce his integral by showing that it is integral at which one arrives if one adopts a somewhat naive approach, one which reveals its connection to the Riemann integral. Namely, given elements X and Y of $\mathcal{S}(\mathbb{P}; \mathbb{R})$, we want to define the *Stratonovich integral*[2] $\int_0^t Y(\tau) \circ dX(\tau)$ of Y with respect to X as the limit of the Riemann integrals

$$\lim_{N \to \infty} \int_0^t Y^N(\tau) \, dX^N(\tau),$$

where, for $\alpha : [0, \infty) \longrightarrow \mathbb{R}$, we use α^N to denote the polygonal approximation of α obtained by linear interpolation on each interval $[m2^{-N}, (m+1)2^{-N}]$. That is,

$$\alpha^N(t) = \alpha(m2^{-N}) + 2^N(t - m2^{-N})\Delta_m^N \alpha$$
$$\text{where } \Delta_m^N \alpha \equiv \alpha\big((m+1)2^{-N}\big) - \alpha\big(m2^{-N}\big)$$

---

[2] This notation is only one of many which are used to indicate it is Stratonovich's, as opposed to Itô's , sense in which an integral is being taken. Others include the use of $\delta X(\tau)$ in place of $\circ dX(\tau)$. The lack of agreement about the choice of notation reflects the inadequateness of all the variants.

for all m, $N \in \mathbb{N}$ and $t \in [m2^{-N}, (m+1)2^{-N}]$.

Of course, before we can adopt this definition, we are obliged to check that this limit exists. For this purpose, write

$$\int_0^t Y^N(\tau)\, dX^N(\tau) = \int_0^t Y([\tau]_N)\, dX(\tau) + \int_0^t \left(Y^N(\tau) - Y([\tau]_N)\right) dX^N(\tau),$$

where the dX-integral on the right can be interpreted as either an Itô or a Riemann integral. By using Corollary 7.2.3, one can easily check that

$$\left\| \int_0^t Y([\tau]_N)\, dX(\tau) - \int_0^t Y(\tau)\, dX(\tau) \right\|_{[0,T]} \longrightarrow 0 \quad \text{in } \mathbb{P}\text{-probability.}$$

At the same time, by (8.1.3),

$$\int_0^t \left(Y^N(\tau) - Y([\tau]_N)\right) dX^N(\tau)$$

$$= 4^N \sum_{m \leq 2^N t} \left(\Delta_m^N X\right)\left(\Delta_m^N Y\right) \int_{m2^{-N}}^{t \wedge (m+1)2^{-N}} (\tau - m2^{-N})\, d\tau \longrightarrow \frac{1}{2} \langle X, Y \rangle(t)$$

in $\mathbb{P}$-probability uniformly for t in compacts. Hence, for computational purposes, it is best to present the Stratonovich integral as

$$(8.1.4) \qquad \int_0^t Y(\tau) \circ dX(\tau) = \int_0^t Y(\tau)\, dX(\tau) + \frac{1}{2} \langle X, Y \rangle(t),$$

where the integral on the right is taken in the sense of Itô.

So far as I know, the formula in (8.1.4) was first given by Itô in [12]. In particular, Itô seems to have been the one who realized that the problems posed by Stratonovich's theory could be overcome by insisting that the integrands be semimartingales, in which case, as (8.1.4) makes obvious, *the Stratonovich integral of Y with respect to X is again a continuous semimartingale*. In fact, if $X = M + B$ is the decomposition of X into its martingale and bounded variation parts, then

$$\int_0^t Y(\tau)\, dM(\tau) \quad \text{and} \quad \int_0^t Y(\tau)\, dB(\tau) + \frac{1}{2} \langle Y, M \rangle(t)$$

are the martingale and bounded variation parts of $I(t) \equiv \int_0^t Y(\tau) \circ dX(\tau)$, and so $\langle Z, I \rangle(dt) = Y(t) \langle Z, X \rangle(dt)$ for all $Z \in \mathcal{S}(\mathbb{P}; \mathbb{R})$.

To appreciate how clever Itô's observation is, notice that Stratonovich's integral is not really an integral at all. Indeed, in order to deserve being called an *integral*, an operation should result in a quantity which can be estimated in terms of zeroth order properties of the integrand. On the other hand, as (8.1.4) shows, no such estimate is possible. To see this, take $Y(t) = f(Z(t))$, where $Z \in \mathcal{S}(\mathbb{P}; \mathbb{R})$ and $f \in C^2(\mathbb{R}; \mathbb{R})$. Then, because $\langle X, Y \rangle(dt) = f'(Z(t)) \langle X, Z \rangle(dt)$,

$$\int_0^t Y(\tau) \circ dX(\tau) = \int_0^t f(Z(\tau)) \, dX(\tau) + \frac{1}{2} \int_0^t f'(Z(\tau)) \langle X, Z \rangle(d\tau),$$

which demonstrates that there is, in general, no estimate of the Stratonovich integral in terms of the zeroth order properties of the integrand.

Another important application of (8.1.4) is to the behavior of Stratonovich integrals under iteration. That is, suppose that $X, Y \in \mathcal{S}(\mathbb{P}; \mathbb{R})$, and set $I(t) = \int_0^t Y(\tau) \circ dX(\tau)$. Then $\langle Z, I \rangle(dt) = Y(t)\langle Z, X \rangle(dt)$ for all $Z \in \mathcal{S}(\mathbb{P}; \mathbb{R})$, and so

$$\int_0^t Z(\tau) \circ dI(\tau) = \int_0^t Z(\tau) \, dI(\tau) + \frac{1}{2} \int_0^t Y(\tau) \langle Z, X \rangle(d\tau)$$

$$= \int_0^t Z(\tau)Y(\tau) \, dX(\tau) + \frac{1}{2} \int_0^t \left(Z(\tau)\langle Y, X \rangle(d\tau) + Y(\tau)\langle Z, X \rangle(d\tau) \right)$$

$$= \int_0^t Z(\tau)Y(\tau) \, dX(\tau) + \frac{1}{2}\langle ZY, X \rangle(t) = \int_0^t ZY(\tau) \circ dX(\tau),$$

since, by Itô's formula,

$$ZY(t) - ZY(0) = \int_0^t Z(\tau) \, dY(\tau) + \int_0^t Y(\tau) \, dZ(\tau) + \frac{1}{2}\langle Z, Y \rangle(t),$$

and therefore $\langle ZY, X \rangle(dt) = Z(t)\langle Y, X \rangle(dt) + Y(t)\langle Z, X \rangle(dt)$. In other words, we have now proved that

$$(8.1.5) \qquad \int_0^t Z(\tau) \circ d\left(\int_0^\tau Y(\sigma) \circ X(\sigma) \right) = \int_0^t Z(\tau)Y(\tau) \circ dX(\tau).$$

§8.1.3. Ito's Formula and Stratonovich Integration: Because the origin of Stratonovich's integral is in Riemann's theory, it should come as no surprise that Itô's formula looks deceptively like the fundamental theorem of calculus when Stratonovich's integral is used. Namely, let $Z = (Z_1, \ldots, Z_n) \in \mathcal{S}(\mathbb{P}; \mathbb{R})^n$ and $f \in C^3(\mathbb{R}^n; \mathbb{R})$ be given, and set $Y_i = \partial_i f \circ Z$ for $1 \leq i \leq n$. Then $Y_i \in \mathcal{S}(\mathbb{P}; \mathbb{R})$ and, by Itô's formula applied to $\partial_i f$, the local martingale part of Y_i is given by the sum over $1 \leq j \leq n$ of the

Itô stochastic integrals of $\partial_i \partial_j f \circ Z$ with respect to the local martingale part of Z_j. Hence,

$$\langle Y_i, Z_i \rangle(dt) = \sum_{j=1}^{n} \partial_i \partial_j f \circ Z(\tau) \langle Z_i, Z_j \rangle(dt).$$

But, by Itô's formula applied to f, this means that

$$df(Z(t)) = \sum_{i=1}^{n} \left(Y_i(\tau) \, dZ_i(\tau) + \frac{1}{2} \langle Y_i, Z_i \rangle(dt) \right),$$

and so we have now shown that, in terms of Stratonovich integrals, Itô's formula does look like the "the fundamental theorem of calculus"

$$(8.1.6) \qquad f(Z(t)) - f(Z(0)) = \sum_{i=1}^{n} \int_0^t \partial_i f(Z(\tau)) \circ dZ_i(\tau).$$

As I warned above, (8.1.6) is deceptive. For one thing, as its derivation makes clear, it, in spite of its attractive form, is really just Itô's formula (8.1.2) in disguise. In fact, if, as we will, one adopts Itô's approach to Stratonovich's integral, then it is not even that. Indeed, one cannot write (8.1.6) unless one knows that $\partial_i f \circ Z$ is a semimartingale. Thus, in general, (8.1.6) requires us to assume that f is three times continuously differentiable, not just twice, as in the case with (8.1.2). Ironically, we have arrived at a first order fundamental theorem of calculus which applies only to functions with three derivatives. In view of these remarks, it is significant that, at least in the case of Brownian motion, Itô found a way (cf. [13] or Exercise 8.1.8 below) to make (8.1.6) closer to a *true* fundamental theorem of calculus, at least in the sense that it applies to all $f \in C^1(\mathbb{R}^n; \mathbb{R})$.

REMARK 8.1.7. Putting (8.1.6) together with footnote 2 about the notation for Stratonovich integrals, one might be inclined to think the *right* notation should be $\int_0^t Y(\tau)\dot{X}(\tau)\, d\tau$. For one thing, this notation recognizes that the Stratonovich integral is closely related to the notion of generalized derivatives *à la* Schwartz's distribution theory. Secondly, (8.1.6) can be summarized in differential form by the expression

$$df(Z(t)) = \sum_{i=1}^{n} \partial_i f(Z(t)) \circ dZ_i(t),$$

which would take the appealing form

$$\frac{d}{dt} f(Z(t)) = \sum_{i=1}^{n} \partial_i f(Z(t)) \dot{Z}(t).$$

Of course, the preceding discussion should also make one cautious about being too credulous about all this.

§8.1.4. Exercises

EXERCISE 8.1.8. In this exercise we will describe Itô's approach to extending the validity of (8.1.6).

(i) The first step is to give another description of Stratonovich integrals. Namely, given $X, Y \in \mathcal{S}(\mathbb{P}; \mathbb{R})$, show that the Stratonovich integral of Y with respect to X over the interval $[0, T]$ is almost surely equal to

$$(*) \quad \lim_{N \to \infty} \sum_{m=0}^{2^N - 1} \frac{Y\left(\frac{(m+1)T}{2^N}\right) + Y\left(\frac{mT}{2^N}\right)}{2} \left(X\left(\frac{(m+1)T}{2^N}\right) - X\left(\frac{mT}{2^N}\right) \right).$$

Thus, even if Y is not a semimartingale, we will say that $Y : [0, T] \times \Omega \longrightarrow \mathbb{R}$ is *Stratonvich integrable* on $[0, T]$ with respect to X if the limit in $(*)$ exists in $\mathbb{P}$-measure, in which case we will use $\int_0^T Y(t) \circ dX(t)$ to denote this limit. It must be emphasized that the definition here is "T by T" and not simultaneous for all T's in an interval.

(ii) Given an $X \in \mathcal{S}(\mathbb{P}; \mathbb{R})$ and a $T > 0$, set $\check{X}^T(t) = X((T - t)^+)$, and suppose that $(\check{X}^T(t), \check{\mathcal{F}}_t^T, \mathbb{P})$ is a semimartingale relative to some filtration $\{\check{\mathcal{F}}_t^T : t \geq 0\}$. Given a $Y : [0, T] \times \Omega \longrightarrow \mathbb{R}$ with the properties that $Y(\cdot, \omega) \in C([0, T]; \mathbb{R})$ for $\mathbb{P}$-almost every ω and that $\omega \rightsquigarrow Y(t, \omega)$ is $\mathcal{F}_t \cap \check{\mathcal{F}}_t^T$ for each $t \in [0, T]$, show that Y is Stratonovich integrable on $[0, T]$ with respect to X. In fact, show that

$$\int_0^{T} Y(t) \circ dX(t) = \frac{1}{2} \int_0^T Y(t)\, dX(t) - \frac{1}{2} \int_0^T Y(t)\, d\check{X}^T(t),$$

where each of the integrals on the right is the taken in the sense of Itô.

(iii) Let $(\beta(t), \mathcal{F}_t, \mathbb{P})$ be an $\mathbb{R}^n$-valued Brownian motion. Given $T \in (0, \infty)$, set $\check{\beta}^T(t) = \beta((T - t)^+)$, $\check{\mathcal{F}}_t^T = \sigma(\{\check{\beta}^T(\tau) : \tau \in [0, t]\})$, and show that, for each $\xi \in \mathbb{R}^n$, $((\xi, \check{\beta}^T(t))_{\mathbb{R}^n}, \check{\mathcal{F}}_t^T, \mathbb{P})$ is a semimartingale with $\langle \check{\beta}^T, \check{\beta}^T \rangle(t) = t \wedge T$ and bounded variation part $t \rightsquigarrow -\int_0^t \frac{\check{\beta}_\tau^T}{T - \tau}\, d\tau$. In particular, show that, for each $\xi \in \mathbb{R}^n$ and $g \in C(\mathbb{R}^n; \mathbb{R})$, $t \rightsquigarrow g(\beta(t))$ is Stratonovich integrable on $[0, T]$ with respect to $t \rightsquigarrow (\xi, \beta(t))_{\mathbb{R}^n}$. Further, if $\{g_n\}_1^\infty \subseteq C(\mathbb{R}^n; \mathbb{R})$ and $g_n \longrightarrow g$ uniformly on compacts, show that

$$\int_0^T g_n(\beta(t)) \circ d(\xi, \beta(t))_{\mathbb{R}^n} \longrightarrow \int_0^T g(\beta(t)) \circ d(\xi, \beta(t))_{\mathbb{R}^n}$$

in $\mathbb{P}$-measure.

(iv) Continuing with the notation in **(iii)**, show that

$$f\big(\beta(T)\big) - f(0)) = \sum_{i=1}^{n} \int_0^T \partial_i f\big(\beta(t)\big) \circ d\beta_i(t)$$

for every $f \in C^1(\mathbb{R}^n; \mathbb{R})$. In keeping with the comment at the end of **(i)**, it is important to recognize that although this form of Itô 's formula holds for all continuously differentiable functions, it is, in many ways less useful than forms which we obtained previously. In particular, when f is no better than once differentiable, the right hand side is defined only up to a $\mathbb{P}$-null set for each T and not for all T's simultaneously.

(v) Let $\big(\beta(t), \mathcal{F}_t, \mathbb{P}\big)$ be an $\mathbb{R}$-valued Brownian motion, show that, for each $T \in (0, \infty)$, $t \rightsquigarrow \mathrm{sgn}\big(\beta(t)\big)$ is Stratonovich integrable on $[0, T]$ with respect to β, and arrive at

$$\big|\beta(T)\big| = \int_0^T \mathrm{sgn}\big(\beta(t)\big) \circ d\beta(t).$$

After comparing this with the result in (6.1.7), conclude that the local time $\ell(T, 0)$ of β at 0 satisfies

$$\int_0^T \frac{|\beta(t)|}{t} \, dt - \ell(T, 0) - |\beta(T)| = \int_0^T \mathrm{sgn}\big(\beta(T-t)\big) \, d\check{M}^T(t),$$

where $\check{M}^T$ is the martingale part of $\check{\beta}^T$. In particular, the expression on the left hand side is a centered Gaussian random variable with variance T.

EXERCISE 8.1.9. Itô's formula proves that $f \circ Z \in \mathcal{S}(\mathbb{P}; \mathbb{R})$ whenever $Z = (Z_1, \ldots, Z_n) \in \mathcal{S}(\mathbb{P}; \mathbb{R})^n$ and $f \in C^2(\mathbb{R}^n; \mathbb{R})$. On the other hand, as Tanaka's treatment (cf. §6.1) of local time makes clear, it is not always necessary to know that f has two continuous derivatives. Indeed, both (6.1.3) and (6.1.7) provide examples in which the composition of a martingale with a continuous function leads to a continuous semimartingale even though the derivative of the function is discontinuous. More generally, as a corollary of the Doob-Meyer decomposition theorem (alluded to at the beginning of §7.1) one can show that $f \circ Z$ will be a continuous semimartingale whenever $Z \in \mathcal{M}_{\mathrm{loc}}(\mathbb{P}; \mathbb{R}^n)$ and f is a continuous, convex function. Here is a more pedestrian approach to this result.

(i) Following Tanaka's procedure, prove that $f \circ Z \in \mathcal{S}(\mathbb{P}; \mathbb{R})$ whenever $Z \in \mathcal{S}(\mathbb{P}; \mathbb{R})^n$ and $f \in C^1(\mathbb{R}^n; \mathbb{R})$ is convex. That is, if M_i denotes the martingale part of Z_i, show that

$$f\big(Z(t)\big) - \sum_{i=1}^{n} \int_0^t \partial_i f\big(Z(\tau)\big) \, dM_i(\tau)$$

is the bounded variation part of $f \circ Z$.

Hint: Begin by showing that when $A(t) \equiv \left((\langle Z_i, Z_j \rangle(t)) \right)_{1 \leq i,j \leq n}$, $A(t) - A(s)$ is $\mathbb{P}$-almost surely non-negative definite for all $0 \leq s \leq t$, and conclude that if $f \in C^2(\mathbb{R}^n; \mathbb{R})$ is convex then

$$t \rightsquigarrow \sum_{i,j=1}^{n} \int_0^t \partial_i \partial_j f(Z(\tau)) \langle Z_i, Z_j \rangle(d\tau)$$

is $\mathbb{P}$-almost surely nondecreasing.

(ii) By taking advantage of more refined properties of convex functions, see if you can prove that $f \circ Z \in \mathcal{S}(\mathbb{P}; \mathbb{R})$ when f is a continuous, convex function.

EXERCISE 8.1.10. If one goes back to the original way in which we described Stratonovich in terms of Riemann integration, it becomes clear that the only reason why we needed Y to be a semimartingale is that we needed to know that

$$\sum_{m \leq 2^N t} (\Delta_m^N Y)(\Delta_m^N X)$$

converges in $\mathbb{P}$-probability to a continuous function of locally bounded variation uniformly for t in compacts.

(i) Let $Z = (Z_1, \ldots, Z_n) \in \mathcal{S}(\mathbb{P}; \mathbb{R})^n$, and set $Y = f \circ Z$, where $f \in C^1(\mathbb{R}^n; \mathbb{R})$. Show that, for any $X \in \mathcal{S}(\mathbb{P}; \mathbb{R})$,

$$\sum_{m \leq 2^N t} (\Delta_m^N Y)(\Delta_m^N X) \longrightarrow \sum_{i=1}^{n} \int_0^t \partial_i f(Z(\tau)) \langle Z_i, X \rangle(d\tau)$$

in $\mathbb{P}$-probability uniformly for t compacts.

(ii) Continuing with the notation in **(i)**, show that

$$\int_0^t Y(\tau) \circ dX(\tau) \equiv \lim_{N \to \infty} \int_0^t Y^N(\tau) \, dX^N(\tau)$$

$$= \int_0^t Y(\tau) \, dX(\tau) + \frac{1}{2} \sum_{i=1}^{n} \int_0^t \partial_i f(Z(\tau)) \langle Z_i, X \rangle(d\tau),$$

where the convergence is in $\mathbb{P}$-probability uniformly for t in compacts.

(iii) Show that (8.1.6) continues to hold for $f \in C^2(\mathbb{R}^n; \mathbb{R})$ when the integral on the right hand side is interpreted using the extension of Stratonovich integration developed in **(ii)**.

EXERCISE 8.1.11. Let $X, Y \in \mathcal{S}(\mathbb{P}; \mathbb{R})$. In connection with Remark 8.1.7, it is interesting to examine whether it is sufficient to mollify only X when defining the Stratonovich integral of Y with respect to X.

(i) Show that $\int_0^1 Y([\tau]_N) \, dX^N(\tau)$ tends in $\mathbb{P}$-probability to $\int_0^1 Y(\tau) \, dX(\tau)$.

(ii) Define $\psi_N(t) = 1 - 2^N(t - [t]_N)$, set $Z_N(t) = \int_0^t \psi_N(\tau) \, dY(\tau)$, and show that

$$\int_0^1 Y(\tau) \, dX^N(\tau) - \int_0^1 Y([\tau]_N) \, dX^N(\tau) = \sum_{m=0}^{2^N-1} \left(\Delta_m^N X \right) \left(\Delta_m^N Z_N \right).$$

(iii) Show that

$$\sum_{m=0}^{2^N-1} \left(\Delta_m^N X \right) \left(\Delta_m^N Z_N \right) - \int_0^1 \psi_N(t) \langle X, Y \rangle(dt) \longrightarrow 0 \quad \text{in } \mathbb{P}\text{-probability}.$$

(iv) Show that for any Lebesgue integrable function $\alpha : [0, 1] \longrightarrow \mathbb{R}$, $\int_0^1 \psi_N(\tau)\alpha(\tau) \, d\tau$ tends to $\frac{1}{2} \int_0^1 \alpha(\tau) \, d\tau$.

(v) Under the condition that $\langle X, Y \rangle(dt) = \beta(t) \, dt$, where $\beta : [0, \infty) \times \Omega \longrightarrow [0, \infty)$ is a progressively measurable function, use the preceding to see that $\int_0^1 Y(\tau) \, dX^N(\tau)$ tends in $\mathbb{P}$-probability to $\int_0^1 Y(\tau) \circ dX(\tau)$.

EXERCISE 8.1.12. One place where Stratonovich's theory really comes into its own is when it comes to computing the determinant of the solution to a linear stochastic differential equation. Namely, suppose that $A = ((A_{ij}))_{1 \leq i,j \leq n} \in \mathcal{S}(\mathbb{P}; \mathrm{Hom}(\mathbb{R}^n; \mathbb{R}^n))$ (i.e., $A_{ij} \in \mathcal{S}(\mathbb{P}; \mathbb{R})$ for each $1 \leq i, j \leq n$), and assume that $X \in \mathcal{S}(\mathbb{P}; \mathrm{Hom}(\mathbb{R}^n; \mathbb{R}^n))$ satisfies $dX(t) = X(t) \circ dA(t)$ in the sense that

$$dX_{ij}(t) = \sum_{k=1}^n X_{ik}(t) \circ dA_{kj}(t) \quad \text{for all } 1 \leq i, j \leq n.$$

Show that

$$\det \left(X(t) \right) = \det \left(X(0) \right) e^{\mathrm{Trace} \left(A(t) - A(0) \right)}.$$

8.2 Stratonovich Stochastic Differential Equations

Since every Stratonovich integral can be converted into an Itô integral, it might seem unnecessary to develop a separate theory of Stratonovich stochastic differential equations. On the other hand, the replacement of a Stratonovich equation by the equivalent Itô equation removes the advantages, especially the coordinate invariance, of Stratonovich's theory. With this in mind, we devote the present section to a *non-Itô* analysis of Stratonovich stochastic differential equations.

Let $\mathbb{P}^0$ denote the standard Wiener measure for r-dimensional paths $p = (p_1, \ldots, p_r) \in C([0, \infty); \mathbb{R}^r)$. In order to emphasize the coordinate invariance of the theory, we will write our Stratonovich stochastic differential equations in the form

$$
(8.2.1) \qquad dX(t, x, p) = V_0\big(X(t, x, p)\big)\, dt + \sum_{k=1}^{r} V_k\big(X(t, x, p)\big) \circ dp_k(t)
$$

$$
\text{with } X(0, x, p) = x,
$$

where, for each $0 \le k \le r$, $V_k : \mathbb{R}^n \longrightarrow \mathbb{R}^n$ is a smooth function. To see what the equivalent Itô equation is, notice that

$$
\Big\langle (V_k)_i\big(X(\,\cdot\,, x, p)\big), p_k \Big\rangle(dt) = \sum_{j=1}^{n} \partial_j (V_k)_i\big(X(t, x, p)\big) \langle X_j(\,\cdot\,, x, p), p_k \rangle(dt)
$$

$$
= \sum_{j=1}^{n}\sum_{\ell=1}^{r} (V_\ell)_j \partial_j (V_k)_i\big(X(t, x, p)\big) \langle p_\ell, p_k \rangle(dt) = D_{V_k}(V_k)_i\big(X(t, x, p)\big)\, dt,
$$

where $D_{V_k} = \sum_{i=1}^{n} (V_k)_j \partial_j$ denotes the directional derivative operator determined by V_k. Hence, if we think of each V_k, and therefore $X(t, x, p)$, as a column vector, then the Itô equivalent to (8.2.1) is

$$
(8.2.2) \qquad dX(t, x, p) = \sigma\big(X(t, x, p)\big) dp(t) + b\big(X(t, x, p)\big)\, dt
$$

$$
\text{with } X(0, x, p) = x,
$$

where $\sigma(x) = \big(V_1(x), \ldots, V_r(x)\big)$ is the $n \times r$-matrix whose kth column is $V_k(x)$ and $b = V_0 + \frac{1}{2}\sum_1^r D_{V_k} V_k$. In particular, if

$$
L = \frac{1}{2}\sum_{i,j=1}^{n} a_{ij}\partial_i\partial_j + \sum_{i=1}^{n} b_i\partial_i, \quad \text{where } a = \sigma\sigma^\top,
$$

then L is the operator associated with (8.2.1) in the sense that for any $f \in C^{1,2}\big([0, T] \times \mathbb{R}^n\big)$

$$
\left(f\big(t \wedge T, X(t, x, p)\big) - \int_0^{t \wedge T} \big(\partial_\tau f + Lf\big)\big(X(\tau, x, p)\big)\, d\tau, \mathcal{B}_t, \mathbb{P}^0 \right)
$$

is a local martingale. However, a better way to write this operator is directly in terms of the directional derivative operators D_{V_k}. Namely,

$$(8.2.3) \qquad\qquad L = D_{V_0} + \frac{1}{2} \sum_{k=1}^{r} (D_{V_k})^2.$$

Hörmander's famous paper [10] was the first to demonstrate the advantage of writing a second order elliptic operator in the form given in (8.2.3), and, for this reason, (8.2.3) is often said to be the *Hörmander form* expression for L. The most obvious advantage is the same as the advantage that Stratonovich's theory has over Itô's: it behaves well under change of variables. To wit, if $F : \mathbb{R}^n \longrightarrow \mathbb{R}^n$ is a diffeomorphism, then

$$(L\varphi) \circ F = D_{F_* V_0} \varphi + \frac{1}{2} \sum_{k=1}^{r} (D_{F_* V_k})^2 \varphi,$$

where

$$(F_* V_k)_i = D_{V_k} F_i, \quad 1 \le i \le n,$$

is the "pushforward" under F of V_k. That is, $D_{F_* V_k} \varphi = D_{V_k}(\varphi \circ F)$.

§**8.2.1. Commuting Vector Fields:** Until further notice, we will be dealing with vector fields $V_0, \ldots, V_r$ which have two uniformly bounded, continuous derivatives. In particular, these assumptions are more than enough to assure that the equation (8.2.2), and therefore (8.2.1), admits a $\mathbb{P}^0$-almost surely unique solution. In addition, by Corollary 4.2.6 or Corollary 7.3.6, the $\mathbb{P}^0$-distribution of the solution is the unique solution $\mathbb{P}_x^L$ to the martingale for L starting from x.

In this section we will take the first in a sequence of steps leading to an alternative (especially, non-Itô) way of thinking about solutions to (8.2.1). Given $\xi = (\xi_0, \ldots, \xi_r) \in \mathbb{R}^{r+1}$, set $V_\xi = \sum_{k=0}^{r} \xi_k V_k$, and determine $E(\xi, x)$ for $x \in \mathbb{R}^n$ so that $E(0, x) = x$ and

$$\frac{d}{dt} E(t\xi, x) = V_\xi\big(E(t\xi, x)\big).$$

From elementary facts (cf. §4 of Chapter 2 in [4]) about ordinary differential equations, we know that $(\xi, x) \in \mathbb{R}^{r+1} \times \mathbb{R}^n \longmapsto E(\xi, x) \in \mathbb{R}^n$ is a twice continuously differentiable function which satisfies estimates of the form

$$(8.2.4) \quad \begin{aligned} & \big|E(\xi, x) - x\big| \le C|\xi|, \quad \big|\partial_{\xi_k} E(\xi, x)\big| \vee \big|\partial_{x_i} E(\xi, x)\big| \le C e^{\nu|\xi|} \\ & \big|\partial_{\xi_k} \partial_{\xi_\ell} E(\xi, x)\big| \vee \big|\partial_{\xi_k} \partial_{x_i} E(\xi, x)\big| \vee \big|\partial_{x_i} \partial_{x_j} E(\xi, x)\big| \le C e^{\nu|\xi|}, \end{aligned}$$

for some $C < \infty$ and $\nu \in [0, \infty)$. Finally, define

(8.2.5) $V_k(\xi, x) = \partial_{\xi_k} E(\xi, x)$ for $0 \leq k \leq r$ and $(\xi, x) \in \mathbb{R}^{r+1} \times \mathbb{R}^n$.

For continuously differentiable, $\mathbb{R}^n$-valued functions V and W on $\mathbb{R}^n$, we will use the notation $[V, W]$ to denote the $\mathbb{R}^n$-valued function which is determined so that $D_{[V,W]}$ is equal to the commutator, $\left[D_V, D_W\right] = D_V \circ D_W - D_W \circ D_V$, of D_V and D_W. That is, $[V, W] = D_V W - D_W V$.

LEMMA 8.2.6. $V_k(\xi, x) = V_k\big(E(\xi, x)\big)$ for all $0 \leq k \leq r$ and $(\xi, x) \in \mathbb{R}^{r+1} \times \mathbb{R}^n$ if and only if $[V_k, V_\ell] \equiv 0$ for all $0 \leq k, \ell \leq r$.

PROOF: First assume that $V_k(\xi, x) = V_k\big(E(\xi, x)\big)$ for all k and (ξ, x). Then, if e_ℓ is the element of $\mathbb{R}^{r+1}$ whose ℓth coordinate is 1 and whose other coordinates are 0, we see that

$$\frac{d}{dt} E(\xi + te_\ell, x) = V_\ell(\xi + te_\ell, x) = V_\ell\big(E(\xi + te_\ell, x)\big),$$

and so, by uniqueness, $E(\xi + te_\ell, x) = E\big(te_\ell, E(\xi, x)\big)$. In particular, this leads first to

$$E\big(te_\ell, E(se_k, x)\big) = E(te_\ell + se_k, x) = E\big(se_k, E(te_\ell, x)\big),$$

and thence to

$$D_{V_k} V_\ell(x) = \frac{\partial^2}{\partial s \partial t} E\big(te_\ell, E(se_k, x)\big)\Big|_{s=t=0}$$
$$= \frac{\partial^2}{\partial t \partial s} E\big(se_k, E(te_\ell, x)\big)\Big|_{s=t=0} = D_{V_\ell} V_k(x).$$

To go in the other direction, first observe that $V_k(\xi, x) = V_k\big(E(\xi, x)\big)$ is implied by $E(\xi + te_k, x) = E\big(te_k, E(\xi, x)\big)$. Thus, it suffices to show that $E(\xi + \eta, x) = E\big(\eta, E(\xi, x)\big)$ follows from $[V_\xi, V_\eta] \equiv 0$. Second, observe that $E(\xi + \eta, x) = E\big(\eta, E(\xi, x)\big)$ is implied by $E\big(\xi, E(\eta, x)\big) = E\big(\eta, E(\xi, x)\big)$. Indeed, if the second of these holds and $F(t) \equiv E\big(t\xi, E(t\eta, x)\big)$, then $\dot{F}(t) = (V_\xi + V_\eta)\big(F(t)\big)$, and so, by uniqueness, $F(t) = E\big(t(\xi + \eta), x\big)$. In other words, all that remains is to show that $E\big(\xi, E(\eta, x)\big) = E\big(\eta, E(\xi, x)\big)$ follows from $[V_\xi, V_\eta] \equiv 0$. To this end, set $F(t) = E\big(\xi, E(t\eta, x)\big)$, and note that $\dot{F}(t) = E(\xi, \cdot)_* V_\eta\big(E(t\eta, x)\big)$. Hence, by uniqueness, we will know that $E\big(\xi, E(t\eta, x)\big) = F(t) = E\big(t\eta, E(\xi, x)\big)$ once we show that $E(\xi, \cdot)_* V_\eta = V_\eta\big(E(\xi, \cdot)\big)$. But

$$\frac{d}{ds} E(s\xi, \cdot)_*^{-1} V_\eta\big(E(s\xi, \cdot)\big) = [V_\xi, V_\eta]\big(E(s\xi, \cdot)\big),$$

and so we are done. $\square$

THEOREM 8.2.7. *Assume that the vector fields V_k commute. Then the one and only solution to (8.2.1) is $(t, p) \rightsquigarrow X(t, x, p) \equiv E\big((t, p(t)), x\big).$*

PROOF: Let $X(\,\cdot\,, x, p)$ be given as in the statement. By Itô's formula,

$$dX(t, x, p) = V_0\big((t, p(t)), X(t, x, p)\big)dt + \sum_{k=1}^{r} V_k\big((t, p(t)), X(t, x, p)\big) \circ dp_k(t),$$

and so, by Lemma 8.2.6, $X(\,\cdot\,, x, p)$ is a solution. As for uniqueness, simply rewrite (8.2.1) in its Itô equivalent form, and conclude, from Theorem 5.2.2, that there can be at most one solution to (8.2.1). □

REMARK 8.2.8. A significant consequence of Theorem 8.2.7 is that the solution to (8.2.1) is a *smooth* function of p when the V_k's commute. In fact, for such vector fields, $p \rightsquigarrow X(\,\cdot\,, x, p)$ is the unique continuous extension to $C\big([0, \infty); \mathbb{R}^r\big)$ of the solution to the ordinary differential equation

$$\dot{X}(t, x, p) = V_0\big(X(t, x, p)\big) + \sum_{k=1}^{r} V_k\big(X(t, x, p)\big) \dot{p}_k(t)$$

for $p \in C^1\big([0, \infty); \mathbb{R}^r\big)$. This fact should be compared to the examples given in §3.3.

§8.2.2. General Vector Fields: Obviously, the preceding is simply not going to work when the vector fields do not commute. On the other hand, it indicates how to proceed. Namely, the commuting case plays in Stratonovich's theory the role that the constant coefficient case plays in Itô's . In other words, we should suspect that the commuting case is correct locally and that the general case should be handled by perturbation. We continue with the assumption that the V_k's are smooth and have two bounded continuous derivatives.

With the preceding in mind, for each $N \geq 0$, set $X^N(0, x, p) = x$ and

(8.2.9)
$$X^N(t, x, p) = E\big(\Delta^N(t, p), X^N([t]_N, x, p)\big)$$
$$\text{where } \Delta^N(t, p) \equiv \big(t - [t]_N, p(t) - p([t]_N)\big).$$

Equivalently (cf. (8.2.5)):

$$dX^N(t, x, p) = V_0\big(\Delta^N(t, p), X^N([t]_N, x, p)\big) dt$$
$$+ \sum_{k=1}^{r} V_k\big(\Delta^N(t, p), X^N([t]_N, x, p)\big) \circ dp_k(t).$$

Note that, by Lemma 8.2.6, $X^N(t, x, p) = E\big((t, p(t)), x\big)$ for each $N \geq 0$ when the V_k's commute. In order to prove that $\{X^N(\cdot, x, p) : N \geq 0\}$ is $\mathbb{P}^0$-almost surely convergent even when the V_k's do not commute, we proceed as follows. Set $D^N(t, x, p) \equiv X(t, x, p) - X^N(t, x, p)$ and $W_k(\xi, x) \equiv V_k\big(E(\xi, x)\big) - V_k(\xi, x)$, and note that $D^N(0, x, p) = 0$ and

$$
\begin{aligned}
dD^N(t, x, p) = &\Big(V_0\big(X(t, x, p)\big) - V_0\big(X^N(t, x, p)\big)\Big)\, dt \\
&+ \sum_{k=1}^{r} \Big(V_k\big(X(t, x, p)\big) - V_k\big(X^N(t, x, p)\big)\Big) \circ dp_k(t) \\
&+ W_0\big(\Delta^N(t, p), X^N([t]_N, x, p)\big)\, dt \\
&+ \sum_{k=1}^{r} W_k\big(\Delta^N(t, p), X^N([t]_N, x, p)\big) \circ dp_k(t).
\end{aligned}
$$

(8.2.10)

LEMMA 8.2.11. *There exist a $C < \infty$ and $\nu > 0$ such that*

$$
\Big|V_k\big(\eta, E(\xi, x)\big) - V_k(\xi + \eta, x) - \tfrac{1}{2}\big[V_\xi, V_k\big](x)\Big| \leq C\big(|\xi| + |\eta|\big)^2 e^{\nu(|\xi| + |\nu|)}
$$

for all $\xi, \eta \in \mathbb{R}^{r+1}$, $x \in \mathbb{R}^n$, and $0 \leq k \leq r$. In particular, if

$$
\tilde{W}_k(\xi, x) \equiv W_k(\xi, x) - \sum_{\ell \neq k} \xi_\ell \big[V_\ell, V_k\big](x),
$$

then $\tilde{W}_k(0, x) = 0$ and $\big|\partial_\xi \tilde{W}_k(\xi, x)\big| \leq C|\xi| e^{\nu|\xi|}$ for all $(\xi, x) \in \mathbb{R}^{r+1} \times \mathbb{R}^n$ and some $C < \infty$ and $\nu > 0$.

PROOF: In view of the estimates in (8.2.4), it suffices for us to check the first statement. To this end, observe that

$$
E\big(\eta, E(\xi, x)\big) = E(\xi, x) + V_\eta\big(E(\xi, x)\big) + \tfrac{1}{2} D_{V_\eta} V_\eta\big(E(\xi, x)\big) + R_1(\xi, \eta, x),
$$

where $R_1(\xi, \eta, x)$ is the remainder term in the second order Taylor expansion of $\eta \rightsquigarrow E(\cdot, E(\xi, x))$ around 0 and is therefore (cf. (8.2.4)) dominated by constant $C_1 < \infty$ times $|\eta|^3 e^{\nu|\eta|}$. Similarly,

$$
\begin{aligned}
E(\xi, x) &= x + V_\xi(x) + \tfrac{1}{2} D_{V_\xi} V_\xi(x) + R_2(\xi, x), \\
V_\eta\big(E(\xi, x)\big) &= V_\eta(x) + D_\xi V_\eta(x) + R_3(\xi, \eta, x), \\
D_{V_\eta} V_\eta\big(E(\xi, x)\big) &= D_{V_\eta} V_\eta(x) + R_4(\xi, \eta, x), \\
E(\xi + \eta, x) &= x + D_{V_{\xi+\eta}} V_{\xi+\eta}(x) + \tfrac{1}{2} D^2_{V_{\xi+\eta}} V_{\xi+\eta}(x) + R_2(\xi + \eta, x),
\end{aligned}
$$

where

$$
|R_2(\xi, x)| \leq C_2 |\xi|^3 e^{\nu|\xi|}, \quad |R_3(\xi, \eta, x)| \leq C_3 |\xi|^2 |\eta| e^{\nu|\xi|},
$$
$$
|R_4(\xi, \eta, x)| \leq C_4 |\xi| |\eta|^2 e^{\nu|\xi|}.
$$

Hence

$$E\big(\eta, E(\xi, x)\big) - E(\xi + \eta, x) = \tfrac{1}{2}\big[V_\xi, V_\eta\big](x) + R_5(\xi, \eta, x),$$
$$\text{where } |R_5(\xi, \eta, x)| \le C_5\big(|\xi| + |\eta|\big)^3 e^{\nu(|\xi|+|\eta|)},$$

and so the required estimate follows. $\square$

Returning to (8.2.10), we now write

$$\begin{aligned}
dD^N(t, x, p) &= W_0\big(\Delta^N(t, p), X^N([t]_N, x, p)\big)\, dt \\
&\quad + \frac{1}{2} \sum_{1 \le k \ne \ell \le r} \big[V_\ell, V_k\big]\big(X^N([t]_N, x, p)\big)\Delta_\ell^N(t, p)\, dp_k(t) \\
&\quad + \sum_{k=1}^{r} \tilde{W}_k\big(\Delta^N(t, p), X^N([t]_N, x, p)\big) \circ dp_k(t) \\
&\quad + \Big(V_0\big(X(t, x, p)\big) - V_0\big(X^N(t, x, p)\big)\Big)\, dt \\
&\quad + \sum_{k=1}^{r} \Big(V_k\big(X(t, x, p)\big) - V_k\big(X^N(t, x, p)\big)\Big) \circ dp_k(t),
\end{aligned}$$

where, for $1 \le \ell \le r$, we have used $\Delta_\ell^N(t, p) = p_\ell(t) - p_\ell([t]_N)$ to denote the kth coordinate of $\Delta^N(t, p)$. Also, it is important to observe that, because $\langle p_k, p_\ell \rangle \equiv 0$ for $\ell \ne k$, we were able to replace the Stratonovich stochastic integral $\Delta_\ell^N(t, p) \circ dp_k(t)$ by the Itô stochastic integral $\Delta_\ell^N(t, p)\, dp_k(t)$. In fact, it is this replacement which makes what follows possible.

To proceed, first use an obvious Gaussian computation to see that there exists for each $q \in [1, \infty)$ a $A_q < \infty$ such that

$$\mathbb{E}^{\mathbb{P}^0}\Big[\big|\Delta^N(t, p)\big|^{2q} e^{2q\nu|\Delta^N(t, p)|}\Big]^{\frac{1}{q}} \le A_q\big(t - [t]_N\big).$$

Hence, there exist constants $C_q < \infty$ such that

$$\mathbb{E}^{\mathbb{P}^0}\left[\left\|\int_0^{\cdot} W_0\big(\Delta^N(t, p), X^N([t]_N, x, p)\big)\, dt\right\|_{[0,T]}^{2q}\right]^{\frac{1}{q}} \le C_q T^2 2^{-N}$$

$$\mathbb{E}^{\mathbb{P}^0}\left[\left\|\int_0^{\cdot} \big[V_k, V_\ell\big]\big(X^N([t]_N, x, p)\big)\Delta_\ell^N(t, p)\, dp_k(t)\right\|_{[0,T]}^{2q}\right]^{\frac{1}{q}} \le C_q T 2^{-N}$$

$$\mathbb{E}^{\mathbb{P}^0}\left[\left\|\int_0^{\cdot} \tilde{W}_k\big(\Delta^N(t, p), X^N(t, x, p)\big) \circ dp_k(t)\right\|_{[0,T]}^{2q}\right]^{\frac{1}{q}} \le C_q\big(T + T^2\big)2^{-N}.$$

In the derivation of the last two of these, we have used Burkholder's inequality (cf. Exercise 5.1.27 or 7.2.15). Of course, in the case of the last, we had to first convert the Stratonovich integral to its Itô equivalent and then also had to apply the final part of Lemma 8.2.11. Similarly, we obtain

$$
\mathbb{E}^{\mathbb{P}^0}\left[\left\|\int_0^{\cdot}\Big(V_k\big(X(t,x,p)\big)-V_k\big(X^N(t,x,p)\big)\Big)\circ dp_k(t)\right\|_{[0,T]}^{2q}\right]^{\frac{1}{q}}
$$
$$
\leq C_q(1+T)\int_0^T \mathbb{E}^{\mathbb{P}^0}\left[\left\|X(\,\cdot\,,x,p)-X^N(\,\cdot\,,x,p)\right\|_{[0,T]}^{2q}\right]^{\frac{1}{q}}\,dt.
$$

After combining the preceding with Gronwall's inequality, we arrive at the goal toward which we have been working.

THEOREM 8.2.12. *For each $T > 0$ and $q \in [1,\infty)$ there exists a $C(T,q) < \infty$ such that*

$$
\mathbb{E}^{\mathbb{P}^0}\left[\left\|X(\,\cdot\,,x,p)-X^N(\,\cdot\,,x,p)\right\|_{[0,t]}^{2q}\right]^{\frac{1}{q}}\leq C(T,q)t2^{-N}\quad\text{if }t\in[0,T].
$$

§8.2.3. **Another Interpretation:** In order to exploit the result in Theorem 8.2.12, it is best to first derive a simple corollary of it, a corollary containing an important result which was proved, in a somewhat different form and by entirely different methods, originally by Wong and Zakai [42]. Namely, when p is a locally absolutely continuous element of $C\big([0,\infty);\mathbb{R}^n\big)$, then it is easy to check that, for each $T > 0$,

$$
\lim_{N\to\infty}\left\|X^N(\,\cdot\,,x,p)-X(\,\cdot\,,x,p)\right\|_{[0,T]}=0,
$$

where here $X(\,\cdot\,,x,p)$ is the unique locally absolutely continuous function such that

$$
(8.2.13)\quad X(t,x,p)=x+\int_0^t\left(V_0\big(X(\tau,x,p)\big)+\sum_{k=1}^r V_k\big(X(\tau,x,p)\big)\dot{p}_k(\tau)\right)d\tau.
$$

THEOREM 8.2.14. *For each $N \geq 0$ and $p \in C\big([0,\infty);\mathbb{R}^r\big)$, let $p^N \in C\big([0,\infty);\mathbb{R}^r\big)$ be determined so that*

$$
p^N(m2^{-N})=p(m2^{-N})\text{ and }p\restriction I_{m,N}\equiv[m2^{-N},(m+1)2^{-N}]\text{ is linear}
$$

for each $m \geq 0$. Then, for all $T > 0$ and $\epsilon > 0$,

$$
\lim_{N\to\infty}\mathbb{P}^0\left(\left\|X(\,\cdot\,,x,p)-X(\,\cdot\,,x,p^N)\right\|_{[0,T]}\geq\epsilon\right)=0
$$

where $X(\,\cdot\,,x,p^N)$ is the unique, absolutely continuous solution to (8.2.13).

PROOF: The key to this result is the observation that $X^N(m2^{-N}, x, p) = X(m2^{-N}, x, p^N)$ for all $m \geq 0$. Indeed, for each $m \geq 0$, both the maps $t \in (m2^{-N}, (m+1)2^{-N}) \longmapsto X(t, x, p^N)$ and

$$t \in (m2^{-N}, (m+1)2^{-N})$$
$$\longrightarrow E\Big(t(2^{-N}, p((m+1)2^{-N}) - p(m2^{-N})), X(m2^{-N}, x, p)\Big) \in \mathbb{R}^n$$

are solutions to the same ordinary differential equation. Thus, by induction on $m \geq 0$, the asserted equality follows from the standard uniqueness theory for the solutions to such equations.

In view of the preceding and the result in Theorem 8.2.12, it remains only to prove that

$$\lim_{N \to \infty} \mathbb{P}^0 \Bigg(\sup_{t \in [0,T]} \big| X(t, x, p^N) - X([t]_N, x, p^N) \big|$$

$$\vee \big| X^N(t, x, p) - X^N([t]_N, x, p) \big| \geq \epsilon \Bigg) = 0$$

for all $T \in (0, \infty)$ and $\epsilon > 0$. But

$$\big| X(t, x, p^N) - X([t]_N, x, p^N) \big| \leq C2^{-N} \vee \big| p([t]_N + 2^{-N}) - p([t]_N) \big| \quad \text{and}$$
$$\big| X^N(t, x, p) - X^N([t]_N, x, p) \big| \leq Ce^{\nu \|p\|_{[0,T]}} 2^{-N} \vee \big| p(t) - p([t]_N) \big|,$$

and so there is nothing more to do. $\square$

An important dividend of the preceding is best expressed in terms of the *support* in $C([0, \infty); \mathbb{R}^n)$ of the solution $\mathbb{P}^L_x$ to the martingale problem for L (cf. (8.2.3)) starting from x. Namely, take (cf. Exercise 8.2.16 below for more information)

$$S(x; V_0, \dots, V_r) = \big\{ X(\cdot, x, p) : p \in C^1([0, \infty); \mathbb{R}^r) \big\}.$$

COROLLARY 8.2.15. *Let L be the operator described by (8.2.3), and (cf. Corollary 4.2.6) let $\mathbb{P}^L_x$ be the unique solution to the martingale problem for L starting at x. Then the support of $\mathbb{P}^L_x$ in $C([0, \infty); \mathbb{R}^n)$ is contained in the closure there of the set $S(x; V_0, \dots, V_r)$.*

PROOF: First observe that $\mathbb{P}^L_x$ is the $\mathbb{P}^0$-distribution of $p \rightsquigarrow X(\cdot, x, p)$. Thus, by Theorem 8.2.14, we know that

$$\mathbb{P}^L_x\big(\overline{S(x; V_0, \dots, V_r)}\big) \geq \lim_{N \to \infty} \mathbb{P}^0\big(\{p : X(\cdot, x, p^N) \in \overline{S(x; V_0, \dots, V_r)}\}\big).$$

But, for each $n \in \mathbb{N}$ and $p \in C([0, \infty); \mathbb{R}^r)$, it is easy to construct $\{p_\epsilon^N : \epsilon > 0\} \subseteq C^\infty([0, \infty); \mathbb{R}^r)$ so that $p_\epsilon^N(0) = p(0)$ and

$$\lim_{\epsilon \searrow 0} \int_0^T \left| \dot{p}_\epsilon^N(t) - \dot{p}^N(t) \right|^2 dt = 0$$

and to show that $\lim_{\epsilon \searrow 0} \| X(\,\cdot\,, x, p_\epsilon^N) - X(\,\cdot\,, x, p^N) \|_{[0,T]} = 0$ for all $T > 0$. Hence, $X(\,\cdot\,, x, p^N) \in \overline{S(x; V_0, \ldots, V_r)}$ for all $n \in \mathbb{N}$ and $p \in C([0, \infty); \mathbb{R}^n)$. □

§8.2.4. Exercises

EXERCISE 8.2.16. Except when $\mathrm{span}(\{V_1(x), \ldots, V_r(x)\}) = \mathbb{R}^n$ for all $x \in \mathbb{R}^n$, in which case $\overline{S(x; V_0, \ldots, V_r)}$ is the set of $p \in C([0, \infty); \mathbb{R}^n)$ with $p(0) = x$, it is a little difficult to get a feeling for what paths are and what paths are not contained in $\overline{S(x; V_0, \ldots, V_r)}$. In this exercise we hope to give at least some insight into this question. For this purpose, it will be helpful to introduce the space $\mathcal{V}(V_0, \ldots, V_r)$ of bounded $V \in C^1([0, \infty); \mathbb{R}^n)$ with the property that for all $\alpha \in \mathbb{R}$ and $x \in \mathbb{R}^n$ the integral curve of $V_0 + \alpha V$ starting at x is an element of $\overline{S(x; V_0, \ldots, V_r)}$. Obviously, $V_k \in \mathcal{V}(V_0, \ldots, V_r)$ for each $1 \leq k \leq r$.

(i) Given $V, W \in \mathcal{V}(V_0, \ldots, V_r)$ and $(T, x) \in (0, \infty) \times \mathbb{R}^n$, determine $X \in C([0, \infty); \mathbb{R}^n)$ so that

$$X(t) = \begin{cases} x + \int_0^t V(X(\tau))\, d\tau & \text{if } t \in [0, T] \\ X(T) + \int_T^t W(X(\tau))\, d\tau & \text{if } t > T, \end{cases}$$

and show that $X \in \overline{S(x; V_0, \ldots, V_r)}$.

(ii) Show that if $V, W \in \mathcal{V}(V_0, \ldots, V_r)$ and $\varphi, \psi \in C_{\mathrm{b}}^1(\mathbb{R}^n; \mathbb{R})$, then $\varphi V + \psi W \in \mathcal{V}(V_0, \ldots, V_r)$.

Hint: Define $X : \mathbb{R}^3 \times \mathbb{R}^n \longrightarrow \mathbb{R}^n$ so that $X((0,0,0), x) = x$ and

$$\frac{d}{dt} X(t(\alpha, \beta, \gamma), x) = (\alpha V_0 + \beta V + \gamma W)(X(t(\alpha, \beta, \gamma), x)).$$

Given $N \in \mathbb{N}$, define $Y_N : [0, \infty) \times \mathbb{R}^n \longrightarrow \mathbb{R}^n$ so that $Y_N(t, x)$ equals

$$\begin{cases} X(t(1, 2\varphi(x), 0), x) & \text{for } 0 \leq t \leq 2^{-N-1} \\ X((t - 2^{-N-1})(1, 0, 2\psi(x)), Y_N(2^{-N-1})) & \text{for } 2^{-N-1} \leq t \leq 2^{-N} \\ Y_N(t - [t]_N, Y_M([t]_N, x)) & \text{for } t \geq 2^{-N}. \end{cases}$$

Show that $Y_N(\,\cdot\,, x) \longrightarrow Y(\,\cdot\,, x)$ in $C([0, \infty); \mathbb{R}^n)$, where $Y(\,\cdot\,, x)$ is the integral curve of $V_0 + \varphi V + \psi W$ starting at x.

(iii) If $V, W \in \mathcal{V}(V_0, \ldots, V_r)$ have two bounded continuous derivatives, show that $[V, W] \in \mathcal{V}(V_0, \ldots, V_r)$.

Hint: Use the notation in the preceding. For $N \in \mathbb{N}$, define $Y_N : [0, \infty) \times \mathbb{R}^n \longrightarrow \mathbb{R}^n$ so that $Y_N(t, x)$ is equal to

$$
\begin{aligned}
& X\big((t, 2^{N/2+2}t, 0), x\big) \\
& X\big((t - 2^{-N-2}, 0, 2^{N/2+2}(t - 2^{N-2}), Y_N(2^{-N-2}, x)\big) \\
& X\big((t - 2^{-N-1}, -2^{N/2+2}(t - 2^{-N-1}), 0), Y_N(2^{-N-1}, x)\big) \\
& X\big((t - 32^{-N-2}, 0, -2^{N/2+2}(t - 32^{N-2})), Y_N(32^{-N-2}, x)\big) \\
& Y_N\big(t - [t]_N, Y_N([t]_N, x)\big)
\end{aligned}
$$

according to whether $0 \le t \le 2^{-N-2}$, $2^{-N-2} \le t \le 2^{N-1}$, $2^{-N-1} \le t \le 32^{-N-2}$, $32^{-N-2} \le t \le 2^{-N}$, or $t \ge 2^{-N}$. Show that $Y_N(\,\cdot\,, x) \longrightarrow Y(\,\cdot\,, x)$ in $C([0, \infty); \mathbb{R}^n)$, where $Y(\,\cdot\,, x)$ is the integral curve of $V_0 + [V, W]$ starting at x.

(iv) Suppose that M is a closed submanifold of $\mathbb{R}^n$ and that, for each $x \in M$, $\{V_0(x), \ldots, V_r(x)\}$ is a subset of the tangent space $T_x M$ to M at x. Show that, for each $x \in M$, $\overline{S(x; V_0, \ldots, V_r)} \subseteq C([0, \infty); M)$ and therefore that $\mathbb{P}_x^L\big(\forall t \in [0, \infty)\ p(t) \in M\big) = 1$. Moreover, if $\{V_1, \ldots, V_r\} \subseteq C^\infty(\mathbb{R}^n; \mathbb{R}^n)$ and $\mathrm{Lie}(V_1, \ldots, V_r)$ is the smallest Lie algebra of vector fields on $\mathbb{R}^n$ containing $\{V_1, \ldots, V_r\}$, show that

$$
\overline{S(x; V_0, \ldots, V_r)} = \big\{p \in C([0, \infty); M) : p(0) = x\big\}
$$

if $M \ni x$ is a submanifold of $\mathbb{R}^n$ with the property that

$$
V_0(y) \in T_y M = \big\{V(y) : V \in \mathrm{Lie}(V_1, \ldots, V_r)\big\} \quad \text{for all } y \in M.
$$

8.3 The Support Theorem

Corollary 8.2.15 is the easier half of the following result which characterizes the support of the measure $\mathbb{P}_x^L$. In its statement, and elsewhere, we use the notation (cf. Theorem 8.2.14) $\|q\|_{M,I} \equiv \|\dot{q}^M\|_{L^2(I; \mathbb{R}^r)}$ for $M \in \mathbb{N}$, $q \in C^1([0, \infty); \mathbb{R}^r))$, and closed intervals $I \subseteq [0, \infty)$.

THEOREM 8.3.1. *The support of $\mathbb{P}_x^L$ in $C([0, \infty); \mathbb{R}^n)$ is the closure there of (cf. Corollary 8.2.15) $S(x; V_0, \ldots, V_r)$. In fact, if for each smooth $g \in C^1([0, \infty); \mathbb{R}^n)$ $X(\,\cdot\,, x, g)$ is the solution to (8.2.13) with $p = g$, then*

(8.3.2)
$$
\mathbb{P}^0\Big(\big\|X(\,\cdot\,, x, p) - X(\,\cdot\,, x, g)\big\|_{[0,T]} < \epsilon \,\big|\, \|p - g\|_{M,[0,T]} \le \delta\Big)
$$
$$
\longrightarrow 1 \quad \text{as first } \delta \searrow 0 \text{ and then } M \to \infty.
$$

for all $T \in (0, \infty)$ and $\epsilon > 0$.[3]

Notice that, because we already know $\mathrm{supp}(\mathbb{P}_x^L) \subseteq \overline{S(x; V_0, \ldots, V_r)}$, the first assertion will follow as soon as we prove (8.3.2). Indeed, since

$$p(0) = 0 \implies \|p - g\|_{M, [0,T]}^2 \leq 2^{M+1} \sum_{m=1}^{[2^M T]+1} |p(m2^{-M}) - g(m2^{-M})|^2$$

and, for any $\ell \in \mathbb{Z}^+$, the $\mathbb{P}^0$-distribution of

$$p \in C([0, \infty); \mathbb{R}^n) \longmapsto (p(2^{-M}), \ldots, p(\ell 2^{-M})) \in (\mathbb{R}^n)^\ell$$

has a smooth, strictly positive density, we know that

(8.3.3) $\mathbb{P}^0 \big(\|p - g\|_{M, [0,T]} \leq \delta \big) > 0$ for all $\delta > 0$.

Hence, (8.3.2) is much more than is needed to conclude that

$$\mathbb{P}^0 \big(\|X(\cdot, x, p) - X(\cdot, x, g)\|_{[0,T]} < \epsilon \big) > 0$$

for all $\epsilon > 0$.

To begin the proof of (8.3.2), first observe that, by Theorem 8.2.12 and (8.3.3),

$$\mathbb{P}^0 \Big(\|X(\cdot, x, p) - X(\cdot, x, g)\|_{[0,T]} > \epsilon \,\Big|\, \|p - g\|_{M, [0,T]} \leq \delta \Big)$$

$$\leq \varlimsup_{N \to \infty} \mathbb{P}^0 \Big(\|X^N(\cdot, x, p) - X(\cdot, x, g)\|_{[0,T]} > \epsilon \,\Big|\, \|p - g\|_{M, [0,T]} \leq \delta \Big)$$

$$\leq \sup_{N > M} \mathbb{P}^0 \Big(\|X^N(\cdot, x, p) - X^M(\cdot, x, g)\|_{[0,T]} > \tfrac{\epsilon}{2} \,\Big|\, \|p - g\|_{M, [0,T]} \leq \delta \Big)$$

$$+ \mathbb{P}^0 \Big(\|X^M(\cdot, x, p) - X(\cdot, x, g)\|_{[0,T]} > \tfrac{\epsilon}{2} \,\Big|\, \|p - g\|_{M, [0,T]} \leq \delta \Big).$$

---

[3] This sort of statement was proved for the first time in [39]. However, the conditioning there was somewhat different. Namely, the condition was that $\max_{1 \leq i \leq n} \|p_i - g_i\|_{[0,T]} < \delta$. Ikeda and Watanabe [16] follow the same basic strategy in their treatment, although they introduce an observation which not only greatly simplifies the most unpleasant part of the argument in [39] but also allows them to use the more natural condition $\|p - g\|_{[0,T]} < \delta$. The strategy of the proof followed below was worked out in [38]. Finally, if all that one cares about is the support characterization, [40] shows that one need not assume that the operator L can be expressed in Hörmander's form.

Hence, since

$$\left\|X^M(\,\cdot\,,x,p) - X(\,\cdot\,,x,g)\right\|_{[0,T]}$$

$$\leq \left\|X^M(\,\cdot\,,x,p) - X(\,\cdot\,,x,p^M)\right\|_{[0,T]} + \left\|X(\,\cdot\,,x,p^M) - X(\,\cdot\,,x,g^M)\right\|_{[0,T]}$$

$$+ \left\|X(\,\cdot\,,x,g^M) - X(\,\cdot\,,x,g)\right\|_{[0,T]}$$

$$\leq C\Bigg(2^{-M} + e^{\nu\|p\|_{[0,T]}} \sup_{\substack{0\leq s<t\leq T \\ t-s\leq 2^{-M}}} |p(t) - p(s)|$$

$$+ \|p-g\|_{M,[0,T]} + 2^{-\frac{M}{2}}\|g\|_{M,[0,T]}\Bigg),$$

it follows that we need only show that, for each $\epsilon > 0$,

$$(8.3.4) \quad \lim_{M\to\infty}\sup_{\delta\in(0,1]} \mathbb{P}^0\left(\sup_{\substack{0\leq s<t\leq T \\ t-s\leq 2^{-M}}} |p(t)-p(s)| \geq \epsilon \,\middle|\, \|p-g\|_{M,[0,T]}\leq \delta\right) = 0$$

and that
(8.3.5)

$$\sup_{\substack{N>M \\ \delta\in(0,1]}} \mathbb{P}^0\left(\left\|X^N(\,\cdot\,,x,p) - X^M(\,\cdot\,,x,g)\right\|_{[0,T]} \geq \epsilon \,\middle|\, \|p-g\|_{M,[0,T]}\leq\delta\right)$$

$$\longrightarrow 0 \quad \text{as } M\to\infty.$$

§8.3.1. The Support Theorem, Part I: The key to our proof of both (8.3.4) and (8.3.5) is the following lemma about Wiener measure. As the astute reader will recognize, this lemma is a manifestation of the same property of Gaussian measures on which (6.2.5) is based.

LEMMA 8.3.6. *For $M\in\mathbb{N}$ and $p\in C\big([0,\infty);\mathbb{R}^n\big)$, set $\tilde{p}^M = p - p^M$. Then $\sigma\big(\{\tilde{p}^M(t) : t\geq 0\}\big)$ is $\mathbb{P}^0$-independent of $\mathcal{B}^M \equiv \sigma\big(\{p(m2^{-M}) : m\in\mathbb{N}\}\big)$. Hence, if, for some Borel measurable $F : C\big([0,\infty);\mathbb{R}^n\big) \longrightarrow [0,\infty)$,*

$$\tilde{F}^M(q) \equiv \int F\big(\tilde{p}^M + q^M\big)\,\mathbb{P}^0(dp), \quad q\in C\big([0,\infty);\mathbb{R}^r\big),$$

then $\tilde{F}^M = \mathbb{E}^{\mathbb{P}^0}[F|\mathcal{B}^M]$ $\mathbb{P}^0$-almost surely. In particular,

$$\mathbb{E}^{\mathbb{P}^0}\big[F(p)\,\big|\,\|p-h\|_{M,[0,T]}\leq\delta\big] = \frac{\mathbb{E}^{\mathbb{P}^0}\big[\tilde{F}^M(p),\,\|p-h\|_{M,[0,T]}\leq\delta)\big]}{P^0(\|p-h\|_{M,[0,T]}\leq\delta)}.$$

PROOF: Clearly, the second part of this lemma is an immediate consequence of the first part. Thus (cf. part **(i)** in Exercise 4.2.39 in [36]) because all elements of $\text{span}(\{p(t) : t \geq 0\})$ are centered Gaussian random variables, it suffices to prove that $\mathbb{E}^{\mathbb{P}^0}\big[\tilde{p}^M(s)p(m2^{-M})\big] = 0$ for all $s \in [0, \infty)$ and $m \in \mathbb{N}$. Equivalently, what we need to show is that

$$\mathbb{E}^{\mathbb{P}^0}\big[p(s)p^M(m2^{-M})\big] = \mathbb{E}^{\mathbb{P}^0}\big[p^M(s)p^M(m2^{-M})\big] \text{ for all } s \in [0, \infty) \ \& \ m \in \mathbb{N}.$$

But, after choosing $\ell \in \mathbb{N}$ so that $\ell 2^{-M} \leq s \leq (\ell+1)2^{-M}$ and using the fact that $\mathbb{E}^{\mathbb{P}^0}\big[p(u)p(v)\big] = u \wedge v$, this becomes an elementary computation. $\square$

Knowing Lemma 8.3.6, one gets (8.3.4) easily. Namely, given $M \in \mathbb{N}$ and $(p, h) \in C\big([0, \infty); \mathbb{R}^r\big)$, set

$$\tilde{p}_h^M(t) \equiv \tilde{p}^M(t) + h^M(t).$$

Then, we will have proved (8.3.4) once we show that

$$\sup_{\|h-g\|_{M,[0,T]} \leq 1} \mathbb{E}^{\mathbb{P}^0}\left[\sup_{0 \leq s < t \leq T} \frac{|\tilde{p}_h^M(t) - \tilde{p}_h^M(s)|}{(t-s)^{\frac{1}{4}}}\right] < \infty.$$

But, $|h^M(t) - h^M(s)| \leq |t - s|^{\frac{1}{2}}\|h\|_{M,[0,T]}$,

$$\sup_{0 \leq s < t \leq T} \frac{|\tilde{p}^M(t) - \tilde{p}^M(s)|}{(t-s)^{\frac{1}{4}}} \leq 4 \sup_{0 \leq s < t \leq T+1} \frac{|p(t) - p(s)|}{(t-s)^{\frac{1}{4}}},$$

and so the required estimate follows immediately from part **(ii)** of Exercise 2.4.16.

§8.3.2. The Support Theorem, Part II: Unfortunately, the proof of (8.3.5) requires much more effort. To get started, fix $x \in \mathbb{R}^n$ and set $D^{M,N}(t, x, p) = X^N(t, x, p) - X^M(t, x, p)$ for $N > M$. Then, in view of Lemma 8.3.6, (8.3.5) will follow once we show that

$$(8.3.7) \qquad \lim_{M \to \infty} \sup_{\substack{N > M \\ \|h-g\|_{M,[0,T]} \leq 1}} \mathbb{E}^{\mathbb{P}^0}\Big[\big\|D^{M,N}(\,\cdot\,, x, \tilde{p}_h^M)\big\|_{[0,T]}^2\Big] = 0.$$

Because

$$X^N(t, x, p) = X^N\big(t - [t]_M, X^N([t]_M, x, p), \delta_{[t]_M}p\big)$$
$$\text{and } X^M(t, x, p) = X^M\big(t - [t]_M, X^M([t]_M, x, p), \delta_{[t]_M}p\big),$$

where, for each $s \geq 0$, $\delta_s : C\big([0, \infty); \mathbb{R}^r\big) \longrightarrow C\big([0, \infty); \mathbb{R}^r\big)$ is defined so that $\delta_s p(t) = p(s+t) - p(s)$, we have that $D^{M,N}(t, x, p)$ can be decomposed into

$$D^{M,N}\big(t - [t]_M, X^N([t]_M, x, p), \delta_{[t]_M}p\big)$$
$$+ E\big(\Delta^M(t, p), X^N([t]_M, x, p)\big) - E\big(\Delta^M(t, p), X^M([t]_M, x, p)\big)$$
$$= D^{M,N}([t]_M, x, p) + D^{M,N}\big(t - [t]_M, X^N([t]_M, x, p), \delta_{[t]_M}p\big)$$
$$+ \bar{E}\big(\Delta^M(t, p), X^N([t]_M, x, p)\big) - \bar{E}\big(\Delta^M(t, p), X^M([t]_M, x, p)\big),$$

where $\bar{E}(\xi, x) \equiv E(\xi, x) - x$. Hence, we can write

$$
D^{M,N}(t, x, p) = \int_0^t W_0^{M,N}(\tau, x, p)\, d\tau + \sum_{k=1}^r \int_0^t W_k^{M,N}(\tau, x, p) \circ dp_k(\tau)
$$

(8.3.8)
$$
+ \tilde{D}^{M,N}(t, x, p),
$$

where $W_k^{M,N}(t, x, p)$ is used to denote

$$
V_k\big(\Delta^M(t, p), X^N([t]_M, x, p)\big) - V_k\big(\Delta^M(t, p), X^M([t]_M, x, p)\big)
$$

and $\tilde{D}^{M,N}(t, x, p)$ is given by

$$
\sum_{m=0}^{[2^M t]} D^{M,N}\big((t - m2^{-M}) \wedge 2^{-M}, X^N(m2^{-M}, x, p), \delta_{m2^{-M}} p\big).
$$

The terms involving the $W_k^{M,N}$'s are relatively easy to handle. Namely, because (cf. (5.1.25) and use Brownian scaling) for any $\lambda \geq 0$ and $T \in (0, \infty)$

(8.3.9) $\mathbb{E}^{\mathbb{P}^0}\big[e^{\lambda T^{-\frac{1}{2}}\|p\|_{[0,T]}}\big] = \mathbb{E}^{\mathbb{P}^0}\big[e^{\lambda\|p\|_{[0,1]}}\big] \equiv K(\lambda) < \infty,$

and, for any $(m, M) \in \mathbb{N}^2$ (cf. the notation in Theorem 8.2.14),

$$
\|\Delta^M(\cdot, \tilde{p}_h^M)\|_{I_{m,M}} \leq 2\|\Delta^M(\cdot, p)\|_{I_{m,M}} + 2^{-\frac{M}{2}}\|h\|_{M, I_{m,M}},
$$

we have that

(8.3.10) $\mathbb{E}^{\mathbb{P}^0}\Big[e^{\lambda 2^{M/2}\|\Delta^M(\cdot, \tilde{p}_h^M)\|_{I_{m,M}}} \,\Big|\, \mathcal{B}_{m2^{-M}}\Big] \leq K(2\lambda)e^{\lambda\|h\|_{M, I_{m,M}}}.$

Hence, since (cf. (8.2.4))

$$
\left\| \int_0^{\cdot} W_0^{M,N}(\tau, x, p)\, d\tau \right\|_{[0,T]}^2 \leq T \int_0^T \big|W_0^{M,N}(\tau, x, p)\big|^2\, d\tau
$$

$$
\leq CT \int_0^T e^{2\nu|\Delta^M(\tau, p)|}\big|D^{M,N}([\tau]_M, x, p)\big|^2\, d\tau,
$$

it is clear that there are finite constants K and K' such that

$$
\mathbb{E}^{\mathbb{P}^0}\left[\left\| \int_0^{\cdot} W_0^{M,N}(\tau, x, \tilde{p}_h^M)\, d\tau \right\|_{[0,T]}^2\right]
$$

$$
\leq KT \int_0^T \mathbb{E}^{\mathbb{P}^0}\Big[e^{2\nu|\Delta^M(\tau, \tilde{p}_h^M)|}\big|D^{M,N}([\tau]_M, x, \tilde{p}_h^M)\big|^2\Big]\, d\tau
$$

$$
\leq K'Te^{2\nu\|h\|_{M,[0,T]}} \int_0^T \mathbb{E}^{\mathbb{P}^0}\Big[\big|D^{M,N}([\tau]_M, x, \tilde{p}_h^M)\big|^2\Big]\, d\tau.
$$

Next, suppose that $1 \leq k \leq r$. Then, after converting Stratonovich integrals to their Itô equivalents, we have

$$\int_0^t W_k^{M,N}(\tau, x, \tilde{p}_h^M) \circ d(\tilde{p}_h^M)_k(\tau)$$

$$= \int_0^t W_k^{M,N}(\tau, x, \tilde{p}_h^M)\, dp_k(\tau) + \frac{1}{2} \int_0^t W_{k,k}^{M,N}(\tau, x, \tilde{p}_h^M)\, d\tau$$

$$- \int_0^t W_k^{M,N}(\tau, x, \tilde{p}_h^M)\dot{p}_k^M(\tau)\, d\tau + \int_0^t W_k^{M,N}(\tau, x, \tilde{p}_h^M)\dot{h}_k^M(\tau)\, d\tau,$$

where $W_{k,k}^{M,N}(t, x, p)$ denotes

$$V_{k,k}\big(\Delta^M(t,p), X^N([t]_M, x, p)\big) - V_{k,k}\big(\Delta^M(t,p), X^M([t]_M, x, p)\big)$$

when $V_{k,k}(\xi, x) \equiv \big[\partial_k V_k(\,\cdot\,, x)\big](\xi)$. The second and fourth terms on the right are handled in precisely the same way as the one involving $W_0^{M,N}$ and satisfy estimates of the form

$$\mathbb{E}^{\mathbb{P}^0}\left[\left\|\int_0^\cdot W_{k,k}^{M,N}(\tau, x, \tilde{p}_h^M)\, d\tau\right\|_{[0,T]}^2\right]$$

$$\leq KTe^{2\nu\|h\|_{M,[0,T]}} \int_0^T \mathbb{E}^{\mathbb{P}^0}\left[\big|D^{M,N}([\tau]_M, x, \tilde{p}_h^M)\big|^2\right] d\tau$$

and

$$\mathbb{E}^{\mathbb{P}^0}\left[\left\|\int_0^\cdot W_k^{M,N}(\tau, x, \tilde{p}_h^M)\dot{h}_k^M(\tau)\, d\tau\right\|_{[0,T]}^2\right]$$

$$\leq K\|h\|_{M,[0,T]}^2 e^{2\nu\|h\|_{M,[0,T]}} \int_0^T \mathbb{E}^{\mathbb{P}^0}\left[\big|D^{M,N}([\tau]_M, x, \tilde{p}_h^M)\big|^2\right] d\tau.$$

In addition, because the first term is an Itô integral, we can apply Doob's inequality followed by the above reasoning to obtain the estimate

$$\mathbb{E}^{\mathbb{P}^0}\left[\left\|\int_0^\cdot W_k^{M,N}(\tau, x, \tilde{p}_h^M)\, dp_k(\tau)\right\|_{[0,T]}^2\right]$$

$$\leq Ke^{2\nu\|h\|_{M,[0,T]}} \int_0^T \mathbb{E}^{\mathbb{P}^0}\left[\big|D^{M,N}([\tau]_M, x, \tilde{p}_h^M)\big|^2\right] d\tau.$$

As for the third term on the right, it is best to first decompose it into

$$\int_0^t W_k^{M,N}([\tau]_M, x, \tilde{p}_h^M)\, dp_k(\tau)$$

$$+ \int_0^t \big(W_k^{M,N}(\tau, x, p) - W_k^{M,N}([\tau]_M, x, p)\big)\dot{p}_k^M(\tau)\, d\tau.$$

Since the first of these is an Itô integral, it presents no new problem. To deal with the second, observe that (cf. (8.2.4))

$$\left\| \int_0^\cdot \left(W_k^{M,N}(\tau, x, p) - W_k^{M,N}([\tau]_M, x, p) \right) \dot{p}_k^M(\tau) \, d\tau \right\|_{[0,T]}^2$$

$$\leq C^2 4^M T \sum_{m=0}^{[2^M T]} \left(p_k \left((m+1)2^{-M} \right) - p_k(m2^{-M}) \right)^2$$

$$\times \int_{I_{m,M}} e^{2\nu |\Delta^M(\tau, \tilde{p}_h^M)|} |\Delta^M(\tau, \tilde{p}_h^M)|^2 \, d\tau \, |D^{M,N}(m2^{-M}, x, \tilde{p}_h^M)|^2,$$

and pass from this to the same sort of estimate at which we arrived for the other terms. Hence, by combining these, we can now replace (8.3.8) by

$$\mathbb{E}^{\mathbb{P}^0} \left[\left\| D^{M,N}(\,\cdot\,, x, \tilde{p}_h^M) \right\|_{[0,T]}^2 \right]$$

(8.3.11)
$$\leq C(T, \|h\|_{M,[0,T]}) \int_0^T \mathbb{E}^{\mathbb{P}^0} \left[\left\| D^{M,N}(\,\cdot\,, x, \tilde{p}_h^M) \right\|_{[0,t]}^2 \right] dt$$

$$+ 2\mathbb{E}^{\mathbb{P}^0} \left[\left\| \tilde{D}^{M,N}(\,\cdot\,, x, \tilde{p}_h^M) \right\|_{[0,T]}^2 \right],$$

where $(s,t) \in [0,\infty) \longmapsto C(s,t) \in (0,\infty)$ is nondecreasing in each variable separately.

§8.3.3. **The Support Theorem, Part III:** After applying Gronwall's inequality to (8.3.11), we know that, there is a $(s,t) \in [0,\infty)^2 \longmapsto K(s,t) \in (0,\infty)$ which is nondecreasing in each variable separately, such that

$$\sup_{\|h-g\|_{M,[0,T]} \leq 1} \mathbb{E}^{\mathbb{P}^0} \left[\left\| D^{M,N}(\,\cdot\,, x, \tilde{p}_h^M) \right\|_{[0,T]}^2 \right]$$

$$\leq K(T,g) \sup_{\|h-g\|_{M,[0,T]} \leq 1} \mathbb{E}^{\mathbb{P}^0} \left[\left\| \tilde{D}^{M,N}(\,\cdot\,, x, \tilde{p}_h^M) \right\|_{[0,T]}^2 \right].$$

Hence, what remains is to show that

(8.3.12)
$$\lim_{M \to \infty} \sup_{\substack{N > M \\ \|h-g\|_{M,[0,T]} \leq 1}} \mathbb{E}^{\mathbb{P}^0} \left[\left\| \tilde{D}^{M,N}(\,\cdot\,, x, \tilde{p}_h^M) \right\|_{[0,T]}^2 \right] = 0,$$

and this turns out to be quite delicate.

To get started on the proof of (8.3.12), we again break the computation into parts. Namely, because

$$X^N\big(t - [\tau]_M, X^N([\tau]_M, x, p), \delta_{[\tau]_M} p\big) = X^N\big(t - [\tau]_N, X^N([\tau]_N, x, p), \delta_{[\tau]_N} p\big),$$

$\tilde{D}^{M,N}(t, x, \tilde{p}_h^M)$ can be written as

$$\int_0^t \tilde{W}_0^{M,N}(\tau, x, \tilde{p}_h^M)\, d\tau$$

$$+ \sum_{k=1}^r \int_0^t \tilde{W}_k^{M,N}(\tau, x, \tilde{p}_h^M)\, dp_k(\tau) + \frac{1}{2}\sum_{k=1}^r \int_0^t \tilde{W}_{k,k}^{M,N}(\tau, x, \tilde{p}_h^M)\, d\tau$$

$$- \sum_{k=1}^r \int_0^t \tilde{W}_k^{M,N}(\tau, x, \tilde{p}_h^M)\dot{p}_k^M(\tau)\, d\tau + \sum_{k=1}^r \int_0^t \tilde{W}_k^{M,N}(\tau, x, \tilde{p}_h^M)\dot{h}_k^M(\tau),$$

where $\tilde{W}_k^{M,N}(t, x, p)$ denotes

$$V_k\big(\Delta^N(t,p), X^N([t]_N, x, p)\big) - V_k\big(\Delta^M(t,p), X^N([t]_M, x, p)\big)$$

and $\tilde{W}_{k,k}^{M,N}(t, x, p)$ is equal to

$$V_{k,k}\big(\Delta^N(t,p), X^N([t]_N, x, p)\big) - V_{k,k}\big(\Delta^M(t,p), X^N([t]_M, x, p)\big).$$

With the exception of those involving $\dot{p}_k^M(\tau)$, none of these terms is very difficult to estimate. Indeed,

$$\mathbb{E}^{\mathbb{P}^0}\left[\left\|\int_0^\cdot \tilde{W}_0^{M,N}(\tau, x, \tilde{p}_h^M)\, d\tau\right\|_{[0,T]}^2\right] \leq T \int_0^T \mathbb{E}^{\mathbb{P}^0}\left[\big|\tilde{W}_0^{M,N}(\tau, x, \tilde{p}_h^M)\big|^2\right]\, d\tau,$$

$$\mathbb{E}^{\mathbb{P}^0}\left[\left\|\int_0^\cdot \tilde{W}_k^{M,N}(\tau, x, \tilde{p}_h^M)\, dp_k(\tau)\right\|_{[0,T]}^2\right] \leq 4 \int_0^T \mathbb{E}^{\mathbb{P}^0}\left[\big|\tilde{W}_k^{M,N}(\tau, x, \tilde{p}_h^M)\big|^2\right]\, d\tau,$$

$$\mathbb{E}^{\mathbb{P}^0}\left[\left\|\int_0^\cdot \tilde{W}_{k,k}^{M,N}(\tau, x, \tilde{p}_h^M)\, d\tau\right\|_{[0,T]}^2\right] \leq T \int_0^T \mathbb{E}^{\mathbb{P}^0}\left[\big|\tilde{W}_{k,k}^{M,N}(\tau, x, \tilde{p}_h^M)\big|^2\right]\, d\tau,$$

and

$$\mathbb{E}^{\mathbb{P}^0}\left[\left\|\int_0^\cdot \tilde{W}_k^{M,N}(\tau, x, \tilde{p}_h^M)\dot{h}_k^M(\tau)\right\|_{[0,T]}^2\right]$$

$$\leq \|h\|_{M,[0,T]}^2 \int_0^T \mathbb{E}^{\mathbb{P}^0}\left[\big|\tilde{W}_k^{M,N}(\tau, x, \tilde{p}_h^M)\big|^2\right]\, d\tau.$$

At the same time, by (8.2.4), $\tilde{W}_k^{M,N}(t, x, \tilde{p}_h^M)$ is dominated by a constant times

$$\Big(\big|\tilde{p}_h^M([\tau]_N) - \tilde{p}_h^M([\tau]_M)\big|$$

$$+ \big|X^N([\tau]_N, x, \tilde{p}_h^M) - X^N([\tau]_M, x, \tilde{p}_h^M)\big|\Big)e^{\nu\|\Delta^M(\cdot, \tilde{p}_h^M)\|_{I_{m,M}}},$$

and so, in view of (8.3.9) and (8.3.10), all the above terms will have been handled once we check that, for each $q \in [1, \infty)$, there exists a $C_q < \infty$ such that

(8.3.13)
$$\mathbb{E}^{\mathbb{P}^0}\left[\left\|X^N(\cdot, x, \tilde{p}_h^M) - X^N(m2^{-M}, x, \tilde{p}_h^M)\right\|_{I_{m,M}}^q\right]^{\frac{1}{q}}$$
$$\leq C_q\left(1 + \|h\|_{M, I_{m,M}}\right)2^{-\frac{M}{2}}$$

for all $x \in \mathbb{R}^n$, $(m, M) \in \mathbb{N}^2$, and $N \geq M$. To this end, first note that, because $X^N(\cdot + m2^{-M}, x, p) = X^N(\cdot, X^N(m2^{-M}, x, p), \delta_{m2^{-M}}p)$, it suffices to treat the case when $m = 0$. Second, write $X^N(t, x, \tilde{p}_h^M) - x$ as

$$\int_0^t V_0\left(\Delta^N(\tau, \tilde{p}_h^M), x, \tilde{p}_h^M\right) d\tau$$

$$+ \sum_{k=1}^r \left(\int_0^t V_k\left(\Delta^N(\tau, \tilde{p}_h^M), x, \tilde{p}_h^M\right) dp_k(\tau) + \frac{1}{2}\int_0^t V_{k,k}\left(\Delta^N(\tau, \tilde{p}_h^M), x, \tilde{p}_h^M\right) d\tau\right)$$

$$- \sum_{k=1}^r \int_0^t \left(V_k\left(\Delta^N(\tau, \tilde{p}_h^M), x, \tilde{p}_h^M\right)\dot{p}_k^M(\tau) - V_k\left(\Delta^N(\tau, \tilde{p}_h^M), x, \tilde{p}_h^M\right)\dot{h}_k^M(\tau)\right) d\tau,$$

and apply (8.2.4) and standard estimates to conclude from this that

$$\mathbb{E}^{\mathbb{P}^0}\left[\left\|X^N(\cdot, x, \tilde{p}_h^M) - x\right\|_{I_{0,M}}^q\right]^{\frac{1}{q}}$$

is dominated by an expression of the form on the right hand side of (8.3.13).

§8.3.4. The Support Theorem, Part IV: The considerations in §8.3.3 reduce the proof of (8.3.12) to showing that, for each $T \in [0, \infty)$,

(8.3.14) $$\lim_{M \to \infty} \sup_{\substack{N > M \\ \|h - g\|_{M,[0,T]} \leq 1}} \mathbb{E}^{\mathbb{P}^0}\left[\left\|\int_0^\cdot \tilde{W}_k^{M,N}(\tau, x, \tilde{p}_h^M)\dot{p}_k^M(\tau)\, d\tau\right\|_{[0,T]}^2\right] = 0,$$

and for this purpose it is best to begin with yet another small reduction. Namely, dominate $\left\|\int_0^\cdot \tilde{W}_k^{M,N}(\tau, x, \tilde{p}_h^M)\dot{p}_k^M(\tau)\, d\tau\right\|_{[0,T]}$ by

$$\max_{0 \leq m \leq 2^M T} \int_{I_{m,M}} \left|\tilde{W}_k^{M,N}(\tau, x, \tilde{p}_h^M)\right|\left|\dot{p}_k^M(\tau)\right| d\tau$$

$$+ \max_{0 \leq m \leq 2^M T} \left|\int_0^{m2^{-M}} \tilde{W}_k^{M,N}(\tau, x, \tilde{p}_h^M)\dot{p}_k^M(\tau)\, d\tau\right|.$$

Again, the first of these is easy, since, by lines of reasoning which should be familiar by now:

$$\mathbb{E}^{\mathbb{P}^0}\left[\max_{0\leq m\leq[2^M T]}\left(\int_{I_{m,M}}|\tilde{W}_k^{M,N}(\tau,x,\tilde{p}_h^M)||\dot{p}_k^M(\tau)|\,d\tau\right)^2\right]$$

$$\leq C\mathbb{E}^{\mathbb{P}^0}\left[\sum_{m=0}^{[2^M T]-1}e^{4\nu\|\Delta^M\tilde{p}_h^M\|_{I_{m,M}}}\left|p_k\big((m+1)2^{-M}\big)-p_k(m2^{-M})\right|^4\right]^{\frac{1}{2}}$$

$$\leq C'T^{\frac{1}{2}}e^{2^{1-\frac{M}{2}}\nu}\|h\|_{M,[0,T]}^2\,2^{-\frac{M}{2}}$$

for appropriate, finite constants C and C'. Hence, the proof of (8.3.14), and therefore (8.3.12), is reduced to showing that

$$(8.3.15) \quad \lim_{M\to\infty}\sup_{N>M}\mathbb{E}^{\mathbb{P}^0}\left[\max_{m\leq 2^M T}\left|\int_0^{m2^{-M}}\tilde{W}_k^{M,N}(\tau,x,\tilde{p}_h^M)\dot{p}_k^M(\tau)\,d\tau\right|^2\right]=0$$

uniformly with respect to h with $\|h-g\|_{M,[0,T]}\leq 1$.

To carry out the proof of (8.3.15), we decompose $\tilde{W}_k^{M,N}(\tau,x,p)$ into the sum

$$\sum_{L=M+1}^N\left(\hat{V}_k^{M,L}\big(\tau,X^N([\tau]_M,x,p),p\big)+\hat{W}_k^{M,L}\big(\tau,X^N([\tau]_M,x,p),p\big)\right),$$

where

$$\hat{V}_k^{M,L}(\tau,x,p)\equiv V_k\big(\Delta^L(\tau,p),X^L([\tau]_L-[\tau]_M,x,\delta_{[\tau]_M}p)\big)$$
$$-V_k\big(\Delta^L(\tau,p),X^{L-1}([\tau]_L-[\tau]_M,x,\delta_{[\tau]_M}p)\big),$$

and

$$\hat{W}_k^{M,L}(\tau,x,p)\equiv V_k\big(\Delta^L(\tau,p),X^{L-1}([\tau]_L-[\tau]_M,x,\delta_{[\tau]_M}p)\big)$$
$$-V_k\big(\Delta^{L-1}(\tau,p),X^{L-1}([\tau]_{L-1}-[\tau]_M,x,\delta_{[\tau]_M}p)\big).$$

After making this decomposition, it is clear that (8.3.15) will be proved once we show that

$$(8.3.16)\quad \lim_{M\to\infty}\sup_{\substack{N>M\\\|h-g\|_{M,[0,T]}\leq 1}}\mathbb{E}^{\mathbb{P}^0}\left[\left(\sum_{m=0}^{[2^M T]-1}|p_k\big((m+1)2^{-M}\big)-p_k(m2^{-M})|\right.\right.$$
$$\left.\left.\times F_k^{M,N}\big(m,x,\tilde{p}_h^M\big)\right)^2\right]=0,$$

where $F_k^{M,N}(m, x, p)$ is given by

$$\sum_{L=M+1}^{N} 2^M \int_{I_{m,M}} \left| \hat{V}_k^{M,L}\left(\tau, X^N(m2^{-M}, x, p), p\right) \right| d\tau,$$

and that

(8.3.17) $$\lim_{M \to \infty} \sup_{\substack{N > M \\ \|h-g\|_{M,[0,T]} \le 1}} \mathbb{E}^{\mathbb{P}^0} \left[\max_{m < [2^M T]} \left| J_k^{M,N}(m, x, (p, h)) \right|^2 \right] = 0,$$

where $J_k^{M,N}(m, x, (p, h))$ is equal to

$$\sum_{m'=0}^{m} \sum_{L=M+1}^{N} \int_{I_{m',M}} \hat{W}_k^{M,L}\left(\tau, X^N(m'2^{-M}, x, \tilde{p}_h^M), \tilde{p}_h^M\right) \dot{p}_k(\tau) \, d\tau.$$

By Schwarz's inequality, the expectation value in (8.3.16) is dominated by

$$2^M T \sum_{m=0}^{[2^M T]-1} \mathbb{E}^{\mathbb{P}^0} \left[\left(p_k\left((m+1)2^{-M}\right) - p_k(m2^{-M}) \right)^2 F_k^{M,N}(m, x, \tilde{p}_h^M)^2 \right]$$

$$\le 2T \sum_{m=0}^{[2^M T]-1} \mathbb{E}^{\mathbb{P}^0} \left[F_k^{M,N}(m, x, \tilde{p}_h^M)^2 \right].$$

In addition, by Minkowski's and Jensen's inequalities,

$$\mathbb{E}^{\mathbb{P}^0} \left[F_k^{M,N}(m, x, \tilde{p}_h^M)^2 \right]^{\frac{1}{2}}$$

$$\le \sum_{L=M+1}^{N} \sup_{x \in \mathbb{R}^n} \mathbb{E}^{\mathbb{P}^0} \left[\left(2^M \int_{I_{m,M}} \left| \hat{V}^{M,L}(\tau, x, \tilde{p}_h^M) \right| d\tau \right)^2 \right]^{\frac{1}{2}}$$

$$\le 2^{M/2} \sum_{L=M+1}^{N} \left(\sup_{x \in \mathbb{R}^n} \int_{I_{m,M}} \mathbb{E}^{\mathbb{P}^0} \left[\left| \hat{V}^{M,L}(\tau, x, \tilde{p}_h^M) \right|^2 \right] d\tau \right)^{\frac{1}{2}}.$$

Thus, (8.3.16) will be proved once we show that

(8.3.18) $$\mathbb{E}^{\mathbb{P}^0} \left[\int_{I_{m,M}} \left| \hat{V}_k^{M,L}(\tau, x, \tilde{p}_h^M) \right|^2 d\tau \right] \le C\left(\|h\|_{M,[0,T]} \right) 2^{-L-\frac{3M}{2}}$$

for some nondecreasing $r \in [0, \infty) \longmapsto C(r) \in [0, \infty)$ and all $x \in \mathbb{R}^n$ and $0 \le m < [2^M T]$. To this end, first observe that the left hand side of (8.3.18) is dominated by

$$C\left(\|h\|_{M,[0,T]} \right) \sup_{x \in \mathbb{R}^n} \int_{I_{0,M}} \mathbb{E}^{\mathbb{P}^0} \left[\left| D^{L-1,L}([\tau]_L, x, \delta_{m2^{-M}} \tilde{p}_h^M) \right|^2 \right] d\tau.$$

With the help of the following lemma, we will be able to use Theorem 8.2.12 to control this last expression.

LEMMA 8.3.19. *Given* $0 \leq \tau < 2^{-M}$ *and a* $\mathcal{B}_\tau$-*measurable, continuous* $\Psi : C([0,\infty); \mathbb{R}^r) \longrightarrow [0,\infty)$,

$$\mathbb{E}^{\mathbb{P}^0} \left[\Psi \left(\delta_{m2^{-M}} \tilde{p}_h^M \right) \right] \leq \left(\frac{2^M}{1 - 2^M \tau} \right)^{\frac{n}{2q}} e^{\frac{\|h\|_{M,I_m,M}^2}{2q}} \|\Psi\|_{L^q(P^0)}$$

for every $q \in [1,\infty]$.

PROOF: First observe that (cf. Lemma 8.3.6)

$$\mathbb{E}^{\mathbb{P}^0} \left[\Psi \left(\delta_{m2^{-M}} \tilde{p}_h^M \right) \right] = \mathbb{E}^{\mathbb{P}^0} \left[\Psi \left(\tilde{p}_{\delta_{m2^{-M}}h}^M \right) \right] = \mathbb{E}^{\mathbb{P}^0} \left[\Psi \left(\tilde{p}^M + h_m^M \right) \right] = \tilde{\Psi}^M (h_m^M),$$

where $h_m^M(t) = (1 \wedge 2^M t)\left(h((m+1)2^M) - h(m2^M) \right)$. Indeed, the first of these is just the time shift-invariance of Brownian increments, the second comes from the fact that $\delta_{m2^{-M}} h^M$ coincides with h_m^M on $[0, 2^{-M}]$, and the last is the definition of $\tilde{\Psi}^M$ in Lemma 8.3.6.

The next step is to show that another expression for $\tilde{\Psi}^M (h_m^M)$ is

$$\left(\frac{2^M}{1 - 2^M \tau} \right)^{\frac{n}{2}} e^{2^{M-1} |h_m^M(2^{-M})|^2} \mathbb{E}^{\mathbb{P}^0} \left[\Psi(p) \exp \left(- \frac{2^M |p(\tau) - h_m^M(2^{-M})|^2}{2(1 - 2^M \tau)} \right) \right],$$

and, while doing this, we may and will assume that Ψ is not only continuous and non-negative but also bounded. But, by Lemma 8.3.6, we know that, for any bounded continuous $f : \mathbb{R}^r \longrightarrow \mathbb{R}$,

$$\mathbb{E}^{\mathbb{P}^0} \left[\tilde{\Psi}^M (p) f \left(p(2^{-M}) \right) \right] = \mathbb{E}^{\mathbb{P}^0} \left[\Psi(p) f \left(p(2^{-M}) \right) \right].$$

Further, if $y \in \mathbb{R}^r \longmapsto \ell_y^M \in C([0,\infty); \mathbb{R}^r)$ is given by $\ell_y^M(t) = (t \wedge 2^{-M})y$, then

$$\mathbb{E}^{\mathbb{P}^0} \left[\tilde{\Psi}^M (p) f \left(p(2^{-M}) \right) \right] = \int_{\mathbb{R}^r} \tilde{\Psi}^M (\ell_y^M) f(y) \gamma_{2^{-M}}(y) \, dy$$

while

$$\mathbb{E}^{\mathbb{P}^0} \left[\Psi(p) f \left(p(2^{-M}) \right) \right] = \int_{\mathbb{R}^r} \mathbb{E}^{\mathbb{P}^0} \left[\Psi(p) \gamma_{2^{-M} - \tau} (y - p(\tau)) \right] f(y) \, dy,$$

where $\gamma_t(y) = (2\pi t)^{-\frac{r}{2}} \exp \left(- \frac{|y|^2}{2t} \right)$. Hence, because Ψ, and therefore also $\tilde{\Psi}^M$, is bounded and continuous, the asserted equality follows by choosing a sequence of f's which form an approximate identity at $h((m+1)2^{-M}) - h(m2^{-M})$.

Given the preceding, there are two ways in which the proof can be completed. One is to note that, trivially,

$$\mathbb{E}^{\mathbb{P}^0} \left[\Psi \left(\tilde{p}^M + h_m^M \right) \right] \leq \|\Psi\|_{L^\infty(\mathbb{P}^0)},$$

whereas, by the preceding,

$$\mathbb{E}^{\mathbb{P}^0}\left[\Psi\left(\tilde{p}^M + h_m^M\right)\right] \le \|\Psi\|_{L^1(\mathbb{P}^0)} \left(\frac{2^M}{1 - 2^M\tau}\right)^{\frac{n}{2}} e^{2^{M-1}|h_m^M(2^{-M})|^2}.$$

Hence interpolation provides the desired estimate. Alternatively, one can apply Hölder's inequality to get

$$\mathbb{E}^{\mathbb{P}^0}\left[\Psi(p)\exp\left(-\frac{2^M|p(\tau) - h_m^M(2^{-M})|^2}{2(1 - 2^M\tau)}\right)\right]$$

$$\le \|\Psi\|_{L^q(\mathbb{P}^0)}\left(\int_{\mathbb{R}^r}\exp\left(-\frac{q'2^M|y - h_m^M(2^{-M})|^2}{2(1 - 2^M\tau)}\right)\gamma_\tau(y)\,dy\right)^{\frac{1}{q'}},$$

perform an elementary Gaussian integration, and arrive at the desired result after making some simple estimates. $\square$

If we now apply Lemma 8.3.19 with

$$\Psi(p) = \left|X^L([\tau]_L, x, p) - X^{L-1}([\tau]_L, x, p)\right|^2$$

and $q = n$, we see from Theorem 8.2.12 that

$$\mathbb{E}^{\mathbb{P}^0}\left[\left|X^L([\tau]_L, x, \tilde{p}_h^M) - X^{L-1}([\tau]_L, x, \tilde{p}_h^M)\right|^2\right]$$

$$\le C\left(\|h\|_{M, I_{m,M}}\right)\left(2^{-L}\tau\right)\left(\frac{2^M}{1 - 2^M\tau}\right)^{\frac{1}{2}},$$

for an appropriate choice of nondecreasing $r \rightsquigarrow C(r)$. In view of the discussion prior to Lemma 8.3.19, (8.3.18), and therefore (8.3.16), is now an easy step.

§8.3.5. **The Support Theorem, Part V:** What remains is to check (8.3.17). For this purpose, we have to begin by rewriting $\hat{W}_k^{M,L}(\tau, x, p)$ as

$$V_k\left(\Delta^L(\tau, p), E\left(\xi^L(\tau, p), X^{L-1}([\tau]_{L-1} - [\tau]_M, x, \delta_{[\tau]_M}p)\right)\right)$$

$$- V_k\left(\Delta^L(\tau, p) + \xi^L(\tau, p), X^{L-1}([\tau]_{L-1} - [\tau]_M, x, \delta_{[\tau]_M}p)\right),$$

where $\xi^L(\tau, p) \equiv \left([\tau]_L - [\tau]_{L-1}, p([\tau]_L) - p([\tau]_{L-1})\right)$. Having done so, one sees that Lemma 8.2.11 applies and shows that $\hat{W}_k^{M,L}(\tau, x, p)$ can be written as

$$\frac{1}{2}\sum_{\ell \ne k}[V_\ell, V_k]\left(X^{L-1}([\tau]_{L-1} - [\tau]_M, x, \delta_{[\tau]_M}p)\right)\xi_\ell^L(\tau, p) + R_{k,\ell}^{M,L}(\tau, x, p),$$

where

$$\left|R_{k,\ell}^{M,L}(\tau,x,p)\right| \le Ce^{2\nu\|\Delta^{L-1}(\cdot\,,p)\|_{I_{m,M}}}\|\Delta^{L-1}(\cdot\,,p)\|_{I_{m,M}}^2 \quad \text{for } \tau \in I_{m,M}.$$

Thus, since

$$\mathbb{E}^{\mathbb{P}^0}\left[\left(\sum_{m=0}^{[2^M T]-1}\left|p_k((m+1)2^{-M}) - p_k(m2^{-M})\right|\right.\right.$$

$$\left.\left.\times \sum_{L=M+1}^{N} e^{2\nu\|\Delta^{L-1}(\cdot\,,\tilde{p}_h^M)\|_{I_{m,M}}}\|\Delta^{L-1}(\cdot\,,\tilde{p}_h^M)\|_{I_{m,M}}^2\right)^2\right]$$

$$\le 2^M T \sum_{m=0}^{[2^M T]-1}\left(\sum_{L=M+1}^{N}\mathbb{E}^{\mathbb{P}^0}\left[\left|p_k((m+1)2^{-M}) - p_k(m2^{-M})\right|^2\right.\right.$$

$$\left.\left.\times e^{4\nu\|\Delta^{L-1}(\cdot\,,\tilde{p}_h^M)\|_{I_{m,M}}}\|\Delta^{L-1}(\cdot\,,\tilde{p}_h^M)\|_{I_{m,M}}^4\right]^{\frac{1}{2}}\right)^2$$

$$\le 2^M T^2 C\left(\sum_{L=M+1}^{N}\mathbb{E}^{\mathbb{P}^0}\left[e^{8\nu\|\Delta^{L-1}(\cdot\,,\tilde{p}_h^M)\|_{I_{0,M}}}\|\Delta^{L-1}(\cdot\,,\tilde{p}_h^M)\|_{I_{0,M}}^8\right]^{\frac{1}{4}}\right)^2$$

$$\le C(\|h\|_{M,[0,T]})T^2 2^{-M},$$

we are left with showing that, for each $1 \le k \le r$ and $\ell \ne k$,

$$(8.3.20) \quad \lim_{M\to\infty}\sup_{\substack{N>M \\ \|h-g\|_{M,[0,T]}\le 1}}\mathbb{E}^{\mathbb{P}^0}\left[\max_{m<[2^M T]}\left|\sum_{m'=0}^{m} H_{k,\ell}^{M,N}(m',x,(p,h))\right|^2\right] = 0,$$

where $H_{k,\ell}^{M,N}(m,x,(p,h))$ equals

$$2^M\left(p_k((m+1)2^{-M}) - p_k(m2^{-M})\right)$$

$$\times \int_{I_{m,M}}\sum_{L=M+1}^{N}\alpha_{k,\ell}^{M,L,N}([\tau]_{L-1},x,\tilde{p}_h^M)\xi_\ell^L(\tau,\tilde{p}_h^M)\,d\tau$$

with

$$\alpha_{k,\ell}^{M,L,N}(\tau,x,p) \equiv [V_\ell, V_k]\left(X^{L-1}(\tau - [\tau]_M, X^N([\tau]_M,x,p), \delta_{[\tau]_M}p)\right).$$

Since

$$\left|H_{k,0}^{M,N}(m,x,(p,h))\right| \le C2^{-M}\left|p_k((m+1)2^{-M}) - p_k(m2^{-M})\right|,$$

the case when $\ell = 0$ is easy. Thus, we will restrict our attention to $1 \leq k \neq \ell \leq r$, in which case we decompose $H_{k,\ell}^{M,N}(m, x, (p, h))$ into the sum

$$\left(p_k\big((m+1)2^{-M}\big) - p_k(m2^{-M})\right) A_{k,\ell}^{M,N}(m, x, (p, h))$$
$$+ \left(p_k\big((m+1)2^{-M}\big) - p_k(m2^{-M})\right)$$
$$\times \left((h-p)_\ell\big((m+1)2^{-M}\big) - (h-p)_\ell(m2^{-M})\right) B_{k,\ell}^{M,N}(m, x, (p, h))$$
$$+ \left(p_k\big((m+1)2^{-M}\big) - p_k(m2^{-M})\right)$$
$$\times \left((h-p)_\ell\big((m+1)2^{-M}\big) - (h-p)_\ell(m2^{-M})\right) C_{k,\ell}^{M,N}(m, x, (p, h)),$$

where $A_{k,\ell}^{M,N}(m, x, (p, h))$ equals

$$2^M \int_{I_{m,M}} \sum_{L=M+1}^{N} \alpha_{k,\ell}^{M,L,N}\big([\tau]_{L-1}, x, \tilde{p}_h^M\big) \big(p_\ell([\tau]_L) - p_\ell([\tau]_{L-1})\big) \, d\tau,$$

$B_{k,\ell}^{M,N}(m, x, (p, h))$ equals

$$4^M \int_{I_{m,M}} \sum_{L=M+1}^{N} \alpha_{k,\ell}^{M,L,N}\big(m2^{-M}, x, \tilde{p}_h^M\big) \big([\tau]_L - [\tau]_{L-1}\big) \, d\tau,$$

and $C_{k,\ell}^{M,N}(m, x, (p, h))$ equals

$$4^M \int_{I_{m,M}} \sum_{L=M+1}^{N} \Big(\alpha_{k,\ell}^{M,L,N}\big([\tau]_{L-1}, x, \tilde{p}_h^M\big)$$
$$- \alpha_{k,\ell}^{M,L,N}\big(m2^{-M}, x, \tilde{p}_h^M\big)\Big) \big([\tau_L] - [\tau]_{L-1}\big) \, d\tau.$$

Because $\big|C_{k,\ell}^{M,N}(m, x, (p, h))\big|$ is dominated by a constant times

$$\left\| X^{L-1}\big(\cdot, X^N(m2^{-M}, x, \tilde{p}_h^M), \delta_{m2^{-M}}\tilde{p}_h^M\big) - X^N(m2^{-M}, x, \tilde{p}_h^M)\right\|_{I_{0,M}},$$

a simple application of (8.3.13) leads to

$$\mathbb{E}^{\mathbb{P}^0}\left[\max_{m<[2^M T]}\left|\sum_{m'=0}^{m}\Big(p_k\big((m'+1)2^{-M}\big)-p_k(m'2^{-M})\Big)\right.\right.$$
$$\times\Big((h-p)_\ell\big((m'+1)2^{-M}\big)-(h-p)_\ell(m'2^{-M})\Big)C_{k,\ell}^{M,N}\big(m',x,(p,h)\big)\Big|^2\Big]$$

$$\leq 2^M TC\sup_x \mathbb{E}^{\mathbb{P}^0}\left[\sum_{m=0}^{[2^M T]-1}\left|\Big(p_k\big((m+1)2^{-M}\big)-p_k(m2^{-M})\Big)\right.\right.$$
$$\times\Big((h-p)_\ell\big((m+1)2^{-M}\big)-(h-p)_\ell(m2^{-M})\Big)$$
$$\times\big\|X^{L-1}\big(\cdot,X^N(m2^{-M}x,\tilde{p}_h^M),\delta_{m2^{-M}}\tilde{p}_h^M\big)$$
$$-X^N(m2^{-M},x,\tilde{p}_h^M)\big\|_{I_{0,M}}\Big|^2\Big]$$

$$\leq \big(1+\|h\|_{M,[0,T]}^2\big)CT^2\sup_{\substack{x\in\mathbb{R}^n\\ N\geq M}}\mathbb{E}^{\mathbb{P}^0}\big[\|X^N(\cdot,x,\tilde{p}_{h_m^M}^M)-x\|_{I_{0,M}}^4\big]^{\frac{1}{2}}$$

$$\leq C\big(\|h\|_{M,[0,T]}\big)T^2 2^{-M}.$$

The key to handling the other terms is the observation that, because $\ell\neq k$,

$$\mathbb{E}^{\mathbb{P}^0}\left[\Big(p_k\big((m+1)2^{-M}\big)-p_k(m2^{-M})\Big)A_{k,\ell}^{M,N}\big(m,x,(p,h)\big)\Big|\mathcal{B}_{m2^{-M}}\right]=0$$

and

$$\mathbb{E}^{\mathbb{P}^0}\left[\Big(p_k\big((m+1)2^{-M}\big)-p_k(m2^{-M})\Big)\right.$$
$$\times\Big((h-p)_\ell\big((m+1)2^{-M}\big)-(h-p)_\ell(m2^{-M})\Big)\Big|\mathcal{B}_{m2^{-M}}\right]=0.$$

Thus, since $B_{k,\ell}^{M,N}\big(m,x,(p,h)\big)$ is $\mathcal{B}_{m2^{-M}}$, both

$$\sum_{m'=0}^{m-1}\Big(p_k\big((m'+1)2^{-M}\big)-p_k(m'2^{-M})\Big)A_{k,\ell}^{M,N}\big(m',x,(p,h)\big)$$

and

$$\sum_{m'=0}^{m-1}\Big(p_k\big((m+1)2^{-M}\big)-p_k(m2^{-M})\Big)$$
$$\times\Big((h-p)_\ell\big((m+1)2^{-M}\big)-(h-p)_\ell(m2^{-M})\Big)B_{k,\ell}^{M,N}\big(m,x,(p,h)\big)$$

are $\mathbb{P}^0$-martingales relative to $\{\mathcal{B}_{m2^{-M}} : m \geq 0\}$. In particular, by Doob's inequality, the $\mathbb{P}^0$-expected square of the maximum over $m < [2^M T]$ of each of these are dominated by the sum from $m = 0$ to $[2^M T] - 1$ of, respectively,

$$\mathbb{E}^{\mathbb{P}^0}\left[\left(p_k\big((m+1)2^{-M}\big) - p_k(m2^{-M})\right)^2 \big|A_{k,\ell}^{M,N}(m,x,(p,h))\big|^2\right]$$

$$\leq C2^{-M}\mathbb{E}^{\mathbb{P}^0}\left[\big|A_{k,\ell}^{M,N}(m,x,(p,h))\big|^4\right]^{\frac{1}{2}}$$

and

$$\mathbb{E}^{\mathbb{P}^0}\left[\left(p_k\big((m+1)2^{-M}\big) - p_k(m2^{-M})\right)^2\right.$$

$$\times \left((h-p)_\ell\big((m+1)2^{-M}\big) - (h-p)_\ell(m2^{-M})\right)^2 \big|B_{k,\ell}^{M,N}(m,x,(p,h))\big|^2\right]$$

$$\leq C4^{-M}\big(1 + \|h\|_{M,[0,T]}^2\big)\mathbb{E}^{\mathbb{P}^0}\left[\big|B_{k,\ell}^{M,N}(m,x,(p,h))\big|^2\right].$$

Because $\big|B_{k,\ell}^{M,N}(m,x,(p,h))\big|$ is uniformly bounded, we will be done if we can show that

$$\mathbb{E}^{\mathbb{P}^0}\left[\big|A_{k,\ell}^{M,N}(m,x,(p,h))\big|^4\right]^{\frac{1}{2}} \leq C2^{-M}.$$

To this end, note that

$$\left|\int_{I_{m,M}} \sum_{L=M+1}^{N} \alpha_{k,\ell}^{M,L,N}\big([\tau]_{L-1},x,\tilde{p}_h^M\big)\big(p_\ell([\tau]_L) - p_\ell([\tau]_{L-1})\big)\, d\tau\right|^4$$

$$\leq 8^{-M}\int_{I_{m,M}}\left|\sum_{L=M+1}^{N} \alpha_{k,\ell}^{M,L,N}\big([\tau]_{L-1},x,\tilde{p}_h^M\big)\big(p_\ell([\tau]_L) - p_\ell([\tau]_{L-1})\big)\right|^4 d\tau.$$

Thus

$$\mathbb{E}^{\mathbb{P}^0}\left[\big|A_{k,\ell}^{M,N}(m,x,(p,h))\big|^4\right] \leq 2^M \int_{I_{m,M}} \mathbb{E}^{\mathbb{P}^0}\left[\big|G_{k,\ell}^{M,N}(\tau,x,(p,h))\big|^4\right] d\tau,$$

where

$$G_{k,\ell}^{M,N}(\tau,x,(p,h)) \equiv \sum_{L=M+1}^{N} \alpha_{k,\ell}^{M,L,N}\big([\tau]_{L-1},x,\tilde{p}_h^M\big)\big(p_\ell([\tau]_L) - p_\ell([\tau]_{L-1})\big)$$

$$= \int_{[\tau]_M}^{[\tau]_N} \hat{\alpha}_{k,\ell}^{M,N}(\sigma,\tau,x,\tilde{p}_h^M)\, dp_\ell(\sigma)$$

when $\hat{\alpha}_{k,\ell}^{M,N}(\sigma,\tau,x,p) = \alpha_{k,\ell}^{M,N}([\tau]_{L-1},x,p)$ for $[\tau]_{L-1} \leq \sigma < [\tau]_L$. Finally, by Burkholder's inequality,

$$\mathbb{E}^{\mathbb{P}^0}\left[\big|G_{k,\ell}^{M,N}(\tau,x,(p,h))\big|^4\right] \leq C\mathbb{E}^{\mathbb{P}^0}\left[\left(\int_{[\tau]_M}^{[\tau]_N} |\hat{\alpha}_{k,\ell}^{M,N}(\sigma,\tau,x,\tilde{p}_h^M)|^2\, d\sigma\right)^2\right],$$

which is dominated by a constant times 4^{-M}. Hence, we have at last completed the proof of Theorem 8.3.1.

§8.3.6. Exercises

EXERCISE 8.3.21. It should be pointed out that the characterization of supp$(\mathbb{P}_x^L)$ is *much* easier when L is *strictly elliptic* in the sense that $\{V_1(x), \ldots, V_r(x)\}$ spans $\mathbb{R}^n$ at each $x \in \mathbb{R}^n$. Of course, in this case, the statement is that supp$(\mathbb{P}_x^L)$ is the space of all continuous, $\mathbb{R}^n$-valued paths which start at x. What follows is an outline giving steps for a simple proof of this statement.

For the considerations here, Itô's theory is preferable to Stratonovich's. Thus, we will consider the solution $p \rightsquigarrow X(\,\cdot\,, x, p)$ to the Itô stochastic integral equation (8.2.2). Our ellipticity assumption becomes the assumption that the matrix $a(x) = \sigma(x)\sigma^\top(x)$ is strictly positive definite for each $x \in \mathbb{R}^n$.

(i) If $\big(M(t), \mathcal{F}_t, \mathbb{P}\big)$ is an $\mathbb{R}^n$-valued continuous martingale with $M(0) = 0$ satisfying

$$\sum_{j=1}^{n} \langle M_j \rangle(dt) \le K\,dt$$

for some $K < \infty$, show that

$$\mathbb{P}\big(\|M\|_{[0,T]} < R\big) \ge e^{-\frac{KT}{R^2}}$$

for all $T \in [0, \infty)$ and $R \in (0, \infty)$.

Hint: Set

$$u_R(t, x) = e^{\frac{Kt}{R^2}}\left(1 - \frac{|x|^2}{R^2}\right),$$

and show that $\big(u(t, M(t)), \mathcal{F}_t, \mathbb{P}\big)$ is a submartingale. In particular, by Doob's stopping time theorem, conclude that

$$e^{\frac{KT}{R^2}}\mathbb{P}(\zeta_R > T) \ge \mathbb{E}^{\mathbb{P}}\Big[u\big(T \wedge \zeta_R, M(t \wedge \zeta_R)\big)\Big] \ge 1$$

where $\zeta_R = \inf\{t \ge 0 : |M(t)| \ge R\}$.

(ii) Given $h \in C^1\big([0, \infty); \mathbb{R}^n\big)$ with $h(0) = x$, set $\rho = \|h - x\|_{[0,T]}$, and choose a bounded, measurable $W : [0, \infty) \times \mathbb{R}^n \longrightarrow \mathbb{R}^n$ so that

$$\sigma(y)W(t, y) = b(y) - \dot{h}(t) \quad \text{for } (t, y) \in [0, \infty) \times B_{\mathbb{R}^n}(x, \rho + 1).$$

Next, take $\theta(t, p) = -W\big(t, X(t, x, p)\big)$ and $\zeta(p) = \inf\{t \ge 0 : |X(t, x, p) - x| \ge \rho\}$, set

$$E_\theta(t, p) = \exp\left(\int_0^t \big(\theta(\tau, p), dp(\tau)\big)_{\mathbb{R}^n} - \frac{1}{2}\int_0^t |\theta(\tau, p)|^2 \, d\tau\right),$$

determine (cf. Theorem 6.2.1) the probability measure $\mathbb{Q}$ so that $d\mathbb{Q} \restriction \mathcal{B}_t = E_\theta(t)d\mathbb{P}^0 \restriction \mathcal{B}_t$ for $t \ge 0$, and apply Corollary 6.2.2 together with Doob's stopping time theorem to see that $\big(X(t \wedge \zeta_\rho, x, p) - h(t \wedge \zeta_\rho), \mathcal{B}_t, \mathbb{Q}\big)$ is an $\mathbb{R}^n$-valued martingale which satisfies the conditions in **(i)** with $K = \sup_{|y-x| \le \rho} \text{Trace}\big(\sigma\sigma^\top(y)\big)$.

(iii) By combining (i) with (ii), show that $\mathbb{P}^0\big(\|X(\,\cdot\,,x,p)-h\|_{[0,T]}<\epsilon\big)>$ 0 for all $\epsilon>0$.

EXERCISE 8.3.22. Without much more effort, it is possible to refine (8.3.2) a little. Namely, given $\alpha\in(0,\frac{1}{2})$ and $T\in(0,\infty)$, set

$$\|p\|_{[0,T]}^{(\alpha)}=\sup_{0\le s<t\le T}\frac{|p(t)-p(s)|}{(t-s)^\alpha}\quad\text{for }p\in C\big([0,\infty);\mathbb{R}^n\big).$$

The purpose of this exercise is to show that (8.3.2) can be replaced by the statement that

$$\lim_{M\to\infty}\lim_{\delta\searrow0}\mathbb{P}^0\left(\|X(\,\cdot\,,x,p)-X(\,\cdot\,,x,g)\|_{[0,T]}^{(\alpha)}<\epsilon\,\Big|\,\|p-g\|_{M,[0,T]}\le\delta\right)=1$$

for all $g\in C^1\big([0,\infty);\mathbb{R}^r\big)$ with $g(0)=0$ and $T>0$.

(i) Begin by showing that for each $q\in[1,\infty)$ and $T>0$ there is a nondecreasing $r\rightsquigarrow K_q(T,r)$ such that

$$\mathbb{E}^{\mathbb{P}^0}\left[\big|X^N(t,x,\tilde{p}_h^M)-X^N(s,x,\tilde{p}_h^M)\big|^{2q}\right]^{\frac{1}{2q}}\le K_q\big(T,\|h\|_{M,[0,T]}\big)(t-s)^{\frac{1}{2}}$$

for all $0\le s<t\le T$, $N\ge M$, $x\in\mathbb{R}^n$, and $h\in C\big([0,\infty);\mathbb{R}^n\big)$.

(ii) Using the preceding, Lemma 8.3.6, and Exercise 2.4.17, show that for each $\alpha\in(0,\frac{1}{2})$, $q\in[1,\infty)$, and $T\in[0,\infty)$,

$$\sup_{\substack{x\in\mathbb{R}^n\\\delta\in(0,1]}}\mathbb{E}^{\mathbb{P}^0}\left[\Big(\|X(\,\cdot\,,x,p)\|_{[0,T]}^{(\alpha)}\Big)^q\,\Big|\,\|p-g\|\le\delta\right]<\infty.$$

(iii) Given $0<\alpha<\beta$, show that

$$\|p\|_{[0,T]}^{(\alpha)}\le\big(2\|p\|_{[0,T]}\big)^{1-\frac{\alpha}{\beta}}\big(\|p\|_{[0,T]}^{(\beta)}\big)^{\frac{\alpha}{\beta}};$$

and use this, part (ii) above, and (8.3.2) to complete the program.

EXERCISE 8.3.23. Perhaps the single most important application of Theorem 8.3.1 is to the *strong minimum principle* for degenerate elliptic operators. Namely, let $\mathcal{G}$ be an open subset of $\mathbb{R}\times\mathbb{R}^n$, and, given $(s,x)\in\mathcal{G}$, define $\mathcal{G}_L(s,x)$ to be the set of $\big(t,X(t-s)\big)$ where $t\ge s$ and $X\in S(x;V_0,\ldots,V_r)$ with $\big(\tau,X(\tau-s)\big)\in\mathcal{G}$ for $\tau\in[s,t]$. The minimum principle for L is the statement that if $u\in C^{1,2}(\mathcal{G};\mathbb{R})$ is a (cf. (8.2.3)) (∂_t+L)-supersolution in $\mathcal{G}_L(s,x)$ (i.e., $(\partial_t+L)u\le0$ on $\mathcal{G}_L(s,x)$), then

$$u\restriction\mathcal{G}_L(s,x)\ge u(s,x)\implies u\restriction\mathcal{G}_L(s,x)=u(s,x).$$

Here are some steps which lead to this conclusion.

(i) Suppose that $(t, y) \in \mathcal{G}_L(s, x)$ and that $u(t, y) > u(s, x)$. Choose $X \in S(x; V_0, \ldots, V_r)$ so that $X \upharpoonright [0, t - s] \subseteq \mathcal{G}$ and $y = X(t - s)$. Next, choose $\rho > 0$ so that

$$\{(\tau, z) : \tau \in [s, t] \text{ and } |z - X(\tau - s)| \leq \rho\} \subseteq \mathcal{G}$$

and there exists an $\epsilon > 0$ such that $u(t, z) \geq u(s, x) + \epsilon$ when $|z - y| \leq \rho$. Set

$$\zeta(p) = \inf\{\tau \geq 0 : |p(\tau) - X(\tau)| \geq \rho\} \wedge (t - s),$$

and show that $(u(s + \tau \wedge \zeta(p), p(\tau \wedge \zeta)), \mathcal{B}_\tau, \mathbb{P}_x^L)$ is a supermartingale. In particular, conclude that

$$u(s, x) \geq \mathbb{E}^{\mathbb{P}_x^L}\left[u(s + \zeta(p), p(\zeta))\right].$$

(ii) Show that $\alpha \equiv \mathbb{P}_x^L(\zeta = t - s) > 0$, and combine this with the conclusion reached in (i) to arrive at the contradiction $u(s, x) \geq u(s, x) + \epsilon\alpha > u(s, x)$.

Notation

Notation	Description	See†
$A\complement$	The complement of the set A	
$\mathbf{1}_A$	The indicator function of the set A.	
$a \wedge b \,\&\, a \vee b$	The minimum and the maximum of $a, b \in \mathbb{R}$.	
$\langle \varphi, \mu \rangle$	Alternative notation for $\int \varphi \, d\mu$	§3.1
$\displaystyle\int Y(t) \circ dX(t)$	Stratonovich integral of Y with respect to X	§8.1.2
$\mu_n \longrightarrow \mu$	$\{\mu_n\}$ converges weakly to μ	§2.1
$\|\cdot\|_{\mathrm{u}}$	The uniform or "sup" norm on functions	
$\mathcal{B}_s$	The σ-algebra generated by paths $p \upharpoonright [0,s]$.	Cor. 1.5.9
$\mathcal{B}_E$	The Borel field over the topological space E.	
$C_{\mathrm{b}}(\mathbb{R}^n; \mathbb{R})$	Space of bounded continuous functions	
$C_{\mathrm{c}}(\mathbb{R}^n; \mathbb{R})$	Space of compactly supported continuous functions	
$C_{\mathrm{b}}^k(\mathbb{R}^n; \mathbb{R})$	Space of k-times, continuously and boundedly differentiable functions.	§2.1.1
$C^{1,2}([0,\infty) \times \mathbb{R}^N; \mathbb{R})$	Space of functions $(t, x) \in [0, \infty) \times \mathbb{R}^n \longrightarrow \mathbb{R}$ which are continuously differentiable once in t and twice in x.	
$D([0,\infty); \mathbb{Z}_n)$	Space of right-continuous, piecewise constant, $\mathbb{Z}_n$-valued paths	§1.4.2
$D([0,\infty); \mathbb{R}^n)$	Space of right-continuous, $\mathbb{R}^n$-valued paths with left limits	§2.3.1
δ_s	The time increment map on pathspace	§4.1.2

† This column points to the subsection in the text where the notation is introduced.

$\mathbb{E}^{\mathbb{P}}[X, A]$	To be read *the expectation value of X with respect to $\mathbb{P}$ on A*. Equivalent to $\int_A X\, d\mu$. When A is unspecified, it is assumed to be the whole space.	
$\mathbb{E}^{\mathbb{P}}[X \mid \mathcal{F}]$	To be read *the conditional expectation value of X given the σ-algebra $\mathcal{F}$*.	
$L^{(a,b,M)}$	Operator corresponding to Lévy system (a,b,M)	§2.1.2
$\lambda^{(a,b,M)}$	Infinitely divisable law with Lévy system (a,b,M)	Thm. 2.1.9
$\mathcal{M}^2(\mathbb{P};\mathbb{R})$	Space of square integrable, continuous $\mathbb{R}$-valued, $\mathbb{P}$-martingales	§5.1.1
$\mathcal{M}_{\mathrm{loc}}(\mathbb{P};\mathbb{R})$	Space of continuous, $\mathbb{R}$-valued, local $\mathbb{P}$-maringales	§5.1.3
$\mathbf{M}_1(E)$	Space of Borel probability measures on E.	
$\mu \star \nu$	The convolution product of measures μ and ν.	§§1.2.1 & 2.1.2
$\hat{\mu}$	The characteristic function (Fourier transform) of μ.	§§1.2.1 & 2.1.2
$\mathbb{P}^{(a,b,M)}$	Distribution of Lévy process with Lévy system (a,b,M).	Sec. 2.4
$\mathbb{P}^0$	Wiener measure, equivalent to $\mathbb{P}^{(I,0,0)}$	
$\mathbb{P}^{M'}$	Equivalent to $\mathbb{P}^{(I,0,M')}$	§3.1.1
$\langle M_1, M_2\rangle$	Bounded variation part of the product of continuous martingales M_1 and M_2.	§7.2.1
$\langle M\rangle$	Same as $\langle M, M\rangle$.	§7.1.2
$\mathbb{N}$	Set of non-negative integers	
$\mathbf{N}(t)$	The simple Poisson process.	§1.4.2
$\mathbb{P}_x^L$	Solution to the martingale problem for L starting from x	§§1.5.1 & 2.3.2
$\|p\|_{[a,b]}$	The uniform norm of $p \restriction [a,b]$	
$\|p\|_{M,[a,b]}$	Equivalent to $\sqrt{\int_a^b \|\dot{p}^M(\tau)\|^2\, d\tau}$.	§8.3
$\mathbb{S}^{n-1}$	The unit sphere in $\mathbb{R}^n$.	
Σ_s	The time shift transformation on pathspace.	§4.1.2
$\mathcal{S}(\mathbb{R}^n;\mathbb{C})$	Schwartz test function class	§2.1.2
$[t]$	Integer part of $t \in \mathbb{R}$	
$[t]_N$	Diadic integer part $2^{-N}[2^N t]$.	

$\Theta^2(\mathbb{P}; \mathbb{R}^n)$	Basic space of Itô integrands for Brownian stochastic integrals	§5.1.2
$\Theta^2_{\text{loc}}(\mathbb{P}; \mathbb{R}^n)$	Local version of $\Theta^2(\mathbb{P}; \mathbb{R}^n)$	§5.1.3
$\Theta^2_{\text{loc}}(\langle M \rangle, \mathbb{P}); \mathbb{R})$	Itô integrands for $(M(t), \mathcal{F}_t, \mathbb{P})$	§7.2
$\mathbb{Z}^+$	Set of positive integers	
$\mathbb{Z}_n$	The group $\{0, 1, \ldots, n-1\}$	

REFERENCES

1. Bachelier, L., *Théorie de la speculation*, Ann. Sci. École Normale Superior **3** (1900), 21–86.

2. Choquet, G., *Lectures in Analysis, vol. 2*, Benjamin, NY, 1969.

3. Chung, K.L. & Williams, R.J., *Introduction to Stochastic Integration, 2nd ed.*, Probability & its Applications, Birkhäuser, Boston, 1997.

4. Coddinton, E. & Levinson, N, *Ordinary Differential Equations*, International Series in Pure & Appl. Math., McGraw Hill, NYC, 1955.

5. Dellacherie, C. and Meyer, P.A., *Probabilités et Potentiel*, Chapters I-IV in 1975, V-VIII in 1980, IX-XI in 1983, XII-XVI in 1987, XVII-XXIV in 1992, Hermann, Paris; *Chapitres V à VIII of Théorie des martingales, Revised edition*, Actualités Scientifiques et Industrielles, Hermann, Paris, 1980.

6. Doob, J.L., *Stochastic Processes*, J. Wiley, NY, 1953.

7. Dudley, R., *Real Analysis and Probability*, Mathematics Series, Wadsworth & Brooks Cole, Pacific Grove, CA, 1989.

8. Dynkin, E.B., *Markov Processes, I*, translation from the 1962 Russian edition by D.E. Brown, Springer–Verlag, 1965.

9. Einstein, A., *Zur Theorie der Brownschen Bewegung*, Ann. Phs. **IV** #19, 371–381.

10. Hörmander, L., *Hypoelliptic second order differential equations*, Acta Math. **119** (1967), 147–171.

11. Itô, K., *On stochastic differential equations*, Mem. Amer. Math. Soc. #4, A.M.S., Providence, RI, 1951.

12. _____, *Stochastic differentials*, Appl. Math. Optim. **1** #4 (1974/75), 374–381.

13. _____, *Extension of stochastic integrals*, Proceedings of the International Symposium on Stochastic Differential Equations (Res. Inst. Math. Sci., Kyoto Univ., Kyoto, 1976, Wiley, New York-Chichester-Brisbane, 1978, pp. 95–109.

14. _____, *Kiyosi Itô, Selected Papers* (Stroock, D. & Varadhan S.R.S., eds.), Springer–Verlag, 1987.

15. Itô, K. & McKean, H.P., Jr., *Diffusion Processes and Their Sample Paths, 2nd printing*, Grundlehen Series # 125, Springer–Verlag, NY#, 1974.

16. Ikeda, N. & Watanabe, S., *Stochastic Differential Equations and Diffusion processes*, North-Holland Mathematical Library, 24, North-Holland Publishing Co. & Kodansha, Ltd., Amsterdam-New York & Tokyo, 1981.

17. Jacob, N. & Schilling, R., *Lévy–Type processes and pseudodifferential operators*, Lévy Processes, Theory and Applications (Barndorff–Nielsen, O., Mikosch, T., and Resnik, S., eds.), Birkhäuser, 2001.

18. Kinney, J.R., *Continuity Properties of Markov Processes*, TAMS **74** (1953), 280–302.

19. Kolmogorov, A.N., *Uber der Analytischen Methoden in der Wahrsheinlictkeitsrechnung*, Math. Ann. **104** (1931), 415–458.

20. Kunita, H., *Stochastic Flows and Stochastic Differential Equations*, Cambridge Studies in Advanced Mathematics, vol. 24, Cambridge Univ. Press, Cambridge, UK, 1990.

21. Kunita, H. & Watanabe, S., *On square integrable martingales*, Nagoya Math. J. **30** (1967), 209–245.

22. McKean, H.P., Jr., *A Hölder condition for Brownian local time*, J. Math. Kyoto Univ. **1** (1961/1962), 195–201.

23. _____, *Stochastic Integrals*, Prob. & Math. Stat. # 5, Academic Press, NY & London, 1969.

24. Nualart, D. & Pardoux, E., *Stochastic Calculus with Anticipating Integrands*, Prob. Th. & Related Fields **78** (1988), 535–581.

25. Øksendal, B., *Stochastic Differential Equations, 5th ed.*, Universitext Series, Springer–Verlag, NY, 1998.

26. Parthasarathy, K. R., Ranga Rao, R., & Varadhan, S.R.S., *Probability distributions on locally compact abelian groups*, Illinois J. Math. **7** (1963), 337–369.

27. Revuz, D. & Yor, M., *Continuous martingales and Brownian motion*, Grundlehren Series # 293, Springer–Verlag, Berlin, 1991, 1994, 1999.

28. Skorohod, A.V., *Limit theorems for stochastic processes*, Theory of Prob. & its Appl.'s **I** #3 (1956), 261–289.

29. _____, *Studies in the Theory of Random Processes*, tranlation of 1961 Russian edition, Addison-Wesley, Reading, MA, USA, 1965.

30. _____, *On homogeneous Markov processes that are martingales*, Theory of Prob. & its Appl.'s **VIII** #4 (1963), 355–365.

31. _____, *On the local structure of continuous Markov processes*, Theory of Prob. & its Appl.'s **XI** #3 (1966), 336–372.

32. _____, *On a generalization of a stochastic integral*, Th. of Prob. & its Appl.'s **XX** #4 (1974), 219–234.

33. Strassen, V., *An invariance principle for the law of the iterated logarithm*, Zeitschrift f. Wahrsch. verw. Geb. **3** (1964), 211–226.

34. Stratonovich, R.L., *A new form of representing stochastic integrals and equations*, Vestnik Moscow Univ. Ser. I Mat. Meh. #1 (1964), 3–12.

35. Stroock, D., *A Concise Introduction to the Theory of Integration, 3rd Ed.*, Bürkhauser, Cambridge, MA, USA, 1999.

36. _____, *Probability Theory, an Analytic View*, Cambridge U. Press, NY & Cambridge, UK, 1993 & 1998.

37. _____, *An Introduction to the Analysis of Paths on a Riemannian Manifold*, Mathematical Surveys and Monographs # 74, A.M.S., Providence, RI, 2000.

38. Stroock, D. & Taniguchi, S., *Diffusions as integral curves, or Stratonovich without Itô*, in Progress in Prob. # 34, ed. by M. Freidlin, Birkhäuser, 1994, pp. 331–369.

39. Stroock, D. & Varadhan, S.R.S., *On the support of diffusion processes, with applications to the strong maximum principle*, Proceedings of the Sixth Berkeley Symposium on Probability and Statistics, Berkeley Univ., 1970, pp. 333–359.

40. _____, *On degenerate elliptic-parabolic operators of the second order and their associated diffusions*, Comm. Pure Appl. Math. **XXV** (1972), 651–713.

41. _____, *Multidimensional Diffusion Processes*, Grundlehren Series #233, Springer-Verlag, 1979 & 1998.

42. Wong, E. & Zakai, M., *On the relation between ordinary and stochastic differential equations and applications to stochastic problems in control theory*, Automatic and remote control III (Proc. Third Congr. Internat. Fed. Automat. Control (IFAC), London, 1966), Vol. 1, Inst. Mech. Engrs., London, 1967, pp. 5–13.

Index